Decision Theory with a Human Face

When making decisions, people naturally face uncertainty about the potential consequences of their actions due in part to limits in their capacity to represent, evaluate or deliberate. Nonetheless, they aim to make the best decisions possible. In *Decision Theory with a Human Face*, Richard Bradley develops new theories of agency and rational decision making, offering guidance on how 'real' agents who are aware of their bounds should represent the uncertainty they face, how they should revise their opinions as a result of experience and how they should make decisions when lacking full awareness of, or precise opinions on, relevant contingencies. He engages with the strengths and flaws of Bayesian reasoning, and presents clear and comprehensive explorations of key issues in decision theory, from belief and desire to semantics and learning. His book draws on philosophy, economics, decision science and psychology and will appeal to readers in all these disciplines.

Richard Bradley is Professor of Philosophy at the London School of Economics and Political Science. He is an editor of the journal *Economics and Philosophy*, and his work on decision theory, semantics and epistemology has been published in numerous leading journals.

Decision Theory with a Human Face

RICHARD BRADLEY

London School of Economics and Political Science

CAMBRIDGE
UNIVERSITY PRESS

CAMBRIDGE
UNIVERSITY PRESS

University Printing House, Cambridge CB2 8BS, United Kingdom

One Liberty Plaza, 20th Floor, New York, NY 10006, USA

477 Williamstown Road, Port Melbourne, VIC 3207, Australia

314-321, 3rd Floor, Plot 3, Splendor Forum, Jasola District Centre, New Delhi - 110025, India

79 Anson Road, #06-04/06, Singapore 079906

Cambridge University Press is part of the University of Cambridge.

It furthers the University's mission by disseminating knowledge in the pursuit of education, learning and research at the highest international levels of excellence.

www.cambridge.org
Information on this title: www.cambridge.org/9781108793612
DOI: 10.1017/9780511760105

First published 2017
First paperback edition 2020

A catalogue record for this publication is available from the British Library

Library of Congress Cataloging in Publication data
Names: Bradley, Richard, 1964 December 13– author.
Title: Decision theory with a human face / Richard Bradley, London School of Economics and Political Science.
Description: Cambridge : Cambridge University Press, 2018. |
Includes bibliographical references and index.
Identifiers: LCCN 2017032958 | ISBN 9781107003217 (hardback : alk. paper)
Subjects: LCSH: Bayesian statistical decision theory. | Decision making–Mathematical models. | Uncertainty–Mathematical models.
Classification: LCC QA279.5 .B73 2018 | DDC 519.5/42–dc23
LC record available at https://lccn.loc.gov/2017032958

ISBN 978-1-107-00321-7 Hardback
ISBN 978-1-108-79361-2 Paperback

For Shura

Contents

Figures

Tables

Preface

The aim of this book is to develop a decision theory that is tailored for 'real' agents – i.e. agents who, like us, are uncertain about a great many things and are limited in their capacity to represent, evaluate and deliberate, but who nonetheless want to get things right to the extent that they can. The book is motivated by two broad claims. The first is that Bayesian decision theory provides an account of the rationality requirements on 'unbounded' agents that is essentially correct and is applicable in circumstances in which agents are aware of all the options available to them and are able to form precise judgements about all relevant contingencies. The second is that there are many circumstances in which these conditions are not satisfied and hence in which classical Bayesian theory is not applicable. A normative decision theory, adequate to such circumstances, would provide guidance on how bounded agents should represent the uncertainty they face, how they should revise their opinions as a result of experience and how they should make decisions when lacking full awareness or precise opinions (that they have confidence in) on relevant contingencies. The book tries to provide such a theory.

So many people have helped me with this project over the many years it has taken to complete it that I fear that I will forget to mention many of them. Its origins lie in my PhD dissertation completed under the supervision of Richard Jeffrey, David Malament and Dan Garber. Their support and that of Philippe Mongin, Dorothy Edgington and John Broome in the early years after the start of my career was crucial to my sticking with it. The influence of Dick Jeffrey on my thinking is hard to overestimate. The title of this book mirrors that of a paper of his – 'Bayesianism with a human face' – in which he espoused the kind of heterodox Bayesianism that pervades my writing. To me, he *was* the human face of Bayesianism.

Almost as much of an influence has been Jim Joyce, whom I first met at a workshop on Jeffrey's work some 20 years ago. Big chunks of this book can be read as a dialogue with *The Foundations of Causal Decision Theory* and his subsequent work on Imprecise Bayesianism. Parts of it are based on ideas developed with coauthors on papers: in particular, Christian List and Franz Dietrich (Chapter 10), Mareile Drechsler (Chapter 3), Casey Helgeson and Brian Hill (Chapter 14) and Orri Stefánsson (Chapters 8 and 9). As with many others with whom I have worked over the years (including both PhD students and colleagues), I have largely lost my grip on which ideas are mine and which are theirs, if indeed such a separation can meaningfully be made. It is an unfortunate irony that the ideas that you most thoroughly absorb are often the ones whose origins you forget.

I have been at the LSE for most of my career and it has provided the best possible intellectual environment for writing the book. The weekly seminars of LSE Choice Group have provided an invaluable forum for presenting ideas and acquiring new ones and its members a source of support. A number of people read parts of the book manuscript at various points in its development and gave helpful feedback, including Sven Ove Hansson, Jean Baccelli, Magda Osman, Seamus Bradley, Jay Hodges, Conrad Heilmann, Alex Voorhoeve, Hykel Hosni, Susanne Burri, Philippe van Basshuysen, Casey Helgeson, Aron Vallinder and Silvia Milano. Orri Stefánsson not only read an entire draft, but has been a wonderful interlocutor on its contents over many years. Katie Steele, Anna Mahtani, Jim Joyce, Wlodek Rabinowicz and Christian List provided valuable feedback on individual chapters at a workshop organised by Christian.

I am grateful to the Arts and Humanities Research Council for its support, in the form of a grant (AH/I003118/1) to work on the book and a grant for a project on Managing Severe Uncertainty (AH/J006033/1), the fruits of which are contained in the last part of the book.

Finally, I am deeply grateful to my family, and especially my wife Shura, for putting up with me over the last few years. There have been a good number of 'holidays' and weekends lost to book writing, not to mention grumpiness when nothing seemed to progress, but their patience and support have been undiminished by it all.

Introduction

Decision problems abound. Consumers have to decide what products to buy, doctors what treatments to prescribe, hiring committees which candidates to appoint, juries whether to convict or acquit a defendant, aid organisations what projects to fund, monetary policy committees what interest rates to set, and legislatures what laws to make. The problem for descriptive decision theory is to explain how such decisions are made. The problem that normative decision theory faces, on the other hand, is what to advise individuals and groups facing choices such as these. How should they evaluate the alternatives before them? What criteria should they employ? What procedures should they follow?

As these examples illustrate, decisions have to be made in a wide variety of contexts and by different kinds of decision makers. A pervasive feature, however, is the uncertainty that decision makers face: uncertainty about the state of the world, about the alternatives available to them, about the possible consequences of making one choice rather than another and, indeed, about how to evaluate these consequences. Dealing with this uncertainty is perhaps the most fundamental challenge we face in making a decision.

In the last 100 years or so an impressive body of work on this issue has emerged. At its core stands Bayesian decision theory, a mathematical and philosophical theory of both rational belief formation and decision making under uncertainty. The influence of Bayesian thinking pervades this book, to which the amount of space devoted to examining and criticising it will attest. Indeed, I regard Bayesian decision theory as essentially the correct theory for certain classes of idealised decision problems. But many real problems fall outside its scope. This is for two main reasons.

Firstly, Bayesian theory assumes that decision makers are 'unbounded': rational, logically omniscient and maximally opinionated. Rational in that their attitudes – beliefs, desires and preferences – are consistent both in

1

themselves and with respect to one another; logically omniscient because they believe all logical truths and endorse all the logical consequences of their attitudes; and opinionated because they have determinate beliefs, desires and preferences regarding all relevant prospects. All these assumptions can be criticised on grounds of being unrealistic: human decision makers, for instance, are unlikely to satisfy them for anything but a very small sets of prospects. Some of them can also be criticised on normative grounds. It is surely not required of us, for instance, that we have opinions about everything, nor that we are aware of all possibilities.

Secondly, by formulating the notion of a decision problem in a particular way, Bayesian decision theory excludes many of the kinds of uncertainty mentioned before. Indeed, it essentially restricts uncertainty to our knowledge of the state of the world, leaving out the uncertainty we face in judging how valuable consequences of actions are and the uncertainty we face as to the effect of our interventions in the world. Furthermore, it assumes that all uncertainty is of the same kind or severity – one that can be captured by a (single) probability measure on the possible states of the world. But we are often so unsure of things that we cannot even assign probabilities to them. It follows that Bayesianism is incomplete as a theory of rationality under uncertainty.

The main aim of the book, as its title suggests, is to develop a decision theory that is tailored for 'real' agents facing uncertainty that comes in many forms and degrees of severity. By 'real agents' I mean those who, like us, have limited skills and restricted time and other computational resources that make it impossible and undesirable that they should form attitudes to all contingencies or to think through all the logical consequences of what they believe, but who nonetheless get things right to the best of their ability and who employ quite sophisticated reasoning to this end. Humans, for instance, are capable not just of representing their environment as they find it but also of reflecting prospectively about how it might and would be if certain contingencies turned out to be true or if they were to perform particular actions. And of reflecting retrospectively on experience, and in particular on the outcomes of past actions, enabling them to improve their understanding of the world and the effect of their interventions in it. An examination of these abilities takes us into areas neglected by Bayesianism, such as the study of hypothetical reasoning and of reasoned preference change.

The desirability of moving in the direction of greater realism is, not surprisingly, widely recognised. But the way in which I want to do so is different from the direction taken in, for instance, behavioural economics and the psychology of judgement and choice. The aim is not to describe the way in which we do in fact evaluate prospects and make decisions, but to prescribe how we should, given our limitations and constraints. *The project is thus of giving not a descriptive theory of bounded rationality, but a normative theory of rationality for the bounded.*[1]

A decision theory that aims to play this kind of normative role must address itself to the sorts of agents that we are and the sorts of decision problems we face, taking as its starting point the resources and judgements that are available to us to deal with them. And the guidance that it provides on forming and revising judgements, as well as on making decisions, should be appropriate to the kind of uncertainty we face. This book tries to provide such a normative theory by doing what philosophers do best: proposing and examining candidate principles of rational belief, desire and choice that bounded agents can use to bring order to their deliberations, both prospective and retrospective. It's an enterprise that is at once very ambitious and quite modest. Ambitious because it aims at finding rationality principles of very general scope, applicable to the deliberations of many different kinds of decision makers in many different decision situations. Modest because these principles impose only conditions of consistency. The theory does not attempt to dictate whether we should believe, value or do any specific thing, but only which patterns of believing, valuing and doing are permitted. Rationality alone cannot decide for us what to think or to do, but it can support us in our attempts to do so.

Book Outline

The book is divided into four parts. The first part introduces the basics of Bayesian decision theory and then looks a range of philosophical questions about its foundations, interpretation and application, including the framing of decision problems (Chapter 1), the nature of rationality and the interpretation of probability and utility (Chapter 2) and the classification of forms of uncertainty (Chapter 3). It also assesses the role of representation theorems in motivating decision theories, looking in detail at Leonard Savage's version by way of illustration (Chapter 4).

Part II of the book is devoted to developing a theory of prospectively rational agency, the kind of rationality characteristic of agents who represent and evaluate not only the current state of their environment but also the state that might obtain or would obtain if they (or others) were to intervene in it in some way. The basic building blocks for the account are provided by Richard Jeffrey's version of Bayesian decision theory and the representation theorems for it due to Ethan Bolker and James Joyce, but the theory is extended to the treatment of conditional attitudes (Chapter 6) and then to conditionals (Chapter 7) by enriching the set of prospects and proposing

---

[1] These projects overlap to some degree, of course. Indeed, Herbert Simon's canonical work (see Simon, 1957, 1986, 1990) addresses both normative and descriptive issues, as does the more recent work of Paul Weirich (2004). This book is complementary to their work, but its focus is much further away from the details of cognitive mechanisms.

rationality conditions on belief and desire appropriate to them. Each set of claims is supported by a representation theorem showing how the quantitative claims under consideration have foundations in rationality constraints on relational attitudes of belief and desire.

Part III considers how a prospectively rational agent interacts with the world, applying the framework developed in Part II. There are three aspects to this. The first is the semantic issue of how the agent represents the prospects that are the objects of her attitudes. The second is the issue of how agents should evaluate their own interventions in the world and make decisions on the basis of such evaluations. The third is the effect of experience on the agent's attitudes – i.e. of how she learns from experience. Chapter 8 deals with the first by explaining how conditional prospects are modelled in multidimensional possible-world semantics and showing that this allows for non-trivial satisfaction of the rationality claims made in the second part of the book. Chapter 9 gives the core account of decision making under risk and uncertainty, showing that the theories of John von Neumann and Oskar Morgenstern and of Savage can be derived within the framework of the book in the presence of the special assumptions about the objects of choice, and deriving a particular formulation of causal decision theory. Chapter 10 develops the Bayesian theory of learning, defending forms of conditionalisation appropriate to a variety of different learning experiences.

Part IV develops an account of the rationality of bounded agents: agents that lack full awareness and who are not maximally opinionated. Chapter 11 defends a version of Imprecise Bayesian, seeking foundations for it in the notion of coherently extendable preferences. But it also raises a number of challenges for Imprecise Bayesianism; challenges that are taken up in subsequent chapters. Chapter 12 examines how an agent with imprecise beliefs and desires changes her mind in response to experience, developing a broadly Bayesian account of attitude formation and withdrawal that complements the standard accounts of attitude change. Chapter 13 looks at how such an agent might make decisions, comparing the strategy of making up one's mind with that of applying an alternative decision rule. Special attention is paid to considerations of caution and to the question of whether and how caution can be rationalised. The final chapter argues that considerations of confidence must be drawn on to handle the challenges to Imprecise Bayesianism, to provide a basis for both belief revision and decision making that is appropriately sensitive to the agent's state of uncertainty both about what to believe and what to desire.

Readers can make their way through the book in different ways, depending on background and interests. Those impatient with philosophical preliminaries and with a good background in decision theory can jump right to Part II. Part II is fairly self-contained, but Parts III and IV depend on it. With Part II under your belt, it suffices to read Chapter 9 in order to read Part IV. Figure I.1 summarises these dependencies between chapters.

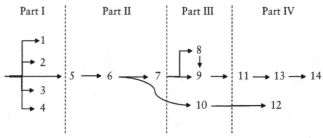

FIGURE I.1. Flow of Chapters

Some parts of the book are more technical than others. When these technical details are essential to the argument, I try to explain them fully. When they are not, I have placed them in a starred section, such as 5.2.2, Boolean Algebras (page 69).

PART I

RATIONALITY, UNCERTAINTY AND CHOICE

1

Decision Problems

I.I MAXIMISATION OF EXPECTED BENEFIT

Decision theory begins with decision problems. Decision problems arise for *agents*: entities with the resources to represent, evaluate and change the world around them in various different ways, typically within the context of ongoing personal and institutional projects, activities or responsibilities. These projects together with the environment, both natural and social, provide the givens for the decision problems agents face: their material resources for acting, their information and often their standards for evaluating outcomes, as well as the source of the problems they must respond to. Social scientists hold very different views about the relative importance of the different aspects of this background and the decisions that are made within it, but few would doubt that choices made by consumers, doctors, policy makers and so on have the power to shape the course of events.

To face a genuine decision problem, agents must have options: actions that they are capable of performing and, equally, of forgoing if they so choose. To get an idea of what sorts of things count as a decision problem, let's look at a few examples.

1. *Take a bus?* You have an appointment that you don't want to miss. If you walk you will arrive late. If you take the bus and the traffic is light, you should arrive ahead of time. On the other hand, if the traffic is heavy then you will arrive very late, perhaps so late that the appointment will be lost. Should you risk your appointment by taking the bus?

2. *Buy health insurance?* You are currently in good health but know that if you were to fall ill you might not be able to continue to earn an income and, in the worst case, you might not be able to afford the health care

TABLE 1.1. *Take a Bus?*

	Heavy traffic	Light traffic
Take a bus	Arrive late Pay for a ticket	Arrive early Pay for a ticket
Walk	Arrive a little late No ticket needed	Arrive a little late No ticket needed

you require. By buying health insurance you can ensure that you have all the care you need. But it is 'expensive' and if your health remains good, the money is wasted. Is it worth insuring yourself?

3. *Free condoms.* By supplying free condoms, rates of transmission of venereal disease can be considerably reduced. But there is the possibility that it will also encourage sexual activity, thereby partially, or perhaps even completely, offsetting the benefits of a decreased transmission rate by virtue of the increase in the number of sexual liaisons. Should they be supplied free of charge?

4. *Vaccinations.* A vaccine has been developed for cervical cancer, a fairly common type of cancer with a high mortality rate. The vaccine is expensive, but if it is developed as part of a large-scale vaccination programme the costs are not exorbitant. The vaccine does, however, have severe side effects in very rare cases (fewer than 1 in 100,000). Should the government offer the vaccine to everyone, actively encourage them to be vaccinated or even introduce compulsory vaccination?

Decision problems such as these can be described in the following way. A decision maker has one or more options before him. The exercise of each option is associated with a number of possible consequences; some of which are desirable from the perspective of the decision maker's goals, others are not. Which consequence will result from the exercise of an option depends on the prevailing features of the environment: whether traffic is light or heavy, how much it is raining, whether you fall ill, and so on.

Let us call the set of environmental features relevant to the determination of the consequences of the exercise of any of the options a 'state of the world'. Then a decision problem can be represented by a matrix showing, for each available option, the consequence that follows from its exercise in each relevant state of the world. In our first example, for instance, taking the bus has the consequence of having to buy a ticket and arriving late in the event of heavy traffic and of paying for a ticket and arriving early in the event of light traffic. This decision problem can be represented in a simple way as in

TABLE I.2. *State–Consequence Matrix*

Options	States			
	State S_1	State S_2	...	State S_n
α	A_1	A_2	...	A_n
β	B_1	B_2	...	B_n
...	...	...	...	...
γ	C_1	C_2	...	C_n

Table 1.1, where the consequences of the available options, for each of the states, are given in the table cells.

More generally, suppose that α, β, ..., and γ are the options open to the decision maker and that S_1 through S_n are n possible states of the world (these must be mutually exclusive and exhaust all the possibilities). For any option γ, let C_1 through C_n be the n consequences that might follow from exercising it. Then a decision problem can be represented by a state-consequence matrix of the kind displayed in Table 1.2.

Given a decision problem of this kind, standard decision theory says that the decision maker should choose the option whose exercise has the *greatest expected benefit*, where benefit is relative to the decision maker's evaluation of the desirability of the possible consequences of her actions. If she knows what the actual state of the world is then she should simply pick the option with the most desirable consequence in that state. Typically, however, the decision maker will be uncertain as to what the actual state is. In this case, she must consider how probable it is that each of the states is the case and pick the option whose expected desirability is greatest given these probability judgements.

For instance, suppose that I consider the probability of heavy traffic to be one-half and the benefit or utility of the various possible consequences to be as below:

	0.5	0.5
Take a bus	3	0
Walk	1	1

Then the expected benefit of taking the bus is a probability weighted average of the benefits of its possible consequences – i.e. $(3 \times 0.5) + (0 \times 0.5) = 1.5$. On the other hand, walking has a certain benefit of 1. So in this case I should take the bus. But had the probability of heavy traffic been a lot greater, walking would have been the better option.

TABLE 1.3. *Probability–Utility Matrix*

	Probabilities of states		
Options	$P(S_1)$	...	$P(S_n)$
α	$U(A_1)$	...	$U(A_n)$
β	$U(B_1)$	...	$U(B_n)$
...	...	...	...
γ	$U(C_1)$	...	$U(C_n)$

The next couple of chapters will be devoted to qualifying, expanding and commenting on the claim illustrated in this simple example, namely that we should pick the option that maximises expected utility. But before we do so, it will be helpful to express it more formally so that the core content is clear. Let P be a probability measure on the states of the world and U a utility measure on consequences (we will say more about what these measures are and where they come from in due course). Then a state–consequence matrix, such as that of Table 1.2, induces another matrix in which options appear as random variables: functions that assign a utility value to each state of the world (intuitively, the utility of the consequence of exercising the option in question in that state). This matrix is given in Table 1.3.

So represented, each option has an expected value that is jointly determined by the functions U and P. For instance, the expected value of option γ, denoted by $\mathbb{E}(\gamma)$, is given by $U(C_1) \cdot P(S_1) + ... + U(C_n) \cdot P(S_n)$. More generally, if the number of possible states of the world is finite:[1]

$$\mathbb{E}(\gamma) = \sum_{i=1}^{n} U(C_i).P(S_i)$$

Now what decision theory says is that rational agents should choose the option with the highest expected value. This is known as the **maximisation of expected utility hypothesis**.

The maximisation hypothesis forms the core of Bayesian decision theory, together with claims about how uncertainty should be represented and resolved through learning (respectively discussed in Chapters 3 and 10). I will argue that this hypothesis is essentially correct for cases in which we can adequately represent the decision problem we face in a manner similar to that of Table 1.2 and Table 1.3 – i.e. when we can display the problem in

---

[1]   The restriction to a finite number of states of the world is made for simplicity, but the expected value will still be well defined even if we drop it.

a state–consequence matrix and can reach probability and utility judgements on all the relevant factors displayed in it. When we cannot (which is quite often the case) then the theory is not false but inapplicable, and much of the last part of this book will be devoted to answering the question as to what we do then. But for the moment our focus will be on understanding what the maximisation of expected utility hypothesis says, examining in this chapter how decision problems should framed and, in the next, how the hypothesis should be interpreted and what notion of rationality it presupposes.

1.2 FRAMING DECISIONS

Decision theory makes a claim about what option(s) it is rational to choose when the decision problem faced by the agent can be represented by a state–consequence matrix of the kind exemplified by Table 1.2. It is very important to stress that the theory does not say that you *must* frame decision problems in this way. Nor does it say that agents *will* always do so. It just says that, *if* they are framed in this way, then only options which maximise expected benefit should be chosen. Nothing precludes the possibility that the same decision situation can be framed in different ways. This is true in more than one sense.

Firstly, it may be that the problem is not naturally represented by a state–consequence decision matrix. As John Broome (1991) points out, the consequences of an action may be distributed across different times or places or people, as well as across states. The desirability of ordering a cold beer or not, for instance, will depend on the location of its consequences: it's good if the beer is served to me, in the evening, with a smile and when I have not had a few too many already; bad when it's for my children, or first thing in the morning or during a philosophy lecture. In this case, my decision problem is better represented by a matrix that associates each action and relevant combination of locations (person, time, place, etc.) with a consequence, rather than by a simple state–consequence one.

Secondly, the problem may not be representable by any kind of decision matrix at all because we are unable to identify the various elements of it: what our options are, what the relevant factors are that determine the consequence of each option, or what the consequences are of exercising one or another of the identified options when these factors are present. In particular, we may not be able to assign a determinate consequence to each state of the world for each option if the world is non-deterministic or if we cannot enumerate all the relevant conditions. This problem, of what I will term option uncertainty, is discussed in detail in Section 3.2.

Finally, it is typically possible to represent the decision problem one faces by any number of different decision matrices that differ in terms of the features of the problem that they explicitly pick out. This is true even if we just confine attention to state–consequence matrices (as I shall do), for our state of uncertainty can be more or less elaborately described by representing more or fewer of the contingencies upon which our decision might depend.

This last point raises the question of whether all such representations are equally good, or whether some are better than others. There are two claims that I want to make in this regard: firstly, that not all representations of a decision problem are equally good; and, secondly, that many representations are nonetheless permissible. This latter point is of some importance because it follows that an adequate decision theory must be 'tolerant' to some degree of the manner in which a problem is represented and that the solution it gives to a decision problem should be independent of the choice of representation

Let us start with the first claim, that some representations of a problem are better than others. A representation of a decision problem should help us arrive at a decision by highlighting certain features of the problem and, in particular, those upon which the decision depends. What makes one way of framing the problem better than another is simply that it is more helpful in this regard. There are at least two considerations that need to be traded off when talking about the usefulness of a representation: the expected quality of the decisions likely to be obtained and the efficiency of obtaining them. Let me say something about them both.

Quality: To make a good decision, a decision maker must give appropriate weight to the factors upon which the decision depends. In deciding whether to take an umbrella or not, for instance, I need to identify both the features of the possible outcomes of doing so that matter to me (e.g. getting wet versus staying dry) and the features of the environment upon which these outcomes depend (e.g. the eventuality of rain). Furthermore, I need to determine how significant these features are: how desirable staying dry is relative to getting wet, how probable it is that it will rain, and so on. If my representation of the decision problem is too sparse, I risk omitting features that are relevant to the decision. If I omit possible weather states from my representation of the umbrella-taking decision, for instance, then I may fail to take into account factors (in particular, the probability of rain) upon which the correctness of the decision depends. So, *ceteris paribus*, a representation that includes more relevant features will be better than one that does not.

Efficiency   One way of ensuring that no relevant features are omitted is simply to list *all* the features of possible outcomes and states of the world. But drawing up and making use of such a list is clearly beyond our human capabilities and those of any real agents. Reaching judgements costs in terms of time and effort. If we try to consider all possible features of the world we will simply run out of time and energy before making a decision. A framing that delivers accuracy but is so complex that it is impossible to specify all the required inputs and to compute the expected utilities is clearly not of much use. More generally, representations that include too many features will result in inefficient decision making requiring more resources than is justified (what level of resources is justified will of course depend on what is at stake). So, *ceteris paribus*, a simpler representation will be better than a more complicated one.

Achieving a good trade-off between quality and efficiency is not just a matter of getting the level of complexity right. It is also a matter of identifying the most useful features to represent explicitly. It is useful to represent a feature if it is (sufficiently) relevant to the decision and if we can determine what significance to attach to it. A feature of the state of the world or of a consequence is relevant to a decision problem if the choice of action is sensitive to values that we might reasonably assign to this feature (its probability or utility). More precisely, one feature is more relevant than another just in case the expected values of the various actions or options under consideration are more sensitive to changes in the values of the former than the latter. For instance, whether it is desirable to take an umbrella with me or not will be sensitive to the probability of rain, but not sensitive at all to the probability of a dust storm on Mars. Likewise it is sensitive to the utility of my getting wet but not to my getting hungry, since my getting wet depends causally on the taking of the umbrella but not my getting hungry. So a good representation of my decision problem will include weather states and 'wet/dry' consequences, but not Martian dust storm states or 'hungry' consequences.

The second aspect of usefulness is equally important. A representation should be appropriate to our informational resources and our cognitive capabilities in specifying features of the environment that we are capable of tracking and features of consequences that we are capable of evaluating. If the weather is relevant to my decision as to take an umbrella or not, but I am incapable of reaching a judgement as to whether it is likely to rain or not (perhaps I have no information relevant to the question or I don't understand the information I have been given), then there is little point in framing the decision problem in terms of weather contingencies. A good representation of a problem helps us to bring the judgements we are able to make to bear on the decision problem.

It follows that whether a framing is a useful one or not will depend on properties of the decision maker (and in more than one way). Firstly, whether the features of the problem it represents are relevant depends on what matters to the decision maker and hence what sort of considerations her decisions will be sensitive to. And, secondly, whether a representation facilitates decision making will depend on the cognitive abilities and resources of the decision maker. Both of these will vary from decision maker to decision maker and from one time and context to another.

It is clearly desirable therefore that a decision theory be representation-tolerant to as great a degree as possible, in the sense of being applicable to a decision problem irrespective of how it turns out to be useful for the decision maker to represent it. Not all decision theories are equal in this regard. On the contrary, as we shall see in the next section, some impose quite severe restrictions on how a decision problem must be represented if the theory is to be used and hence make considerable demands on the decision maker in terms of the number and complexity of judgements that he must reach. Given our aim of a decision theory with a human face, this feature will count heavily against such theories.

1.3 SAVAGE'S THEORY

The modern theory of decision making under uncertainty has its roots in eighteenth-century debates over the value of gambles, with Daniel Bernoulli (1954) giving the earliest precise statement of something akin to the principle of maximising expected utility. The first axiomatic derivation of an expected utility representation of preferences is due to Frank Ramsey (1990/1926) whose treatment in many ways surpasses those of later authors. But modern decision theory descends from Savage, not Ramsey, and it is in his book *The Foundations of Statistics* that we find the first simultaneous derivation of subjective probabilities and utilities from what are clearly candidate rationality conditions on preference.

It is to Leonard Savage too that we owe the representation of decision problems faced by agents under conditions of uncertainty that was described at the beginning of the chapter and that is now standard in decision theory. Its cornerstone is a tripartite distinction between states, consequences and actions. Consequences are the features of the world that the agent cares about and seeks to bring about or avoid by acting; they are, he says, 'anything that may happen to an agent' or 'anything at all about which the person could possibly be concerned' (Leonard Savage, 1974/1954, pp. 13–14). States are those features of the world that are outside the agent's control but determine what consequence follows from the choice of action. Actions are the link between the two; formally, for Savage, they are just functions from states to consequences.

Although the distinction between states, consequences and actions is natural and useful, Savage's theory imposes some quite stringent conditions on how these objects are to be conceived. Firstly, in order that decision problems be representable by state-consequence matrices of the kind given in Table 1.2, he requires that the states of the world suffice to determine the consequence of a choice of action (I will discuss this in more detail in the next chapter). Secondly, he requires that the states themselves be causally and probabilistically independent of the action performed. And thirdly, he requires that the desirability of consequences be independent both of the state of the world in which they are realised and of the action. Jointly these assumptions imply that actions differ in value only insofar as they determine different ordered sets of consequences.

To ensure that the second two conditions hold, Savage suggests that consequences be maximally specific with regard to all that matters to the agent so that there be no uncertainty about how beneficial or desirable the consequence

TABLE 1.4. *Insurance Purchase*

	Good health	Poor health
Purchase insurance	Earn full annual income Make policy payments	Reduced income Insurance pays out
Don't purchase	Earn full annual income No policy payments	Reduced income No payout

is that derives from uncertainty about the state of the world. It follows that the states themselves must be maximally specific, 'leaving no relevant aspect undescribed' (ibid. p. 14), for if this were not the case then there could be features of the consequences of actions that matter to the agent but which are not determined by the prevailing state. So when an agent regards a Savage-style action as open to her, she must take it that she can bring it about that a maximally specific consequence will obtain conditional on each maximally specific state of the world prevailing. This is rather different from what we colloquially understand by an action. When I must choose between walking to the shops or taking the bus, as in the decision problem represented by Table 1.1, I do not do so in the light of anything like full knowledge of the consequences, in each possible state of the world, of these actions. My understanding of them is inevitably coarse-grained to some extent. It would seem, then, that Savage's theory is not well suited to agents like us, who cannot typically represent decision problems in the way required for application of his theory.

Savage was perfectly aware of this objection and drew an important distinction between small-world and grand-world decision problems. Grand-world decision problems are ones which have consequences that are maximally specific with regard to all matters of concern to the agent; small-world problems are ones with coarse-grained specifications of states and consequences. Although his theory is designed for grand-world problems, Savage argues that it could nonetheless be applied to small-world problems, so long as we ensure that the coarse-grained representation of the decision problem has sufficiently similar properties to the fine-grained one that it could be given a numerical representation by a probability–utility matrix of the kind exhibited in Table 1.3. For this, the two conditions of probabilistic independence of states from actions and desirabilistic independence of consequences from states are essential.

It is quite easy to fall foul of these constraints. Suppose that we are deciding whether to purchase health insurance for the coming years and that we represent our decision problem by the state–consequence matrix displayed in Table 1.4. So represented, it looks like a purely financial decision, that can be made on the basis of the expected incomes associated with the two possible acts. It is quite conceivable, however, that the value we attach to income depends on our state of health. We might need more money if our health is poor, for instance, in order to buy services that we can no longer provide for ourselves. This would be a reason to value a particular income more highly if it is gained under poor health than if it is gained under good health. Alternatively, we may get less enjoyment from money when our health is poor, and so we would value it less. Either way, in order to use Savage's theory to make a decision as to whether to purchase insurance or not, in a way that appropriately reflects the sensitivity of the desirability of money on health states, the decision problem must be reframed.

The obvious way of doing this is to take the consequences of options to be combinations of outcomes and the states in which they are realised. The act of buying health insurance, for instance, may be said to have the consequence 'Earn full income, make policy payments, enjoy good health' in good-health states and the consequence 'Earn reduced income, make policy payments, enjoy poor health' in ill-health ones. For reasons that we will examine more closely later on, Savage requires, however, that consequences and states be logically independent. So he is forced to insist that decision makers describe the consequences of their actions in terms which eliminate the sensitivity of their value to the state of the world. This is not straightforward. The way in which income varies with health states is likely to be mediated by an enormous number of variables, including the amount of support that can be expected from friends and family, the services provided by the state, charities or other institutions to help those in poor health, and one's level of psychological well-being. All of these would have to be specified in order for the act of purchasing health insurance to have a state-independent consequence in each health state. We are rarely able to do this.

A second problem in our example concerns the description of the relevant states, for the purchase of health insurance can have a causal effect on how much care one takes of one's health, so that the probability of good health is not independent of the purchase of health insurance. This problem of *moral hazard*, as it often called, plagues insurance markets. When you sell someone fire insurance, for example, you change her incentives in such a way as to make it more probable that a fire will occur. Knowing that they will be reimbursed if a fire occurs, individuals may be less careful. In extreme cases, when the value of the policy is high enough, they may even commit arson. Insurance companies have to be very careful when selling fire insurance not to under-estimate their exposure. In order to eliminate the causal dependence of states on actions fuelling moral hazard the decision problem has to be reframed. In our example this would require identifying all those factors (genetic, environmental, historical) mediating the relationship between purchases of health insurance and health states, combinations of which would serve as states in the reframed decision problem. This can be very difficult to do.

The upshot is that Savage's theory is far from being representation-tolerant in the way that I argued was desirable. It is often possible to ensure that for all practical purposes any one of the three conditions required for application of his theory can be met by being careful about how the decision problem is framed. But ensuring that all three are satisfied at the same time is very difficult indeed, since the demands they impose on the description of the decision problem pull in different directions. Ensuring a determinate consequence for each state is most easily achieved by coarsening the description of outcomes, for instance, but ensuring that they have a state-independent utility requires refining them.

This problem provides strong grounds for turning our attention to a rival version of Bayesian decision theory that is due to Richard Jeffrey and Ethan Bolker. Jeffrey (1990/65) makes two modifications to the Savage framework. First, he recognises that the distinction between states and consequences is both context-and agent-dependent: that it will rain is a possible state of interest to a farmer, but a consequence for a shaman with a rain dance repertoire; that there will be flooding in low-lying areas is a possible state of the world from the perspective of a person buying a house, but a consequence from the point of view of environmental policy. So, instead of distinguishing between the objects of belief and those of desire, he takes the contents of all the decision maker's attitudes to be propositions. This small modification has a very important implication. Since events and consequences are logically interrelated in virtue of being the same kind of object, the dependence of the desirabilities of consequences on states is built into Jeffrey's framework. This means that his theory must dispense with the second of the restrictions required for Savage's.

The second modification that Jeffrey makes is more contentious. If he followed Savage in defining actions as arbitrary functions from partitions of events to consequences, the fact that in principle any proposition could serve as a consequence would imply an explosion in the size of the set of actions. But Jeffrey argues that many of the actions so defined would be inconsistent with the causal beliefs of the decision maker. Someone may think she has the option of making it true that if the traffic is light she will arrive on time for her appointment, and if it's heavy she will arrive late, but not believe that it is possible to make it true that if the traffic is light she arrives late, and if it's heavy she arrives on time. So, instead, Jeffrey conceives of actions as simply those propositions that can be made true at will, characterised for decision purposes by the probabilities and utilities (or desirabilities, as Jeffrey calls them) of the possible consequences that might be brought about by the action.

Two features of this treatment are noteworthy. Firstly, it is not required that the consequence of an action in each state be known in order that a decision be made. All that is required is that the agent have conditional probabilities, given the performance of the act, for the possible consequences of interest to her. This relaxes the first constraint on the applicability of Savage's theory.

Secondly, it is no longer required that the states of the world be probabilistically independent of the available actions. On the contrary, as Jeffrey sees it, actions matter precisely because they influence the probabilities of states (if you like, the consequences of acting *are* changed probabilities of states). This dispenses with the third constraint on the applicability of Savage's theory.

The fact that Jeffrey's theory imposes much weaker requirements on the framing of decision problems is my primary reason for preferring his framework to Savage's for developing a decision theory with a human face. There are other advantages too, such as the simplicity and flexibility of working with sets of propositions and the fact that the foundational representation theorems for his theory require much weaker assumptions about rational preference. But such flexibility does not come without cost. In particular, as we shall see, it opens up the question of exactly how acts should be evaluated, a matter of some controversy. So in the next part of the book I will develop a version of Bayesian decision theory that follows Jeffrey's in defining degrees of belief and desire on a common Boolean algebra of prospects (his propositions). But I will show how it is possible to extend the set of prospects in a way that allows for the re-introduction of Savage-style acts and a formulation of a state-dependent version of his theory. This richer theory is the one that I will defend as giving the best account of ideal rational agency.

2

Rationality

2.1 MODERATE HUMEANISM

The maximisation of expected utility hypothesis brings together two separate claims. The first concerns what rationality requires of the relation between the agent's preferences between different prospects and her beliefs and desires. Stripped of mathematical baggage, the claim can be expressed as follows:

> Rationality Hypothesis Rationality requires of an agent that she prefer one prospect over another if and only if the expectation of benefit conditional on the truth of the former is greater than the expectation of benefit conditional on the truth of the latter, relative to her degrees of belief and desire.

The Rationality Hypothesis is generally taken to express nothing more than a consistency requirement on the agent's preferences, akin to the requirements that logic places on her beliefs. Consistency requirements are purely formal in nature and place no substantial constraints on the content of any preference, belief or desire taken in isolation. Moreover, the constraints that they place on sets of such preferences, beliefs and desires are not such as to rule out many that we might be inclined to regard as defective in some way; for instance, because they are immoral, self-destructive or just plain ill-considered.

In ordinary talk we tend to be more demanding and speak of beliefs as irrational, even if they are consistent, because they fail to meet some or other standard of adequacy. For instance, we might be inclined to criticise someone for not taking into account all available evidence or for failing to give the long-term consequences of their choices sufficient weight. Such talk, it seems to me, runs together two types of requirements that are best kept separate. One is the requirement that we recognise all the available evidence and that we give appropriate weight to all the possible consequences of our actions; the other that our beliefs be consistent with all the evidence that we recognise and that our preferences for actions be consistent with the weight that we give to each of their possible consequences. The former is a requirement that our judgements respond in an adequate way to the world as it is presented to us; the latter that they fit together in a coherent way.

21

Let us call requirements of the first kind *external* requirements, and of the second *internal* requirements. A very demanding example of an external requirement is that we form only true beliefs and correct desires. A weaker one would be that we form only beliefs and desires that we have reason to form. Yet another that beliefs and desires be formed by a reliable process, one that tends to produce beliefs that are true and desires that promote some benefit. Whatever their merits, it is doubtful that someone who fails to satisfy any of these requirements is irrational. Possibly deficient in some way, but not irrational. The fact that the public transport is not running gives me reason to take my car to work. But if I don't know about the state of the public transport, then my choosing to use it is not irrational. And if I do know about it, then my irrationality stems from the violation of the consistency requirement encoded in the maximisation of expected utility hypothesis: I fail to do what seems best by my own lights. Could it not be said to be irrational not to find out about the state of public transport? Only if I have beliefs (such as that the system is unreliable) that make it rational by the lights of the maximisation hypothesis to seek more information.

The view that rationality places only formal constraints on our attitudes contrasts with the view that beliefs and desires that are contrary to reason are irrational. A belief or desire is contrary to reason when there is a decisive reason not have it. Derek Parfit (2013) offers the example of the person who cares about his future pains and pleasure except when they will occur on a Tuesday; not because he believes that the pains and pleasures on Tuesday will be less painful or pleasurable, but simply because of the day on which they occur. As he prefers a pain on Tuesday to one on Wednesday even though he has no reason to, his preferences are, according to Parfit, irrational. Now, there is no doubting the odd nature of this person's preferences. But are they irrational? That depends on whether he recognises that pains on Tuesday should weigh equally as pains on Wednesday. If he does, then he is being inconsistent. If not, then he is no more irrational than the person who prefers chocolate to strawberry ice cream, not because of the difference in taste but simply because it's chocolate rather than strawberry. Being Tuesday or being chocolate is, for these people, an intrinsic reason for preference. They may be mistaken, but being wrong is not the same as being irrational.[1]

John Broome (1999) gives the name 'moderate Humeanism' to the view expressed here – that rationality constrains our attitudes only indirectly by disallowing certain combinations of beliefs, desires and preferences – and argues that it is not a viable position to hold. His argument is that consistency conditions cannot constrain our attitudes at all unless rationality sets some

---

[1] All of this underscores the extent to which the project of determining the rationality requirements on agents' attitudes is a truly modest one. Consistency is an important property of attitudes, but not *that* important. It is no doubt sometimes better to be warm-hearted or generous than to be consistent, and a thorough-going consequentialist will not attempt to maximise consistency to the exclusion of all else.

limits to what kinds of distinctions between prospects can support different attitudes to them. There is something right and something wrong about this claim. It is true that without some requirements of indifference, as Broome calls them, consistency cannot constrain our attitudes. But these requirements of indifference do not have to be requirements of *rationality*. Substantial value commitments will do the job – e.g. to treating people impartially or to taking care of oneself. It is no part of the theory of rationality that we should have one value commitment or another, but once we do then formal consistency conditions such as transitivity will work to constrain preferences in all kinds of ways.

The point is more general. To say that rationality qua consistency cannot arbitrate between different sets of beliefs and preferences is not to say that we have no grounds upon which to do so. We certainly can criticise people for failing to take account of the reasons that they have for preferring one prospect to another just as we can criticise people for failing to attend to the evidence adequately in forming their beliefs. In doing so we might appeal to external requirements on their preferences; to the facts (as we see them) about what is worthy of preference. Such appeals need not involve adopting a value standpoint which the agent rejects. Suppose I am a hedonist and regard the consumption of Cassoulet on a cool evening as the greatest pleasure. Another may criticise me, by saying that I had failed to properly attend to the superior qualities of freshly grilled sea bass served on a warm evening by the sea, perhaps because of a cultural bias or insufficient experimentation. They criticise me, not for possessing inconsistent preferences, but for poor judgement or poor application of my own values. So too might one be criticised for failing to live up to one's moral commitments or to appreciate what they require of one.

Moderate Humeanism should also not be confused with other views to which the label 'Humean' has been attached. In particular, it does not entail the Humean theory of motivation, according to which belief is never sufficient to motivate action, requiring the presence of desire. Whether this theory is correct or not is largely a matter of empirical psychology, something on which a theory of rationality cannot legislate. Nor does moderate Humeanism entail either moral non-cognitivism or moral subjectivism, both sometimes attributed to the 'Humean' view (more on this later). Finally, it is frequently said to be part of the Humean view that not only are preferences and desires distinct from beliefs, but beliefs do not constrain preferences or desires at all (or vice versa). In his recent book, Ken Binmore (2008) calls this Aesop's Principle and gives the following statement of it:

> Aesop's Principle Preferences, beliefs and assessments of what is feasible, should all be independent of each other.

The conviction that Aesop's Principle is fundamental to decision theory seems widespread (as we have seen, Savage embraces it). But, as Binmore himself notes, it is easy enough to find objections to the principle. Indeed, not only does decision theory *not* generally require independence of preference from belief, it requires that preferences *be* sensitive to it. What I believe about the weather conditions should influence my preferences over clothing, what I believe about the freshness of the food being served at different restaurants should influence my preferences about which of them to frequent, and so on.

What Binmore really means is that a particular class of preferences is governed by Aesop's Principle, namely fundamental or intrinsic preferences. A preference for one thing over another is intrinsic, according to Binmore, if nothing we can learn would change it.[2] They are thus unconditional in the sense that they do not depend on some or other condition being satisfied or, more exactly, on the belief that the condition is satisfied. In contrast, instrumental preferences are preferences for prospects that do depend on them being a means to some other good. They are thus conditional on one's beliefs about the kinds of things that make them more or less efficacious as a means. I like to eat at a local Italian restaurant because I expect to get a tasty meal there. Its desirability derives from being instrumental to tasty experiences and is, therefore, conditional on the quality of the cooking not declining and fresh ingredients having been delivered that day. Many prospects are both instrumentally and intrinsically desirable. I take the dog for a walk because it gives both of us the exercise that we need and because I like doing it. If the need for exercise were removed, I would still walk, but less frequently, and not when the weather was foul.

Once we restrict Aesop's Principle to intrinsic preferences, it becomes more or less empty, since it seems to be part of the definition of an intrinsic preference that it should satisfy the principle. The substantive issue is whether there are any preferences that we generally hold that are fundamental in the required sense. I don't see any reason to believe that there are. Being wealthy, attractive and in good health are no doubt all things that we might desire under a wide range of circumstances, but not in circumstances when these arouse such envy that others will seek to kill us or when they are brought about at great suffering to ourselves or others. Even rather basic preferences such as for chocolate over strawberry ice cream are contingent on beliefs. In any case, the important point is that the maximisation principle itself does not require that there be prospects that are intrinsically desirable. It requires our preferences for actions to be consistent with the value we attach to their consequences, but not that the value that we attach to these consequences be unconditional, non-revisable or fundamental.

2.2 THE CHOICE PRINCIPLE

The Rationality Hypothesis alone does not say anything about what agents should or will do. For this a second claim, connecting preference to choice, is required.

---

[2] Binmore says nothing that can happen would change it, but this is too strong. Even intrinsic preferences could be changed by a blow to the head or some other non-informational disturbance to mental states.

Choice Principle Of the options available to an agent, she *should/will* select (one of) those that she most prefers.

The Choice Principle states either a descriptive or a normative claim, which together with the Rationality Hypothesis yields descriptive and normative versions of expected utility theory. To assess them, it is essential to be clear about the interpretation of the maximisation hypothesis being worked with. In fact, at least two quite different interpretations of expected utility theory have been doing the rounds for some time. In one usage, maximisation of expected utility means something like doing what is in one's best interest, be this a matter of experiencing pleasure and avoiding pain, or of acquiring wealth, power and reputation or of having a high level of welfare or well-being. In another usage, maximising expected utility is a matter of doing what one thinks is best, all things considered, in the light of one's beliefs and preferences. The two are quite distinct. The act that one thinks best may not be one that is in one's self-interest (e.g. lending money to an unreliable friend), and vice versa.

More generally, two broad classes of interpretations of the utilities and probabilities figuring in the statement of the maximisation hypothesis are to be found in the literature. On *empirical* interpretations, probabilities and utilities are features of the world relevant to the agent's decision. In debates on probability, for instance, the views that they are long-run frequencies, that they are propensities of physical systems and that they are objective chances of events all belong to this group. Important empirical interpretations of utility include the views that they are hedonic states, that they are degrees of preference satisfaction and that they are measures of well-being or welfare.

On *judgemental* interpretations, on the other hand, probabilities and utilities are features of judgements or states of mind. Bayesians, for instance, view them as measures, respectively, of the decision maker's degrees of belief in the various possible states of the world and degrees of preference or desire for the possible consequences. Bayesianism is the predominant view in contemporary decision theory, but other judgemental views have been important in probability theory: in particular, the 'logical' interpretation of conditional probability as a measure of degree of confirmation or entailment between propositions.

For present purposes an equally important distinction is between interpretations, such as the empirical and logical ones, which imply that probability and/or utility is something objective ('in the world'), and hence something that one can be right or wrong about, and interpretations, such as the Bayesian one, which view them as features of subjective judgement ('in the head'). In principle, a subjective interpretation of utility could be combined with an objective interpretation of probability, and vice versa. So even this crude subjective – objective distinction allows for four different values to be attached to an action, at least three of which have figured prominently in applications of decision theory. Mainstream Bayesian decision theory is doubly subjective, but in the von Neumann and Morgenstern theory, probabilities are objective and in the social ethics of John Harsanyi, utilities (qua welfare) are open to an objective interpretation. There has been much debate over the correct interpretation of both probability and of utility, but I see no reason to think that there should only be one correct construal of either notion. It is better to regard probability and utility as formal notions which can in principle admit of more than one interpretation and debate the appropriateness or usefulness of each for particular applications.

In this regard, three questions are of immediate importance. Which interpretations are appropriate to the Rationality Hypothesis? Which interpretations explicate the role played by the Principle of Choice in the description and explanation of action? And which support its application to normative problems of choice? My earlier claim that rationality is a matter of consistency in one's judgements, not of right relation to features of the world, commits me to a subjective interpretation of the first – the Rationality Hypothesis. But before exploring the exact nature of such a subjective interpretation, let me make a few comments about the other two questions, without claiming thereby to do them proper justice.

Both subjective and objective interpretations of the maximisation hypothesis are often applied descriptively, to the explanation of human behaviour. This has been the cause of a good deal of confusion and misdirected critical discussion. The claim that agents maximise the objective expectation of utility is clearly much stronger than the claim that they maximise their subjective expectation of it. Moreover, there is little doubt that the former claim is false, since false belief is an important causal factor in people's choices. This doesn't mean that these applications are of no explanatory use. There may be contexts in which the hypothesis that agents maximise some kind of objective expected utility (e.g. self-interest) yields good approximations of actual behaviour, perhaps because relevant information is easily accessible or because agents have opportunities to correct their judgements. But in these cases the 'deeper' explanation resides in the subjective version of the maximisation hypothesis, which has the resources both to explain why agents sometimes make the choices that cohere with objective criteria and why sometimes they do not. The problem for the subjective version, on the other hand, is that its claims are notoriously difficult to test. Its critics, consequently, are divided into those who claim that it is unfalsifiable and therefore unscientific and those who claim that it has been falsified.

The normative version of the Choice Principle also makes quite different claims, depending on the interpretation given to expected utility. In this case, however, it is the subjective version that faces the most difficulty. The objection is obvious. Why should subjective expected utility serve as the measure of the choice-worthiness of an action and not, for instance, welfare or moral worth? More bluntly: on many standards of what it is best for someone to do, the best action will *not* be the one that maximises subjective expected utility. When people have false beliefs, choice in accordance with expected utility may lead to a very poor outcome for them – e.g. when they mistake the vinegar for wine and drink it. If this is the case, then surely they should not pick the option that maximises subjective expected utility over an alternative that will in fact deliver a better outcome, even by their own value standards.

The claim that agents should maximise the objective expectation of utility is less vulnerable to this objection. Still, it might be argued that we should pick the option that will in fact have the best outcome, not the option with the most favourable expectations. The person who picks the lottery ticket with the greatest expected pay-off may well find herself wishing that she had picked differently. 'I should have chosen the other ticket' is a reasonable thing to say when it turns out that the alternative was the winning ticket. We might say, parroting Frank Ramsey (1990/1926), that, if asked what option we should choose, we should answer: 'The one that will have the best outcome.' But this violates the dictum of 'Ought implies can': in situations of objective uncertainty, we simply cannot know what the outcome will in fact be. The objective probabilities characterising a lottery express the limits of humanly attainable knowledge. The most that we can be asked to do to is to make the best attainable judgements and decide consistently on the basis of those.

But why stop there? If the 'Ought implies can' dictum can be used to defend objective maximisation, it can also be used in defence of subjective maximisation. For, at the time of making a decision, knowledge of the true probabilities and utilities may be impossible. We are where we are, with the judgements that we have arrived at, and at the moment when the decision must be made the best that we can do is act consistently on the basis of those judgements. From the agent's own perspective, to maximise subjective expected utility just *is* to do what is best on her estimation. This is not inconsistent with the possibility that others will have different views about what that agent should do nor with the claim that they should have done more to improve their opinions. When I say to someone 'You should do x' I am saying something like 'If I were the decision maker I should do x'. When I am better informed than them, they would do well to listen to what I recommend; indeed, consistency will demand it of them if they actually believe that I am better informed. No one is more expert than the Truth, so, when it speaks, we should all listen. But this is not the same as saying we are rationally required to hear it.

2.3 SUBJECTIVISM

The view that the probabilities and utilities figuring in the maximisation hypothesis are the agent's degrees of belief and desire is the predominant one in contemporary decision theory. But this view is only partially correct. To get a handle on what is at stake, let us look at how the quantities occurring in a probability–utility matrix should be interpreted. Here a slightly different interpretation is required depending on whether we view the issue from the perspective of the decision maker or from that of an observer. If a decision maker wants to evaluate an action in the manner suggested by the maximisation hypothesis, she must arrive at judgements about the relative likelihoods of the various possible states of the world and the desirability of the various possible consequences of her action. The probabilities and utilities figuring in the calculation of the expected utilities of actions are thus her *judgements*. An assignment of probability x to state S, for example, is a judgement that S is likely to degree x to be the actual state of affairs. Similarly, an assignment of y to consequence C is a judgement that C is desirable to degree y.

When an observer models the choice confronting the decision maker she can do so either from her own point of view or from that of the decision maker. In the former case, she is adopting a first-person perspective on the choice problem, and so once again the probabilities and utilities she employs are her judgements of likelihood and desirability of truth. In the latter, the probabilities and utilities she writes down are (her estimates of) the decision maker's degrees of belief and preference or desire. What makes it appropriate to model an agent in this way – by imputation of degrees of belief and desire of a particular magnitude to the agent – is the fact that the attributed states play the right kind of causal role in the production of her actions. By 'right kind of role' I mean that they explain, on the assumption that she maximises subjective expected utility, the pattern of choices that she makes. To play this role it is not essential that they be formed as a result of a conscious judgement on the part of the agent. They could, for instance, be part of the agent's cultural or biological inheritance encoded as behavioural dispositions. So it is possible to model the decision making of creatures in terms of maximisation of expected utility even if these creatures don't themselves have the cognitive resources to model the choice problem for themselves. That is, we can adopt a third-person perspective on the utility-maximising actions of agents who do not themselves have the corresponding first-person perspective on the decision problem. When we explain an animal's food choices, for instance, we can offer an explanation in terms of its beliefs about what plants are fit to eat, even if the animal doesn't have a concept of 'fit to eat'. But such an explanation is often less satisfactory than one which is couched in terms of the concepts recognised by the agents themselves. If the animal prefers green foods over red ones, and green foods happen to be those that are fit to eat, then it will be possible to explain its choices in terms of what is fit to eat (that is, such an explanation will cohere with the pattern of its choices) even when it is its colour-of-food judgements that are causally responsible for the development of its choice dispositions.[3]

Both of these two essentially subjective interpretations of probability and utility, as judgements and as mental states, offer an appropriate interpretation of the maximisation hypothesis as a claim about rationality. On the judgement interpretation, it says that rationality requires of agents that they judge actions to be desirable to the degree that they can be expected to have desirable consequences, given how likely they judge the possible states of the world to

---

3 I take this to be the heart of the claim of hermeneutic philosophies that explanation of human action requires understanding, glossed here as identifying the categories that the agent herself uses to formulate the decision problem that she faces.

be and how desirable they judge the possible consequences. Similarly, on the mental state interpretation, the hypothesis says that an agent is rational only if the value she attaches to each action is its expected desirability, relative to her degrees of belief and desire.

The two interpretations are quite closely related, and it is perhaps not surprising that they are not clearly separated in Bayesian decision theory, the predominant subjectivist view. Indeed, since making a judgement normally leads to forming a corresponding belief or desire, it is rather natural to think of the judgement interpretation as just a special case of the belief-desire one. But it is a mistake to do so: although the latter view is the correct one to take in regard to modelling other agents' decisions, it is not satisfactory for first-person normative applications. When we try to make up our mind about what action to perform by attaching utilities to consequences and probabilities to states we are not aiming to describe our own attitudes but to determine what the relevant features of the decision problem are: whether some condition is likely to hold, whether one consequence is preferable to another, and so on. We are making a judgement about the *world*, not about ourselves, and it is accuracy with regard to the former, not the latter, that concerns us. For this reason, the right interpretation of notions such as probability and utility, in these applications, is as judgements of a particular kind.

I stress this rather subtle distinction because of its implications for a related issue. Many Bayesians not only adopt a subjective interpretation of probabilities and utilities but also deny the existence of objective probabilities and utilities of any kind – a view that is known as Subjectivism. Subjectivism has had a number of famous advocates, including Bruno de Finetti, Leonard Savage and Richard Jeffrey, but although the arguments for and against their position are quite well known (in probability theory at least) there has been little recognition of an important ambiguity in it. When subjectivists hold that probabilities and utilities are 'in the head' rather than the world, they can mean two quite distinct things. On one (cognitivist) interpretation, a statement such as 'The probability of rain is one half' is true or false depending on what the agent believes. That is to say, probability and desirability statements are truth-susceptible propositions about the mental states of agents. Hence, both refer to that part of the world occupied by the agent's head. On a second (expressivist) interpretation, such statements do not make descriptive claims at all. Rather, they express an evaluative judgement by the agent that is not susceptible to truth or falsity. Judgemental probabilities are not in the (material) world at all.

Our earlier discussion suggested that both views are tenable. When a probability statement is made in the context of describing, from an observer's point of view, the attitudinal state of an agent, then the statement should be read as a description of the agent that is either true or false of her. On the other hand, when the agent herself makes such a statement, say in the context of thinking through a decision problem, then she should not be read as describing her own state of mind but, rather, as making her mind up by reaching an opinion on features of her environment.

Many decision theorists not only fail to recognise these distinct possibilities but, perhaps as a consequence, adopt a rather extreme subjectivist position on value – a view that I will dub Value Subjectivism. Just as expressivist subjectivists about probability argue that an assertion about probability is an expression of partial belief, not a claim about a feature of the world, ethical subjectivists about value view desirability statements as expressions of preference rather than assertions about some objective value. But they go a step further. Subjectivists about probability typically do not deny that there are objective features of the world that are tracked by probability judgements, just that these features are themselves probabilities. They take probability judgements to be subjective judgements on objective but non-probabilistic facts. Ethical subjectivists, on the other hand, deny not only that there are objective utilities but also that any objective feature of the world at all is tracked by utilities. Utility judgements, on their view, are not subjective judgements of the degree to which the world conforms to one or more objective value standard but (bare) expressions of the agent's subjective tastes or emotions.

This is a much stronger view than the kind of moderate Humeanism that I was defending earlier on. Moderate Humeans hold that the only constraints that rationality places on desirability judgements are formal ones. This, I argued, was consistent with the view that these judgements may be better or worse with respect to satisfaction of external requirements of one kind or another. Value Subjectivism implies a denial of this latter view, since the only requirements it recognises are those of consistency. There is no reason why a subjectivist, even of the expressivist variety, should accept this view. One may consider that utilities express a judgement on the part of the agent and at the same time deem that this judgement can be more or less adequate in the extent to which it coheres with, or tracks, some kind of objective value. All that a subjectivist about utility needs to deny is that a utility judgement is a belief that something has a certain objective utility. But utility judgements can be subjective judgements concerning objective properties of the world, so long as these properties are not themselves utilities, just as probability judgements can be subjective judgements on the facts without these facts having the structure of probabilities.

Finally, let me emphasise that the adoption of one or another subjective interpretation of the Rationality Hypothesis does not entail a commitment to Subjectivism in any of its forms. There are conceptions of objective probability (e.g. as frequencies or chances) and of objective desirability or utility (e.g. as well-being or goodness) that play an important role in decision theory, both as properties of states of affairs that agents do in fact care about and perhaps as properties they should take into consideration. In particular, it is hard to deny that we do experience some uncertainty as objective. This fact can, and should, be accommodated by decision theory, even one which adopts a subjective interpretation of the main decision variables.

3

Uncertainty

Uncertainty is a pervasive feature of human life and almost all our decisions must be made without certainty about what the consequences of our actions will be for ourselves or others. Human attitudes such as hope, fear and even regret depend on it. Despite this, philosophy has given little attention to uncertainty, largely treating it as just lack of certainty (apparently the real interest). In both the mathematical and empirical sciences, on the other hand, the emphasis has been on the development of techniques to manage it. So central has the concept of probability been to this enterprise that it has come to seem as if uncertainty was nothing other than the flipside of probability.

The concept of probability emerged surprisingly late in human history, in the seventeenth-century work of Blaise Pascal and Pierre de Fermat.[1] In this work, and much that followed, probability was conceived both in terms of the stochastic properties of chance processes, such as dice rolls and card deals, and the properties of beliefs about events regarding which full knowledge was lacking. In time this hardened into a distinction between two different forms of uncertainty: objective or *aleatory* uncertainty, which derives from features of the world (indeterminacy, randomness), and subjective or *epistemic* uncertainty, which derives from lack of information about it. In modern decision theory, there is a dominant theory of how each form of uncertainty should be quantified (in both cases, by a probability measure) and of how, so quantified, it should weigh in the evaluation of actions. For situations of objective uncertainty (or risk, as it typically called) decision theorists look to the version of expected utility theory originally due to von Neumann & Morgenstern (2007/1944), while for those characterised by epistemic uncertainty they look to subjective expected utility theory, whose classic statement is to be found in the work of Leonard Savage (1974/1954).

This distinction between risk and epistemic uncertainty, important and useful though it may be, does not remotely do justice to the variety of forms

[1] See Hacking (2006) for an unsurpassed history of probability.

and degrees of uncertainty relevant to decision making. Firstly, epistemic uncertainty comes in different degrees of severity that derive from differences in the quantity and quality of information that we hold. There is a significant difference, for instance, between being unsure about when people will arrive because one lacks precise information about their time of departure, traffic conditions, and so on, and having absolutely no idea when they will arrive because you don't know when or whether they have left, whether they are walking or driving or, indeed, whether they even intend to come. In the former case, the information one holds is such as to make it possible to assign reasonable probabilities to the people arriving within various time intervals. In the latter, one has no basis at all for assigning probabilities, a situation of radical uncertainty or *ignorance*. It may be rare for us to be totally ignorant, but situations of partial ignorance, or *ambiguity*, in which the decision maker is unable to assign determinate probabilities to all relevant contingencies, are both common and important.

Secondly, according to some critics, Bayesian theory fails to distinguish between the different levels of confidence we might have, or have reason to have, in our probability judgements. Compare a situation in which we are presented with a coin about which we know nothing to one in which we are allowed to conduct lengthy trials with it. In both situations we might ascribe probability one-half to it landing heads on the next toss: in the first case for reasons of symmetry, in the second because the frequency of heads in the trials was roughly 50 per cent. It seems reasonable, however, to say that our probability ascriptions are more reliable in the second case than the first and hence that we should feel more confident in them.

Both of these issues will be discussed in detail in the second half of the book. The focus of my concern now will be a third issue. Decision makers confront uncertainty not just concerning what is the case (empirical or factual uncertainty) but also what should be the case (evaluative uncertainty), what could be the case (modal uncertainty) and what would be the case if we were to make an intervention of some kind (option uncertainty). Almost all discussion of uncertainty is directed at the first of these only. The others are just as important, however, and so I shall attempt in this chapter to say something about them, exploring the questions of how they should be captured and whether they can be reduced to a form of empirical uncertainty.[2]

3.1   EVALUATIVE UNCERTAINTY

Although the distinction between certainty and uncertainty is typically used only to characterise the agent's state of knowledge of the world, it is equally important to distinguish cases in which consequences have known, or given,

---

[2]   Other typologies of uncertainty can be found in Hansson (2016) and Bradley & Drechsler (2014).

objective values and those in which these values are either unknown and the decision maker must rely on subjective evaluations of them, or do not exist and the decision maker must construct them. The possibility of such evaluative uncertainty is typically ignored by decision theorists, because of their (often unconscious) attachment to Value Subjectivism, the aforementioned view that values are determined by the agent's subjective preferences. If this view were correct, talk of evaluative uncertainty would be misleading as one is not normally uncertain about what one's own judgement on something is (just what it should be). Indeed it makes questions such as 'What utility should I attach to this outcome?' seem barely intelligible. If a prospect's value for an agent is determined by her preferences, she cannot be right or wrong about what value to attach to them; nor can her preferences be criticised on grounds of their failure to adequately reflect one value or another.

There are at least two ways, however, in which one can be uncertain about the value to attach to a particular consequence or whether one consequence is preferable to another. Firstly, one may be uncertain about the factual properties of the consequence in question. If the latest Porsche model is the prize in a lottery, one may be unsure as to how fast it goes, how safe it is, how comfortable, and so on. This is uncertainty of the ordinary factual kind, and, if one wishes, it can be 'transferred' from the consequence to the state of the world by making the description of the consequence more detailed. For example, the outcome of the lottery may be regarded as having one of several possible consequences, each an instantiation of the schema 'Win a car with such and such speed, such and such safety features and of such and such comfort', with the actual consequence of winning depending on the uncertain state of the world.

Secondly, one can be unsure as to the value of a consequence, not because of uncertainty about its factual properties but because of uncertainty about how valuable these properties are. One may know all the specifications, technical or otherwise, of the latest Porsche and Ferrari models, so that they can be compared on every dimension, but be unsure whether speed matters more than safety or comfort. Once all factual uncertainty has been stripped from a consequence by detailed description of its features, one is left with pure value uncertainty of this kind.

When we assume that values are given, we take this uncertainty to have been resolved in some way. This could be because we assume that there is a fact of the matter as to how good a consequence is or as to whether one outcome is better than another. But it could also be because the description of the decision problem itself comes with values 'built in'. For instance, in a problem involving a decision between two courses of medical treatment, it may be that a limited number of value considerations apply in the assessment of these treatments: number of patients saved, amount of discomfort caused, and so on. The decision theorist will be expected in such circumstances to apply only the relevant values to the assessment of the options, and to set aside any other considerations that she might 'subjectively' consider to be of importance.

In many situations, however, values are not given in any of these ways, and the agent may be uncertain as to the value she should attach to the relevant prospects. She may, for this reason, also be willing to revise her evaluations in the face of new considerations or the criticism of others. These facts would seem to render Value Subjectivism unsustainable. But the subjectivist can insist that an agent cannot be wrong about what value to attach to fully specified consequences, and point out that it suffices that there be factual uncertainty for us to be unsure about the desirability of any less than fully specified prospect. Since in practice complete specification is impossible, this means that evaluative uncertainty of the kind that derives from factual uncertainty will be ubiquitous. Furthermore, since we may hold false beliefs, our ethical judgements on incompletely specified prospects are certainly criticisable. All that is ruled out by their view is pure value uncertainty.

Other views take the possibility of irreducible value uncertainty more seriously. There are three that I will mention here. The first is that evaluative uncertainty is just ordinary uncertainty about agents' tastes or, more generally, about the features of agents that are relevant to the utility of the option for them. The thought is this. What value an agent will assign to a commodity depends not just on features of the commodity itself (the speed, safety and comfort of the cars) but also on features of the consumers: their likes and dislikes, their capacities (for instance, their driving skills) and their needs. And one can be just as uncertain about the latter as the former.

This view has some application to decisions with consequences for different people or ones far in the future and when the value we attach to these consequences depends on the attitudes that the different people, or our future selves, take to them. But it is not plausible as a general account of value uncertainty. When we are uncertain about whether it is more important to help a friend or to further our own interests, the difficulty that we have in deciding the question stems not from the fact that we don't know what we in fact prefer but that we don't know what we *should* prefer. Indeed, I doubt that in such cases there really is anything like a set of pre-given preferences waiting to be discovered. For example, consider evaluative uncertainty associated with the decision whether to take up playing the violin or fencing. Can the problem be described as trying to work out what one's tastes are? I think not. One's tastes are likely to be shaped by the decision itself, for in pursuing the violin one will learn to appreciate one set of skills, in taking up fencing one will learn to appreciate another.

On a second cognitivist view, what I am calling value uncertainty is just ordinary uncertainty about normative facts. For instance, the uncertainty I experience about whether or not to help my friend is uncertainty about whether it is in fact good to help one's friends or whether it is true that it is better to help one's friend than further one's own interests. So, on this view the difference between uncertainty about whether it will rain and about whether it is good that it rains is to be located in the type of proposition about which one is uncertain, not in the nature of this uncertainty.[3]

Both these cognitivist views treat value uncertainty as a kind of factual uncertainty, differing only with regard to the kinds of facts that they countenance and consider relevant. They are in that sense reductive views. The last of the views I want to consider holds that evaluative uncertainty is different in kind from factual uncertainty and is directly expressed in utility judgements, rather than in second-order judgements about tastes or first-order probability judgements about normative facts. Making this precise requires some care. Utility judgements are like probability judgements in that they are judgements about the world (and not just expressions of the agent's mental state). But, while we can say that one's probability for rain tomorrow, say, reflects the degree to which one is uncertain as to whether it will rain, it is not the case that one's utility for rain expresses the degree to which one is uncertain as to whether it is true that it is good that it rains. Rather, it expresses one's uncertainty as to how good it would be if it rained. On the reductive views, once we know all the facts – about what will happen when it rains, how much people like getting wet, and so on – all uncertainty is removed and the value of rain is fully determined by either the relevant normative facts or by the agent's subjective degrees of desire for rain, given the facts. On the non-reductive view, even when we know all the facts we can be unsure as to how desirable rain is.

It will not matter to this book which of these views of evaluative uncertainty is adopted, so long as it is consistent with Bayesian decision theory. On any such view, evaluative uncertainty is captured or measured by the agent's value or utility function. Since evaluations generally depend on the facts, it follows that a value function that adequately represents an agent's state of evaluative uncertainty must be revisable in the face of new factual information (and, potentially, new value experiences as well). This desideratum figures prominently in the choice of value function that I make in the second part of the book and which differentiates it from the utility functions standardly employed in decision theory.

3.2 OPTION UNCERTAINTY

In the state-consequence representation of a decision problem that we started with, actions were associated with definite consequences, one for each state of the world. But in real decision problems we are often unsure about the relationship between actions, worlds and consequences, either because we do not know what consequence follows in each possible state of the world from a

---

3 It is an open question whether this position is consistent with Bayesian decision theory. David Lewis (1988, 1996) famously argues that it is not, but others (e.g. Broome, 1991; Bradley & Stefánsson, 2017) that Lewis' argument is mistaken.

TABLE 3.1. *State–Consequence Correspondence*

	States of the world		
Options	S_1	...	S_n
α	$\{a_1^1,...,a_1^j\}$	...	$\{a_n^1,...,a_n^j\}$
...	...	...	...
γ	$\{c_1^1,...,c_1^j\}$	...	$\{c_n^1,...,c_n^j\}$

choice of action, or because we don't know what state of the world is sufficient, for a given action, to bring about that consequence. For instance, we may be uncertain as to whether taking an umbrella will certainly have the consequence of keeping us dry in the event of rain. Perhaps the umbrella has holes, or the wind will blow it inside out or the rain will be blown in from the sides.

We can put this difficulty in slightly different terms. A possible action may be defined by a particular mapping from states to consequences. Then no uncertainty about the mapping itself can arise. But what we will then be unsure about is which actions are actually available to us – i.e. which of the various hypothetical actions are real options. Whether we describe the problem in these terms – as uncertainty about what options we have – or as uncertainty about the consequences, in each state of the world, of exercising any of the options we know we have, is of little substance, and I shall use the same term – option uncertainty – to denote both.

Option uncertainty arises when we are unsure about what would happen if we were to act in some way, or perform some kind of intervention in our environment. When an agent faces option uncertainty, she cannot represent her decision problem with the simple state–consequence matrix represented by Table 1.2. But she can do something quite similar, by replacing the fine-grained consequences that play the role of Savage's 'sure experiences of the deciding person' with *sets* of such fine-grained consequences – intuitively, the set of consequences the agent regards as possible given the act and state in question. This is exhibited schematically in Table 3.1, in which each act γ is represented as a function from each state S_i to a set of associated possible consequences, $\{c_1^1,...,c_1^j\}$. The larger the sets of possible consequences the greater the option uncertainty facing the agent.

There are three strategies that can be pursued in handling option uncertainty. The first is to try to reduce or transform option uncertainty into empirical uncertainty about the state of the world. The second is to reduce it to evaluative uncertainty. And the third is to treat it as a *sui generis* form of uncertainty. Let's consider each turn.

Reduction to Empirical Uncertainty Decision theorists typically attempt to reduce option uncertainty to uncertainty about the state of the world by

TABLE 3.2. *State Functions*

	States of the world		
Options	$S_1(\alpha^i)$	...	$S_n(\alpha^i)$
α^1	C_1^1	...	C_n^1
...	...	...	...
α^m	C_1^m	...	C_n^m

refining their description of the states until all contingencies are taken care of. They will regard a state of the world as insufficiently described by the absence or presence of rain, for instance, and argue that one needs to specify the speed and direction of the wind, the quality of the umbrella, and so forth. There are at least two reasons why this reductive strategy will not work on all occasions. Firstly, because – according to our best scientific theories – the world is not purely deterministic. When the conditions under which a coin is tossed do not determine whether a coin will land heads or tails, for instance, the act of tossing the coin does not have a predictable consequence in each state of the world. Secondly, even if we are in a purely deterministic set-up, it may be subjectively impossible for the decision maker to conceive of and then weigh up all the relevant contingencies or to provide descriptions of the states of the world that are sufficiently fine-grained as to ensure that a particular consequence is certain to follow, in each state, from the choice of any of the options open to them. And, even if one could envisage all the possibilities, one may simply not know what state of the world is sufficient for the act of taking an umbrella to keep one dry.

To get around these difficulties the reductionist can make a two-pronged attack. To handle objective indeterminacy, she can allow consequences to be objective probability distributions (lotteries) over outcomes, and apply von Neumann and Morgenstern's theory to give a measure of their utility. And to handle enumerability-of-states problems, she can draw on descriptions of the states of the world that identify the set of conditions sufficient for the determination of the consequence, given the performance of the action, without actually listing the conditions. For instance, she can turn Savage's theory around and take actions and consequences as the primitives and then define states of the world as consequence-valued functions ranging over actions. This would lead to a decision matrix of the kind exhibited in Table 3.2, in which each $S_j(\alpha^i)$ denotes the state that maps actions α^i to consequences C_j^i.

This descriptive strategy has some notable advocates. Lewis (1981), for instance, treats states as 'dependency hypotheses', these being maximally specific propositions about how consequences depend causally on acts. Similarly, Stalnaker (1981a) suggests that a state of the world be denoted by a conjunction of conditional sentences of the form 'If action A were performed then consequence C would follow; if action A' were performed then consequence C' would follow; if ... '. In this way, option uncertainty is transformed into a particular kind of state uncertainty, namely uncertainty as to the true mapping from actions to consequences or as to the truth of a dependency hypothesis or particular conjunction of conditionals.

Reduction to Evaluative Uncertainty A second strategy for dealing with option uncertainty is to coarsen the description of the consequences to the degree necessary to ensure that we can be certain it will follow from the exercise of an option in a particular state. Leonard Savage (1974/1954, p. 84), for instance, acknowledges the need for 'acts with actually uncertain consequences to play the role of sure consequences in typical isolated decision situations'. Formally, this amounts to treating the sets of possible consequences associated with an action occurring in Table 3.1 as single coarse-grained consequences and giving them a utility value. Pursuit of this strategy converts option uncertainty not into ordinary uncertainty about the state of the world but into evaluative uncertainty about the desirability of the consequence as described. We may be sure that the act of taking an umbrella will have the consequence in a rainy state of being able to shield ourselves against the rain by opening the umbrella. But whether this is a good thing or not depends on contingencies that by assumption we are unable to enumerate or identify. How bad it is to get soaked, for instance, depends on how cold the rainwater is, and rain temperature may be a variable about whose determinants we know very little. Whatever utility value we assign to the coarse-grained consequence of having an umbrella as rain protection will embody this uncertainty.

Non-Reduction The last strategy to consider is to accept the presence of option uncertainty and try and develop a measure of it. We could, for instance, follow Jeffrey (1990/65) in dispensing with the formal distinction between states and consequences and assign probabilities to consequences that depend on the action performed. So, instead of trying to enumerate the features of the state of the world that will ensure that I stay dry if I take an umbrella, I simply assess the probability that I will stay dry if I take the umbrella and the probability that I will get wet anyhow (even if I take it). In making these probability judgements, I may well try and conceive of the various contingencies under which staying dry will be a consequence of my action, but I need not be able to conceive of all of them in order to do so. Having done so I may directly represent the decision problem I face in terms of the probabilities and utilities of the various possible consequences induced by each option in the manner of Table 3.3, in which $P_\alpha(C)$ is the probability on consequence C induced by action α and $U(C)$ its utility. So, while the other strategies led to alternatives to our initial state–consequence matrix representation of a decision problem, this last one leads to an alternative to the initial quantitative representation of it (the one exhibited by Table 1.3).

TABLE 3.3. *Act-Dependent Consequence Matrix*

Options	States		
	C_1	...	C_n
α	$P_\alpha(C_1), U(C_1)$	...	$P_\alpha(C_n), U(C_n)$
...	...	...	...
γ	$P_\gamma(C_1), U(C_1)$	...	$P_\gamma(C_n), U(C_n)$

When the decision problem is adequately represented by a probability–utility matrix of the kind represented by Table 3.3, the principle of maximisation of expected benefit requires choice of the action α that maximises the quantity:

$$V(\alpha) = \sum_{i=1}^{n} P_\alpha(C_i) \cdot U(C_i)$$

In recent years debate has raged between so-called evidential and causal decision theorists as to the nature of the act-dependent probabilities occurring in this equation. Evidentialists such as Richard Jeffrey regard the conditional probabilities of possible states, given that an action is performed, as giving the correct measure of the uncertainty associated with acting (i.e. they take $P_\alpha(S_i)$ to equal $P(S_i|\alpha)$), while causal decision theorists such as James Joyce (1999) argue that what is required is a measure of their probability under the counterfactual supposition that the action is performed. If the evidentialists are correct then a single probability function suffices not only to measure state uncertainty but option uncertainty as well – i.e. reduction of option uncertainty to empirical uncertainty is possible. But the difficulty that evidential decision theory faces in dealing with Newcomb's Paradox and other more homely cases in which probabilistic correlation fails to provide a good guide to causal efficacy suggests that it is not. More on this in the third part of the book.

3.3 MODAL UNCERTAINTY

Empirical uncertainty arises when we are unsure as to what *is* the case and evaluative uncertainty when we are unsure as to what *should* be the case. The last type of uncertainty to be considered – modal uncertainty – arises when we are unsure as to what is possible or about what *could* be the case: what contingencies might arise, what consequences might follow from our actions and what actions are feasible. In Savage's framework, no modal uncertainty can arise as both the state space and the set of possible consequences are exogenously given. In real decision problems, however, agents must grapple with the possibility of *unforeseen contingencies*: eventualities that even the knowledgeable will fail to take account of. If decision makers are aware of the possibility that they may not be aware of all relevant contingencies – a state that Oliver Walker and Simon Dietz (2011) call 'conscious unawareness' – then they face modal uncertainty.

There are two variants of the problem of unforeseen contingencies that should be distinguished. The first arises when the agent is aware that the states that she can conceive of may be too coarse-grained to capture all decision-relevant considerations. For instance, someone planning where to go on holiday may take into account all factors that seem relevant to her enjoyment of it (costs, climate, cultural amenities, and so on) but nonetheless worry that she has omitted something which mitigates these factors. As a result, she either cannot be sure what the exact consequences of her actions are (i.e. she faces option uncertainty) or, if she can, whether the consequences are sufficiently fine-grained as to capture everything relevant to their value (i.e. she faces evaluative uncertainty).

A second variant of modal uncertainty arises when the agent is aware of the possibility that she has entirely omitted consideration of a possible state or possible consequence. For instance, a businessman considering an investment may be unsure as to what new technologies will be available in the future. So he will be unable to exhaustively enumerate all the states determining the return on his investment. And, consequently, he will not be able to determine whether the investment can be expected to benefit him or not.

Modal uncertainty presents Bayesian decision theory with its most difficult challenge.[4] For, if we don't know what all the relevant possibilities are, is it really rational to try and optimise relative to those we are aware of? A course of action that is best relative to a limited set of considerations may turn out to be disastrous once the unforeseen ones reveal themselves. One response to the problem is to try and reduce modal uncertainty to empirical uncertainty by introducing a catch-all state: the 'any other contingencies not yet enumerated' state. But this catch-all state will have completely unknown consequences, so severe option uncertainty now arises. Furthermore, as we have no way of assigning a probability to this state, severe empirical uncertainty is generated. In short, such reduction of modal uncertainty would seem to come at the cost of its severity. But further examination of this issue must await the last part of the book.

---

4 The problem of how to handle the possibility of unforeseen contingencies has received little attention in philosophical decision theory – a notable exception being Hansson (1996a, 2004, 2016), who gives careful consideration to the question of rational responses to the problem.

4

Justifying Bayesianism

What reasons do we have for accepting or rejecting Bayesian decision theory? Empirical theories stand or fall on the basis of their ability to handle the facts; above all, by the quality of their explanations and the accuracy of their predictions. As an empirical theory of judgement and decision making, subjective expected utility theory has endured a good deal of criticism in the last 30 years or so, with a range of experimental results suggesting that it is a poor predictor of people's behaviour. On the other hand, none of the main rival empirical theories seem to do much better when confronted with data other than that used to generate them; according to some studies they do worse.[1] So, while there is every reason to be cautious about the theory's predictive abilities in a wide range of cases, there is as yet no good reason to abandon it entirely.

It is as a normative theory, however, that we are interested in the problem of justifying acceptance of Bayesian decision theory. Although normative theories cannot be refuted by direct appeal to the facts, the general principles of a normative theory can be assessed in a similar way to an empirical one. A scientific theory will have its laws assessed by deriving predictions about concrete cases and then testing to see whether the predictions turn out to be true or not. A normative theory doesn't make predictions, but it will have implications for concrete situations which can be compared with our judgements about what is correct in those cases. A theory of valid inference, for instance, can be tested against concrete instances of inferences that we are inclined (or otherwise) to make: a theory of grammar against sentences that competent speakers find acceptable; and so on.

The fundamental lesson of the Quine–Duhem problem – that the falsity of a scientific hypothesis can rarely be deduced from a set of observations –

---

[1] See Starmer (2000) for an overview of the empirical evidence.

applies equally to normative theories. When general normative principles clash
with our judgements regarding a concrete case, all that follows is that we
cannot coherently hold onto both. We can revise the principles so that they
can accommodate the intuitive judgements or we can revise the judgements
themselves. Frequently we can also do neither and instead revise one of the
numerous other assumptions that are typically needed in order to draw out
the implications of the general principles for concrete cases.

When something must be revised, foundationalists suggest that we retain
those principles or judgements which have some kind of special justification
and use them as an anvil upon which to beat the rest into shape. Such
thinking is plausibly behind the extensive use of axiomatic methods in
decision theory, with the role of foundational principles being played by
propositions concerning the rationality properties of preferences. Indeed, to
the question 'What grounds are there for thinking that rationality requires
us to maximise subjective expected utility?' decision theorists will typically
produce a representation theorem and argue that it shows that the claims of
expected utility theory can be derived from 'self-evident' principles of rational
preference.

I doubt that the rationality claims about preference that are required for
these arguments can bear the full justificatory load typically piled on them. But
there is no doubting the importance of representation theorems to the Bayesian
enterprise. And as I shall be relying quite heavily on them in the next part of
the book it is best to come clean now about their limits as well as their scope.
So this chapter will be devoted to looking carefully at these theorems, the
assumptions they make and philosophical positions that they are supposed to
support.[2] The broad view that I will take is that they are best viewed as moves
within a search for a reflective equilibrium, in which general principles and
judgements on particular cases are brought in line by systematic refinement of
both with the aim of maximising overall coherence, but in which no (class of)
proposition plays the role of final arbiter of truth. Since this method can lead
to quite different outcomes depending on what choices are made about what
to revise, it is perfectly possible that two people pursuing it will end up with
different normative theories that achieve equal overall coherence. Nonetheless,
it seems to me that we cannot get by without relying on this method to a large
extent.

4.1 PRAGMATISM

A representation theorem for a decision theory proves the existence of an
isomorphism between two kinds of structures: a class of preferences satisfying

---

[2] Despite their centrality, almost all philosophical discussion of representation theorems has
been directed at the status of the axioms they invoke, rather than the arguments that they are
supposed to support (recent exceptions are Meacham & Weisberg, 2011, and Zynda, 2000).

a set of conditions and a class of numerical functions with certain properties. An 'ideal-typical' Bayesian representation theorem, for instance, establishes that, if an agent's preferences satisfy a particular set of axioms, then these preferences can be numerically represented by a pair of probability and utility functions measuring her degrees of belief and desire, in the sense that one alternative is preferred to another iff the expectation of utility given the former exceeds that of the latter.

The central primitive of these theorems is the notion of preference, reflecting widespread adherence amongst decision theorists to Pragmatism, a view which accords conceptual and methodological priority to preferences over numerical degrees of belief and desires: methodological priority, because preferences, as revealed in the behaviours that they engender, are the empirical basis for attributions of degrees of belief and desire to agents; and conceptual priority, because it is the properties of rational preference that are said to explain the laws of rationality for partial belief and desire.

These priority claims should not be confused with those emanating from another commonly held view amongst decision theorists: Behaviourism. Behaviourism accords methodological and conceptual priority to observable behaviour, and particularly choice behaviour, over preference: methodological priority, because observations of choice behaviour are said to furnish the empirical basis for ascriptions of preferences and other mental attitudes to agents; and conceptual priority, because it is the properties of observable behaviour that are said to explain those of preference. These claims are supported by another set of representation theorems, linking choice behaviour to preferences, and which serve to characterise (in the ideal case) the conditions on observed choices necessary and sufficient for their representation by a preference relation having certain properties.

The claim of methodological priority for observable behaviour is motivated by the desire to see the human and social sciences rooted in evidence that is intersubjectively verifiable. It is an entirely defensible position and quite in line with practices in many of these disciplines. But the claim of conceptual priority is much less plausible, and its weakness is evidenced by the failure of attempts by behavourists in many different fields to eliminate mentalistic vocabulary from scientific discourse. Indeed, contrary to the view it expresses, it is the nature of preference and its mental determinants that accounts for the properties of behaviour and not the other way around. It is not surprising, therefore, that Behaviourism's conceptual priority thesis has fallen into philosophical disrepute.

Pragmatism implies neither of the priority claims of Behaviourism. It is true that many decision theorists share with it a distrust of introspection as a means of determining an agent's mental states. Frank Ramsey, for instance, firmly dismisses the idea that we could introspect our degrees of belief, on the grounds that 'the beliefs which are held most strongly are often accompanied by practically no feeling at all; no one feels strongly about the things he takes for granted' (Frank Ramsey, 1990/1926, p. 65). Both he and Leonard Savage argue instead that our judgements of belief are really about how we would act in different hypothetical circumstances. But neither are thereby dismissing introspection altogether: since we can't tell how we would act in a hypothetical circumstance by direct observation, this information would have to come by introspecting on what we would do. It is introspection of quantitative degrees of belief and desire that they consider unreliable, not introspection in general.

These observations point to a second qualification. There are in fact two distinct primacy claims that are rolled together in the kind of Pragmatism espoused by decision theorists. The first is that qualitative attitudes such as preference or comparative belief (attitudes of the form 'X is more credible/probable than Y') have primacy over quantitative ones such as degrees of desire or degrees of belief; the second that practical reason has primacy over theoretical reason. Both have methodological and conceptual dimensions.

The thesis of the methodological priority of qualitative attitudes over quantitative ones says that qualitative attitudes have methodological priority over the corresponding quantitative attitudes because our ability to attribute attitudes of the latter kind depends on our ability to determine attitudes of the former kind. That is, the qualitative attitudes provide the *evidence* for the quantitative ones. The corresponding thesis of conceptual priority says that they have conceptual priority because the rationality properties of the quantitative attitudes (such as degrees of belief being probabilities) derive from the rationality properties of the qualitative ones (such as the transitivity of comparative belief). That is, it is the fact that the qualitative attitudes are rationally required to have certain relational properties that explain why the quantitative attitudes are rationally required to have corresponding numerical ones.

The priority claims for practical over theoretical reason run along similar lines. The methodological priority claim says that practical attitudes such as preference have methodological priority over theoretical ones such as belief because our ability to the determine the latter depends on our ability to determine the former. On the other hand, the conceptual priority claims say that the laws of rational preference explain the laws of rational belief, both qualitatively and quantitatively. In particular, it is the rationality properties of preference that explain those of belief.

With the exception of the last of them, I think that these priority claims are, with some qualifications, true. But I won't at this point argue directly for or against them. Rather, I will focus on the motivation that they provide for decision-theoretic representation theorems and, conversely, on the question of the kind of support they derive from these theorems. First, however, the notion of preference requires further clarification.

4.2 INTERPRETATIONS OF PREFERENCE

Two broad classes of interpretations of the notion of preference can be found in the decision-theoretic literature: those that define preference in terms of choice or behaviour and those that define it in terms of judgements or mental states. (Roughly, the behavioural ones dominate in economics and the mentalistic ones in philosophy.) More than one instance of each has been influential but I will simply spell out what seems to me the most viable version of both.

Choice-Theoretic The basic thought underlying this class of interpretations is that preferences can be defined in terms of the choices or behaviour that they engender. Savage, for instance, regards the claim that someone has a preference between two alternatives acts f and g as meaning that, 'if he were required to choose between f and g, no other acts being available, he would decide on f' (Leonard Savage, 1974/1954, p. 17). The view that Savage seems to be expressing has come to be called the Revealed Preference interpretation of preference. What it says, more formally, is that one alternative α is revealed as preferred to another β iff β is never chosen when both are available. Revealed Preference theorists sometimes speak as if they think that preferences are nothing but the choices that reveal them, but this talk is probably a result of a surfeit of positivistic enthusiasm for elimination of all reference to non-observable entities than a carefully thought-out position. For if preferences were nothing but choices then there would be nothing to reveal and certainly no sense in saying that preferences either explain or rationalise choices.

A more considered explication of the relationship between preference and choice captured in Savage's 'definition' is in terms of choice dispositions. On this account, a preference for α over β is a *disposition* to choose α rather than β when both are available. In contrast to the Revealed Preference account, on this view it is not analytic that β will never be chosen when α is available since dispositions have implicit normality conditions attached. Solubility in water is a matter of being disposed to dissolve when placed in water, but this disposition may not be revealed when the water is frozen, for instance. Similarly, a preference for α over β will not invariably eventuate in choice of α over β, for various other factors (error of judgement, unchecked emotions, etc.) might intervene in some contexts. Although preference is revealed in choice, not everything revealed in choice is preference.

The main advantage of the choice-theoretic construal of preference, and which explains its popularity in economics and other behavioural sciences, is that it requires only the bare minimum of psychological assumptions in order to be applied. Preferences can be attributed to any entity which exhibits patterned choice, irrespective of its psychological constitution and complexity.[3] This gives decision theory enormous potential scope; indeed,

---

[3] See Dennett (1971) for a revealing discussion of these explanatory virtues.

there have been fruitful applications of it to animals, plants, machines and groups which exploit the flexibility of this preference concept. A second advantage of the choice disposition approach is that it ties preference very closely to observable behaviour, thereby making it possible for rival models of preference to be tested empirically. This goes some way towards underpinning the methodological role that Pragmatism accords to preferences.

The main difficulty for choice-theoretic approaches, on the other hand, lies in accounting for the rationality properties that are usually attributed to preferences. Either it must be argued that these properties are embedded in the concept of choice itself or it must be granted that 'rational' preference is simply one subset of the kinds of preferences that a chooser might reveal. The behavioural turn in economics is testimony to the difficulty in making the former strategy work since there is evidence of patterned choice that lacks these properties. The latter strategy, on the other hand, leaves the representation theorems for behaviour without any normative role.

Judgementalism On judgementalist or mentalistic construals of preferences, they are a type of judgement or mental attitude.[4] As such, they are the sorts of things that are susceptible to rationality conditions, an advantage of this interpretation over choice-theoretic ones. What kind of judgement are they? In fields such as welfare economics and social choice theory, they are typically taken to be judgements of personal well-being, so that to prefer one thing to another is to judge that it contributes more to one's well-being, all things considered. This restriction to what may be called self-interested preference, while perhaps appropriate in these fields, is not justified in general as one can clearly prefer things that are not in one's self-interest over those that are.

A better judgementalist interpretation of preference, defended by Dan Hausman (2011a,b), is as an all-things-considered subjective comparative judgement. The 'all-things-considered' part is crucial. In ordinary talk, we attribute preferences to describe someone's tastes, likings or favourings, which, together with her moral beliefs, commitments, social norms, and so on, determine which action she chooses. On the all-things-considered notion of preference, all of these constitute reasons for the agent to prefer one action over another and should be incorporated into her preferences. This brings preference closely into line with choice, for, if one's preferences incorporate all the reasons one has for favouring one action over another, then *ceteris paribus* one should choose it rather than the alternative. But it leaves open the possibility that extra-judgemental factors mediate the relationship between preference and choice.

A Hybrid Definition I do not regard the best versions of these two classes of interpretation as rivals. Indeed, I favour a hybrid of them. On this hybrid

---

4 In Chapter 2, I argued that judgementalist and mentalistic interpretations of notions such as probability are quite different. But I shall ignore these differences here.

account, a preference for α over β is best viewed as an all-things-considered comparative judgement that α is better than β that is instantiated in a disposition to choose the former over the latter when both are available (more generally, to have it be true that α rather than β). That preferences are judgements explains why they are subject to considerations of consistency. That they are also dispositions to choose both explains the connection between preference and choice and fixes the sense of betterness (namely as choice-worthiness) characteristic of preference judgements.

Both these elements are essential. It is possible to make a choice-worthiness judgement but lack the disposition to choose, and vice versa. But in neither case would it be appropriate to speak of preference. This is not to deny that some choice dispositions may never be revealed, since one can have preferences over alternatives between which one cannot choose (such as that one's grandchildren have happy lives). But this does not diminish the importance to the concept of preference of this dispositional relation between judgement and choice.

4.3 REPRESENTATION THEOREMS

Representation theorems are the centrepiece of mathematical decision theory. But, elegant mathematics aside, what do representation theorems achieve? In fact, they play two different, but equally pivotal, roles in decision theory. The first is to provide a demonstration of how the values of the variables occurring within a decision theory (degrees of belief and desire, for instance) can be determined from information that can be gleaned from observation of behaviour. This is typically done via the double-representation argument mentioned before. A first representation theorem establishes that observed choices meeting certain conditions determine attributions of preferences with certain properties (such as completeness and transitivity). A second representation theorem shows that preferences having these properties determine a measure of the agent's degrees of belief and desire.

The second role of a representation theorem for a decision theory, the one of more interest here, is to provide a justification of its normative claims regarding the properties of rational belief and desire and the relationship between them. It does so via an argument of the following kind.

- The axioms of preference either express rationality claims that all sensible people should accept or impose some structural conditions of little conceptual significance (but which are required for numerical representation).
- The representation theorem shows that satisfaction of these axioms by an agent's preferences implies the existence of a probability measure P and utility measure U, respectively, of her degrees of belief and her degrees of desire, that jointly represent these preferences.

- Therefore, since rationality requires her preferences to satisfy these axioms, it requires her to have degrees of belief and desire that are, respectively, probabilities and utilities.

Now, an argument like this is not going to convince people who have little sympathy for the Bayesian picture of rational agency even if they accept the premises. For they can simply deny that the numerical functions whose existence are established by the representation theorem are truly measures of the agent's degrees of belief and desire. They might accept that P measures a determinant of the agent's preferences in some formal sense, but deny that this determinant is the agent's real degrees of belief. They might, for instance, hold the view that rational preference is not sensitive to degrees of belief in the way that Bayesians claim and so conclude that P cannot be a measure of them. The Bayesian can retort that it is only the type of belief to which preference is sensitive that she cares to measure, but this is just a way of ending the discussion, not of convincing her opponent.

What this shows is that representation theorems can play their justificatory role only against the background of some shared assumptions about rational belief and desire and how these cohere with each other. They do not, therefore, give support to a decision theory by showing that its claims can be derived from first principles of preference whose motivation is entirely independent of that of the theory itself. Rather, they serve to give it foundations by exhibiting the core qualitative principles upon which the theory depends, freed from the quantitative packaging in which it is wrapped. A well-constructed representation theorem will give a set of independent principles each expressing a feature of a widely shared conception of rationality, and derive implications from it of a much more precise nature. Its target, therefore, will be those who share the background assumptions that motivate the principles but who do not necessarily accept the full corpus of Bayesian decision theory.

Even if we accept this non-foundationalist view of representation theorems, more domestic challenges remain. The main problem is that numerical representations of an agent's preferences are typically not unique. Not just in the sense that preferences do not determine the scaling of the numerical measures whose existence are established by the theorem, but in the more fundamental sense that they do not uniquely determine what form the numerical representation must take. For an agent's preferences may also be numerically representable by other pairs of functions that, though not probabilities and utilities, can be combined in such a way as to determine the preferences in question.[5] How are we to say which of these representations is the 'true' measure of the agent's degrees of belief and desire?

---

5 Or, indeed, by triples of functions, or any other number of them.

I don't think these questions admit of a single answer. Different representations will have different implications for the properties of the attitudes which they putatively measure and some of these implications may be more plausible than others. If this is so then, ideally, we should be able to pack these considerations into the conditions that underlie the representation so that the classes of representations it admits is more constrained. Equally, there may be reasons for preferring one set of representations over another that do not derive from the attitudes we are trying to measure, but have to do with considerations of simplicity or technical convenience or continuity with other accepted theories. For example, the kinds of constraints that typical representation theorems impose on the representations of agent's degrees of belief do not uniquely determine that they should be represented by a probability function. Indeed, as we will see in the second part of the book, any real valued function that implies a certain kind of ordering over prospects will do. The fact that a probability function assigns the value one to tautologous prospects and zero to contradictory ones is just an artefact of the scaling imposed by the numerical representation, a scaling chosen on grounds of convenience. So it would be just plain silly to say that these theorems show that rationality requires degree of belief zero in contradictory prospects. What is not silly is to say is that rationality requires 'full' belief in the former and 'empty' belief in the latter, for these are properties that *are* picked out by their causal role in the determination of choice and which can be expressed as conditions on preference.[6] All of this no doubt feels extremely abstract, however, so let us turn to the examination of a particular instance of a representation theorem to get a better handle on these claims.

4.4 SAVAGE'S REPRESENTATION THEOREM

Although not the first representation theorem of its kind, that given by Leonard Savage in his book *The Foundations of Statistics* is perhaps the most influential. In this section I will present his theorem following his exposition quite closely, though with a few modifications for continuity with other sections.

Recall Savage's distinction between states, consequences and actions. States are complete descriptions of the features of the world that are causally independent of the agent's actions but which are relevant to determining their outcomes. Consequences, on the other hand, are the features of the world that matter to the decision maker, such as that she is in good health or wins first prize in a beauty contest or is allowed to sleep late on a Sunday morning. Actions are the link between the two, the means by which different consequences are brought about in different states of the world.

To reflect our earlier observation that the distinction between states and consequences is pragmatic rather than ontological, I will treat both as elements

---

[6] See Bradley (2008).

of a single background set containing all the possible ways that the world might be – which I call 'prospects' – rather than follow Savage in treating them as logically distinct types of object. Then the central elements of Savage's framework can be specified as follows.

1. $\Omega = \{A, B, C, ...\}$ is the set of prospects. Informally, we can think of Ω as the set of all possibilities with respect to which an agent can have an attitude.
2. $C \subseteq \Omega$ is the set of consequences. Informally, we can think of C as a partition of Ω whose elements are maximally specific with regard to all that matters to the agent.
3. $S = \{s_1, s_2, ...\}$ is the set of states of the world. Elements of $\wp(S)$, the set of all sets of states, are called events. Informally, we can think of S as a partition of Ω whose elements are maximally specific with regard to factors outside of the agent's control, but that are causally relevant to the determination of the consequences of acting.
4. $\mathcal{F} = \{f, g, h, ...\}$ is the set of actions.
5. $\succsim$ is the two-place 'at least as preferred as' relation on $\mathcal{F}$.

The goal of the representation theorem is to establish, from a set of conditions on the preference relation $\succsim$ on the set of actions, the existence of a value function, V, on $\mathcal{F}$ which represents $\succsim$ in the sense that, for all $f, g \in \mathcal{F}$,

$$V(f) \geq V(g) \Leftrightarrow f \succsim g$$

and which takes the form of an expected utility – i.e. for all $f \in \mathcal{F}$, $V(f) = \sum_{i=1}^{n} U(f(s_i)) \cdot P(s_i)$ – for some real valued function U on C and probability function P on $\wp(S)$.

Savage proves the existence of an expected utility representation of preferences in two steps. First he postulates a set of axioms that are sufficient to establish the existence of a unique probability representation of the agent's beliefs. He then shows that probabilities can be used to construct a utility measure on consequences such that preferences amongst gambles cohere with their expected utilities, first on the assumption that the set of consequences is finite and then for the more general case of an infinite set of consequences.

Recall that, for Savage, actions are just functions from the set of states S into C, the set of consequences. In fact, Savage takes the preference relation to be defined over a very rich set of such acts, namely *all* functions from states to consequences. Because of its importance to his theorem, I have 'promoted' the definition of the domain of the preference relation to the status of an additional postulate:

P0 (*Rectangular Field*)[7] $\mathcal{F} = C^S$

---

[7] I take this term from Broome (1991).

Savage's first official postulate requires that the preference relation orders the set of acts:

P1 (*Ordering*) $\succsim$ is (a) complete and (b) transitive

For any consequence $F \in C$, let act $\bar{f}$ be the corresponding constant act defined by, for all states s, $\bar{f}(s) = F$. Given this definition, it is straightforward to induce preferences over consequences from preferences over constant acts by stipulating that $F \succsim G$ iff $\bar{f} \succsim \bar{g}$. Such a stipulation in effect imposes the requirement that any feature of an act relevant to the preferences an agent has for it should be written into the consequences it determines.

Now P1 alone ensures the existence of a numerical representation V of $\succsim$ on $\mathcal{F}$ (when $\mathcal{S}$ is not countable P1 must be supplemented with a continuity condition, but let's just stick to the countable case). Hence a utility representation U of $\succsim$ on C can be induced by setting, for all $F \in C$, $U(F) = V(\bar{f})$. Note that it follows that V restricted to the constant acts trivially has the form of an expected utility since, in this case, the sum of the probabilities of the states determining the constant consequence of any such act is just one.

Savage's next step is to assume that the preference relation is separable across events – i.e. that the desirability of a consequence of an act in one state of the world is ordinally independent of its consequences in other states. He does so by means of his famous Sure-Thing Principle. Consider the acts displayed in the table below.

	Events	
Acts	*E*	*E'*
f	X	Y
g	X*	Y

Intuitively, act *f* should be preferred to act *g* iff consequence X is preferred to consequence X*. This is because *f* and *g* have the same consequence whenever *E* is not the case, and so should be evaluated solely in terms of their consequences when *E* is the case. Consequently, any other actions *f'* and *g'* having the same consequence as *f* and *g*, respectively, whenever *E* is the case, and identical consequences when it is not, should be ranked in the same order as *f* and *g*. More formally:

P2 (*Sure-Thing Principle*) Suppose that actions f, g, f' and g' are such that, for all states $s \in E$, $f(s) = f'(s)$ and $g(s) = g'(s)$ while, for all states $s \notin E$, $f(s) = g(s)$ and $f'(s) = g'(s)$; then $f \succsim g$ iff $f' \succsim g'$

In view of P2 we can coherently define a conditional preference relation 'is at least as preferred to, given E', denoted $\succsim_E$, on the set of acts by, for all $f, g \in \mathcal{F}$:

$$f \gtrsim_E g \text{ iff } f' \gtrsim g'$$

where the acts f' and g' are as defined in P2. Conditional preference relations on consequences can then be induced in the same way as before. Given P1, it follows from this definition that each such conditional preference relation is complete and transitive.

The main role of P2 is to ensure that the consequences of an act in each state of the world can be evaluated separately. To see the implications of this, let the set of events $\{E_i\}$ be a partition of $\mathcal{S}$ and let f be an act that has the same consequence in every state in any E_i (hence $f(s) = f(s')$ if s and s' belong to the same element of the partition). Now in view of P1, P2 and the definition of conditional preference there exist numerical representations V_{E_i} of the preference relations $\gtrsim_{E_i}$ on $\mathcal{F}$ (again sticking to the countable case) and corresponding event-dependent utility measures U_{E_i} on $\mathcal{C}$ induced by setting, for all $F \in \mathcal{C}$, $U_{E_i}(F) = V_{E_i}(f)$. Note that fixing on these measures amounts to the choice of a representation satisfying $V_E(f') = V_E(f) \Leftrightarrow f'(s) = f(s)$ for all $s \in E$ (something not imposed by P2). Now P2, together with this choice, makes $V(f)$ a function of the $V_{E_i}(f)$; hence of the $U_{E_i}(F)$. In fact, under some additional technical conditions it can be shown that it is possible to choose the V_{E_i} in such a way as to make $V = \sum_i V_{E_i}$, and hence such that $V = \sum_i U_{E_i}$ – i.e. to represent the value of an act as the sum of the event-dependent utilities of its consequences (see Krantz *et al.*, 1971, for details).

What is now needed is a decomposition of the measure $U_{E_i}(F)$ into a probability for E_i and a utility for F. Then the additive representation will be revealed to be an expected utility. First an assumption is required to ensure the comparability of the event-dependent utilities. Let us call an event $E \in \Omega$ a null event iff $f \approx_E g$, for all $f, g \in \mathcal{F}$. Then Savage postulates:

P3   (*State Independence*) Let $B \in \Omega$ be non-null; if $f(s) = F$ and $f'(s) = G$ for every $s \in B$, then $f \gtrsim_B f' \Leftrightarrow F \gtrsim G$

The State Independence assumption ensures the *ordinal* uniformity across states of preferences for consequences. But it is not strong enough to ensure the *cardinal* comparability of the state-dependent utilities. In particular, although it implies that, for any events E and E', $U_E(F) \geq U_E(G) \Leftrightarrow U_{E'}(F) \geq U_{E'}(G)$, it does not imply that $U_E(F) = U_{E'}(F)$.

The next step is the crucial one for ensuring this as well as for obtaining a probability representation of the agent's attitudes to events. First Savage defines an 'at least as probable as' relation, $\trianglerighteq$, on the set of events. Consider the following pair of actions:

	Events	
Action	*A*	*A'*
f	X	Y

	Events	
Action	*B*	*B'*
g	X	Y

Actions *f* and *g* have the same two possible consequences, but *f* has consequence *X* whenever *A* is the case and *g* has it whenever *B* is the case. Now suppose that consequence *X* is preferred to consequence *Y*. Then *f* should be preferred to *g* iff *A* is more probable than *B* because the action which yields the better consequence with the higher probability should be preferred to one which yields it with lower probability. This suggests a definition of $\unrhd$.

Qualitative Probability Suppose $A, B \in \Omega$; then $A \unrhd B$ iff $f \succsim g$ for all actions *f* and *g* and consequences *X* and *Y* such that:

    (i) $f(s) = X$ for all $s \in A$, $f(s) = Y$ for all $s \notin A$
    (ii) $g(s) = X$ for all $s \in B$, $g(s) = Y$ for all $s \notin B$
    (iii) $X \succsim Y$

In effect, the circumstances postulated by this definition provides a 'test' for when one event is more probable than another. Since it requires that $f \succsim g$ for *every* *f* and *g* meeting the conditions (i) – (iii), the existence of such a test does not itself guarantee that any pair of events can be compared in terms of their relative probability using this test. For this a further postulate is required.

    P4 (*Probability Principle*) $\unrhd$ is complete

In the presence of the other postulates, P4 ensures that preferences for actions depend on two factors only: preferences for consequences and the qualitative probability relation on events. It is not difficult to see that the definition of this latter relation implies that it is transitive. In fact, together with P4, it also ensures that it is quasi-additive – i.e. that, for all events *C* such that $A \cap C = \varnothing = B \cap C$,

$$A \unrhd B \Leftrightarrow A \cup C \unrhd B \cup C$$

Two further structural axioms are then required to ensure that the qualitative probability relation can be represented numerically.

    P5 (*Non-Triviality*) There exist actions *f* and *g* such that $f \succ g$
    P6 (*Non-Atomicity*) Suppose $f \succ g$; then, for all $X \in \mathcal{C}$, there is a finite partition of *S* such that for all $s \in S$:

    (i) $(f'(s) = X$ for all $s \in A$, $f'(s) = f(s)$ for all $s \notin A)$ implies $f' \succ g$
    (ii) $(g'(s) = X$ for all $s \in B$, $g'(s) = g(s)$ for all $s \notin B)$ implies $f \succ g'$

P6 is quite powerful and implies that there are no consequences which are so good or bad that they swamp the improbability of any given event A. Nonetheless, neither it nor P5 raises any pressing philosophical issues. And using them Savage proves:

Existence of Probability (Leonard Savage, 1974/1954) P1–P6 imply that there exists a unique probability function P on $\wp(S)$ such that, for all $A, B \in \wp(S)$:

$$P(A) \geq P(B) \Leftrightarrow A \unrhd B$$

The rest of Savage's argument for the existence of an expected utility representation of the preference relation applies the von Neumann & Morgenstern (2007/1944) representation theorem for preferences over lotteries. This requires one more postulate extending the Sure-Thing Principle to infinite cases. In essence, what needs to be established is a correspondence between each act f and a lottery which yields each possible consequence C with probability $P(f^{-1}(C))$, such that Savage's postulates for preferences over acts with a finite number of consequences imply that the induced preferences over the corresponding lotteries satisfy the von Neumann and Morgenstern axioms. For then the value of each such act can be equated with that of the expected utility of the corresponding lottery. The proof is far from trivial, however, and I will not go any of the details here: see Kreps (1988) for a very useful exposition.

4.5 EVALUATION OF SAVAGE'S AXIOMS

In evaluating Savage's axioms it is useful to distinguish, in the manner of Patrick Suppes (2002), between those axioms expressing a requirement of rationality and those that play a technical or structural role in the proof of the representation theorem. In Suppes' view, only P5 and P6 are structural axioms and the rest rationality conditions. In support of this classification, he notes that only these two conditions make existential demands and that neither is implied by the existence of an expected utility representation of preference. James Joyce (1999) adds P0, the rectangular field assumption, and P1a, the completeness assumption, to the list of structural axioms. It is clear that neither is a rationality condition. Furthermore, both make existential claims of a kind – respectively about the richness of the action space and about the judgemental state of the agent – and neither is necessary for the existence of a numerical representation, though the latter is implied by standard expected utility representations.

The only axioms that are unambiguously putative principles of rational preference are P1b, the transitivity condition, and P2, the Sure-Thing Principle. P4, the Probability Principle, is plausibly a principle of rationality, but it is not really a fundamental principle of rational *preference*. Rather, it is a coherence

TABLE 4.1. *Allais' Paradox*

	Ticket numbers		
Actions	1	2 – 10	11 – 100
f	$1,000,000	$1,000,000	$1,000,000
g	$0	$5,000,000	$1,000,000
f'	$1,000,000	$1,000,000	$0
g'	$0	$5,000,000	$0

constraint on the relation between belief and preferences. Finally, P3 – State Independence – is best regarded, I will argue, not as a pure rationality claim but as a constraint on the interpretation of consequences and states.

4.5.1 The Sure-Thing Principle

The most discussed of Savage's axioms is undoubtedly the Sure-Thing Principle. The main focus of attention in this regard has been apparent violation of the principle in the so-called Allais Paradox, a thought experiment proposed by Maurice Allais (1953). To illustrate it, consider the two pairs of acts that are displayed in Table 4.1, which yield monetary outcomes conditional on the draw of a numbered ticket from a hat containing 100 different ones. Allais hypothesised that many people, if presented with a choice between actions *f* and *g*, would choose *f*, but if presented with a choice between *f'* and *g'* would choose *g'*. Such a pattern of choice is, on the face of it, in violation of the Sure-Thing Principle, since the choice between each pair should be independent of the common consequences appearing in the third column of possible outcomes. Nonetheless, Allais' conjecture has been confirmed in numerous choice experiments. Moreover, many subjects are not inclined to revise their choices even after the conflict with the Sure-Thing Principle is pointed out to them. So the 'refutation' seems to extend beyond the descriptive interpretation of the axiom to include its normative pretensions.

There are two lines of defence that are worth exploring. The first is to argue that the choice problem is under-described, especially with regard to the specification of the consequences. One common explanation for subjects' choices in these experiments is that they choose *f* over *g* because of the regret they would feel if they chose *g* and landed up with nothing (albeit quite unlikely), but *g'* over *f'* because in this case the fact that it is quite likely that they will not win anything whatever they choose diminishes the force of regret. If this explanation is correct then we should modify the representation of the choice problem faced by agents so that it incorporates regret as one possible outcome of making a choice. The same would hold for any other explanation of the observed pattern of preferences that refers to additional non-monetary outcomes of choices.[8]

The second line of defensive argument points to the gap between preference and choice and the fact that the specification of the choice set can influence the agent's attitudes. In general, the attitude we take to having or receiving a certain amount of money depends on our expectations. If we expect $100, for instance, then $10 is a disappointment. Now the expectation created by presenting agents with two lotteries to choose from is quite different in the case where the choice is between lotteries f and g and the one in which the choice is between lotteries f' and g'. In the first case they are being placed in a situation in which they can expect to gain a considerable amount of money, while in the second they are not. In the first they can think of themselves as being given $1,000,000 and then having the opportunity to exchange it for lottery g. In the second case they can think of themselves as being handed some much lesser amount (say, whatever they would pay for lottery f') and then being given the opportunity to exchange it for lottery g'. Seen this way it is clear why landing up with nothing is far worse in the first case than in the second. It is because of what one has given up for it. In the first case landing up with nothing as a result of choosing g is equivalent to losing $1,000,000 relative to one's expectations, whereas in the second case it is equivalent to losing some much smaller amount.

Both of these defences are unattractive from the point of view of constructing a testable descriptive theory of decision making under uncertainty. The first approach makes it very hard to tell what choice situation the agents face, since the description of the outcomes of the options may contain subjective elements. The second approach makes it difficult to use choices in one situation as a guide to those that will be made in another, since all preferences are in principle choice set relative. But from a normative point of view, they go some way to supporting the claim that the Sure-Thing Principle is a genuine requirement of rationality. (We return to this issue in Chapter 9.)

It is worth drawing attention to one further issue. As is evident from the informal presentation of the Sure-Thing Principle, its intuitive appeal rests on the thought that, since only the consequences of an action matter to its evaluation, if the consequences of one act are as least as good as those of another, and are better in at least one event, then this act is better overall. This application of consequentialist reasoning is not valid in general, however, for it will normally matter not just what consequences an action has but how probable it makes them. Two actions could have identical consequences but if one of them brings about the better consequences with a higher probability than the other then it should be preferred to it. Such an eventuality is ruled out in Savage's framework because actions are construed as nothing but ordered sets of consequences, from which it follows that any two actions with the same

---

[8] See Broome (1991) for an extended defence of this kind.

consequences must have the same value (formally it is P0 which is doing the work here, by identifying the set of actions $\mathcal{F}$ with $\mathcal{C}^S$). So the Sure-Thing Principle is compelling within his framework, but only because of the special form that actions take.

4.5.2 State Independence / Rectangular Field

It is not hard to produce apparent counterexamples to State Independence. Consider an act which has the constant consequence that I receive £100 and suppose I prefer it to an act with the constant consequence that I receive a case of wine. Would I prefer receiving the £100 to the case of wine given any event? Surely not: in the event of high inflation, for instance, I would prefer the case of wine. One could retort that receiving £100 is not a genuine consequence since its description fails to specify features relevant to its evaluation. Perhaps 'receiving £100 when inflation is low' might be closer to the mark. More generally, State Independence is bound to hold if we simply take consequences to be combinations of outcomes and the states in which they occur. But then the Rectangular Field assumption forces us to countenance actions which have such consequences in any state of the world, including those inconsistent with them. For example, it would require the existence of acts yielding £100 when inflation is low, in states of the world in which inflation is high. Such acts seem nonsensical and it is hard to see how anyone could express a reasonable preference regarding them.

An objection of this kind was famously made by Robert Aumann in a letter to Savage in 1971.[9] Savage's reply is interesting: he suggests that 'a consequence is in the last analysis an experience'(Drèze, 1990, p. 79) and a state of the agent rather than of the world. The thought seems to be that experiences screen out the features of the world that cause them and hence have state-independent utilities. This is unpersuasive. On the whole, I prefer that I be amused than saddened (or experiencing amusement to sadness), but I surely do not prefer it, given that a close friend has died. So even the desirability of experiences is contingent on the state of the world.

To the objection that his theory countenances nonsensical or impossible acts, Savage retorts that such acts 'serve something like construction lines in geometry' (ibid. p. 79), and that they need not be available in order for one to say whether they would be attractive or not. But he seems to under-appreciate the problem. Consider the decision whether or not to buy a life insurance policy that pays out some sum of money in the event of one's death. Now, the payout is not a state-independent consequence in Savage's sense, for I am not indifferent between being paid while alive and being paid while dead. The natural refinement of it gives us the consequence of 'payout and dead', however, which patently cannot be achieved in any state of the world in which

---

9 Printed, along with Savage's letter in reply, in Drèze (1990).

I am alive. And, even if I could summon a preference for being paid and alive, when alive, to being paid and dead, when alive, why does rationality require that my preference between these two consequences be the same for states in which I am dead? Perhaps, conditional on being dead, I am indifferent between the two.

In summary, State Independence is not plausibly a rationality constraint on preference. It is better to read it as a constraint on the specification of consequences, requiring in effect that they be sufficiently specific as to screen out, from an evaluative point of view, the state of the world. The difficulty with this interpretation is that it conflicts with the Rectangular Field assumption, since sufficiently specific consequences cannot occur in every possible state.

4.5.3 Probability Principle

Although P4 simply asserts the completeness of the 'more probable than' relation $\unrhd$, it is the rationality claim underlying it that needs to be assessed. It is this: if two actions, such as f and g below, have the same two consequences then your preferences between them should depend only on the relative probability of the events determining the more preferred consequence. From this it follows, as illustrated below, that, if $f \succsim g$, $X \succsim Y$ and $X^* \succsim Y^*$ then $f^* \succsim g^*$, as required by P4.

$$\text{If } \frac{\begin{array}{c|cc} & A & A' \\ \hline f & X & Y \end{array}}{\begin{array}{c|cc} & B & B' \\ \hline g & X & Y \end{array}} > \text{ then } \frac{\begin{array}{c|cc} & A & A' \\ \hline f^* & X^* & Y^* \end{array}}{\begin{array}{c|cc} & B & B' \\ \hline g^* & X^* & Y^* \end{array}} >$$

This rationale for P4 depends on the cardinal uniformity of utilities of consequences in different events because if event B makes X more desirable than does event A then you could prefer g to f even if the probability of A was greater than that of B. It also requires that the utilities of the consequences in one state be cardinally independent of the consequences of the act in other states. For example, assume that A and B are equiprobable, but that whether an act has consequence Y or Y^* in case of A' (and B') affects how much more desirable X is than Y, in case of A (and of B). Then you could prefer f to g but g' to f', because the desirability of X (respectively X^*), when it would have been the case that Y (respectively Y^*) if A had not been the case, is greater (less) than the desirability of X (X^*), when it would have been the case that Y (Y^*) if B had not been the case. Neither condition on the rationale for P4 can plausibly be said to be a purely formal one.

Most of the discussion of the Probability Principle has been directed elsewhere, however, at a thought experiment of Daniel Ellsberg (1961). In Ellsberg's experiment (see Table 4.2), an urn contains 90 balls, 30 of which are red, and the remaining 60 are black or yellow in an unknown proportion. Subjects are asked to choose between two bets. The first, L_1, pays off \$100 if, in a random draw from the urn, a red ball is drawn. The second, L_2, pays

TABLE 4.2. *The Ellsberg Paradox*

	Red	Black	Yellow
L_1	$100	$0	$0
L_2	$0	$100	$0
L_3	$100	$0	$100
L_4	$0	$100	$100

off \$100 if a black ball is drawn. Most subjects express a preference of L_1 over L_2. In a second choice problem, subjects are asked to choose between L_3 and L_4, which pay out \$100 in the events 'red or black' and 'black or yellow', respectively. Here, most subjects express a preference for L_4 over L_3.

It is evident from the fact that the 'Yellow' column displays the same consequences for the two pairs of acts that this pattern of preferences – the 'Ellsberg preferences' hereafter – violates the Sure-Thing Principle. But they are also inconsistent with the way in which Savage uses the Probability Principle to elicit subjective probabilities. For it follows from the definition of the qualitative probability relation that $L_1 \succ L_2$ iff the event 'red' is more probable than the event 'black' and that $L_4 \succ L_3$ iff the event 'black or yellow' is more probable than the event 'red or yellow'. But the laws of probability require that 'red' is more probable than 'black' iff, for any event X disjoint with both, 'red or X' is more probable than 'black or X'. Now, strictly speaking, the Probability Principle is not violated independently of Savage's other axioms. But the combination of the Probability Principle and the requirement that $A \unrhd B$ iff $B' \unrhd A'$ is. And it is hard to see what justification there is for the former that does not extend to the latter.

Most supporters of Savage have argued that the Ellsberg preferences are simply irrational and do not therefore constitute a refutation of his theory. Others have drawn the opposite conclusion: that the Ellsberg preferences are rational and that this fact shows that probabilistic degrees of belief are not rationally required in the kinds of situations of severe uncertainty exhibited in the Ellsberg set-up. In Chapter 9, I will argue for a third position: that the Ellsberg preferences are both rational and consistent with Savage's axioms. But some work is required to defend this position, so I will defer discussion of it.

4.6 EVALUATION OF SAVAGE'S ARGUMENT

We have seen that Savage's representation theorem establishes that if an agent's preferences satisfy his postulates then she can be represented as a maximiser of subjective expected utility relative to a probability measure P on the set of events and a utility function U on the set of consequences. Our task now is to assess the significance, both methodological and normative, of this result. My focus here will be on the role that it plays within a pragmatist argument for

Bayesian decision theory of the kind sketched at the beginning of the chapter. But first let me first consider the plausibility of a behaviourist interpretation of it.

To provide support for the methodological claims of Behaviourism, one might read Savage's postulates as conditions on choice (in concert with a Revealed Preference interpretation) and hence his theorem as showing that if they are satisfied by observed choices then quantitative degrees of belief and desire can be attributed to the chooser. But this idea faces a fundamental problem. Savage's theorem starts with preferences over acts, construed as functions from states to consequences. But, *although we can observe choices amongst acts, we cannot observe what the acts are that the agent is choosing between*. This is because we cannot tell from an agent's choices how she conceives of the objects of choice; in particular what consequences she believes to follow from this choice in each possible state of the world. Choice reveals preferences only within a framework of common representation of the objects of choice. It would be natural to achieve this by verbal descriptions of the objects of choice. But if recourse must be had to verbal communication then why not simply ask the subjects what they prefer and dispense with the pretence of purely behavioural evidence?

Behaviourism is perhaps something of a straw figure here (even though its influence in decision theory is considerable), since it is doubtful that Savage was committed to it, so let us turn to an assessment of his theorem within a pragmatist framework. There are two claims that need to be evaluated: the methodological claim that Savage's theorem establishes sufficient conditions for attribution of degrees of belief and desire to agents on the basis of their preferences, and the normative claim that his theorem establishes that rationality requires agents to have probabilistic degrees of belief and to maximise expected utility relative to them. Both claims depend, first, on the status (empirical or normative) of his axioms of preference and, second, on the import of the demonstration of the existence of a particular kind of numerical representation of preferences satisfying them.

Our brief discussion of the first issue did not entirely settle the question of whether all his axioms are either genuine rationality conditions or else 'harmless' structural ones. It is clear, however, that the combination of State Independence and the Rectangular Field assumption is problematic from both an empirical and a normative point of view and it would be better for the pragmatist argument if it could be dispensed with. Perhaps the same is true for the Sure-Thing Principle, at least empirically, but I shall set aside further consideration of it until later in the book.

More pressing is the status of the Probability Principle. This postulate is clearly not purely a principle of rational preference; rather, it is a rationality condition on the relationship between preferences and qualitative beliefs. Furthermore, it is not a condition that anyone is likely to accept unless they independently adhere to the view about the role of belief in preference formation that it expresses. Given the central role that it plays in the derivation of degrees of belief from preferences, this somewhat vitiates the claim that Savage's representation theorem supports the conceptual priority of practical reason, since it undermines the assertion that the properties of rational belief can be derived from independent properties of rational preference. (This objection does not of course apply to the other priority thesis of Pragmatism: of the qualitative over the quantitative.)

Although much of the literature on Savage's theory is focused on the status of his axioms, a more fundamental problem derives from the restrictive nature of Savage's framework. Recall that states of the world must be probabilistically independent of the acts over which preferences are defined. So degrees of belief can be inferred from preferences only for such states. But there will be many features of the world that are not independent of the agent's actions that she nonetheless has beliefs about; for instance, whether she will perform any particular action or not! Since such beliefs fall outside the scope of Savage's theorem, he cannot be said to have established either that all degrees of belief are measurable from preferences or that all partial belief must be probabilistic. A similar point can be made about the restriction of utilities to maximally specific consequences. In a nutshell, Savage's method doesn't yield measures of degrees of belief and desire for all prospects (the background set Ω), but only of those belonging in either the set of events $\wp(S)$ or the set of consequences C.

There is a second, more subtle, problem. Although Savage shows that preferences satisfying his postulates have an expected utility representation unique up to a choice of scale for the utility function, he does not show that such a form of representation is unique. Indeed, it quite clearly isn't. For any agent who maximises expected utility relative to a probability measure P and utility U also maximises a function V' defined by

$$V'(f) = \sum_{i=1}^{n} \alpha_i U(f(s_i)) \cdot P'(s_i)$$

where $\alpha_i = nP(s_i)$ and $P'(s_i) = \frac{1}{n}$. So the agent can be represented as if she assigns equal probability to each state of the world, but has utilities for consequences that vary with the state in which they occur. The choice of a cardinally state-independent representation is an entirely arbitrary one, however, not required by Savage's axioms (recall that State Independence is a purely ordinal requirement). This being so, we have no axiomatic basis for saying that the agent's preferences determine that her degrees of belief are measured by P rather than P' (see Karni, 1985, for more details).

The problem can be expressed in a slightly different way. The Probability Principle allows a qualitative probability ordering over events to be constructed from preferences over acts. But why should we take the constructed ordering to represent the agent's relational beliefs? The most plausible answer is that the constructed ordering plays just the role in the determination of the agent's preferences that one would expect of them. But, as I have already shown, this argument presupposes that an agent's preferences for the consequences of her actions should be cardinally independent of both the state in which they occur and of the counterfacts regarding what the consequence would have been if some other state had been the actual one. For if they were not then her beliefs would not combine with her desires in the manner required by the Probability Principle. Neither presupposition is guaranteed by Savage's postulates. So, even if we regard the Probability Principle as a rationality constraint, we cannot infer that the qualitative probability relation constructed from the preference relation in accordance with it correctly represents the agent's relational beliefs and so we cannot conclude that she maximises expected utility relative to them. In summary, an agent who satisfies Savage's postulates maximises expected utility relative to some probability P, but not necessarily relative to her actual degrees of belief.

Savage's representation theorem does not, it seems, deliver the goods. To plausibly interpret his theorem in the manner required by the pragmatist argument for Bayesian decision theory, states and consequences must be specified in a way which ensures the cardinal independence of preferences for consequences from both states and the counterfacts. But this then considerably restricts the scope of any conclusions that can be drawn from his theorem, for many prospects with respect to which we have both beliefs and desires will not meet these conditions. Furthermore, his theorem does not establish that rationality requires maximisation of expected utility relative to the agent's beliefs. To give conceptual and normative foundations to Bayesian decision theory, we will have to do better. In the next part of the book, I hope to do so.

PART II

PROSPECTIVE RATIONALITY

5

Rational Belief and Desire

To make decisions, agents must have a range of capabilities. Firstly, they must have an ability to represent, and discriminate between, different possible features or states of their environment. Secondly, they must be able to perceive or infer the presence or absence of these features, i.e. to form beliefs about the world. Thirdly, they must be able to compare these possible features or states according to some standard of value – i.e. be able to judge whether they are more or less desirable. And, finally, if they are to be capable of not just thinking about the world but also acting on it, agents need to be able to determine what options they have, to evaluate them and to exercise them. In summary, they must be able to identify both the possible states of the world and their options for changing them and to form cognitive and evaluative attitudes towards both.

This list of conditions may seem quite demanding, but it should be borne in mind that there is no requirement that these judgements be at all refined or detailed, or even that they be conscious. So even cognitively quite primitive entities may well have most or all of the required capabilities. Consider a thermostat, for instance. It inarguably represents features of the world – the temperature – and judges which of these states in the actual one. It also has a crude value system that can be summarised as: cold is bad, warm is good. The actions available to it – switching the heating on or off or doing nothing – are chosen in the light of information it receives about the state of the world and implicitly in view of the consequences of selecting the action.

On the other hand, the way the thermostat makes decisions, if one is happy to speak in such terms, lacks a number of the essential characteristics of human decision making. Although we might say that the reason why the thermostat switches the heating on is that the room is cold and that by doing so it will warm up, the reason we refer to is not the cause of the thermostat's switching on. The thermostat does not itself make the judgements about what would happen if it performed one or other of the actions available to it. This judgement is made by the person programming the thermostat and adjusting its settings to the environment in which it is employed. The thermostat, unlike the programmer, lacks the ability to make suppositions of this kind and deliberate on the basis of them.

To capture this difference between humans and thermostats, let us speak of two kinds of evaluation of options or actions. *Reactive* evaluation is when an action is selected on the basis of a set of judgements, cognitive and evaluative, concerning only the actual state of the world. *Prospective* evaluation requires something more: the action must be selected in virtue of the anticipated consequences of so doing. This requires not just judgements about the current state of the world but also judgements about possible, and even counterfactual, states. The agent who picks an action because its expected consequences are better than those of the alternatives must be able to represent and evaluate possible states which may or may not be realised at some point, in part depending on what the agent does. In short, they must be able to form both cognitive and evaluative attitudes to possible states of the world on the supposition that they were to perform one or another action.

There is a second salient difference between humans and thermostats: thermostats, unlike humans, cannot learn from experience. Once programmed, a thermostat will switch on and off in response to signals about ambient temperature without regard either for other changes in the environment or for the efficacy of these actions in achieving the goal of creating a comfortable environment. The difference between the range of environmental changes that humans and thermostats are sensitive to does not make for a fundamental difference between them but is, rather, one of degree or complexity. On the other hand, the difference in their ability to re-evaluate the relationship between states of the environment and actions on the basis of experience is fundamental. Thermostats cannot reprogramme themselves if they are failing to achieve their goals. Humans, on the other hand, assess actions and their consequences not only prospectively but also *retrospectively*, with a view to reappraisal of the appropriateness and efficacy of actions in achieving goals, and, indeed, with a view to reappraisal of the goals themselves.

My aim in this part of the book is to investigate the nature of rational prospective evaluation and, in particular, the question of what states of mind, or judgemental states, are rational. This chapter will lay the foundations of the account that I will develop by describing the theory of rational belief and desire developed by Richard Jeffrey (1990/65). In subsequent chapters I will extend this account, first to conditional attitudes and then to attitudes to conditional prospects. In the next part of the book, I will tackle two related questions. What actions is it rational to choose, given one's judgemental state? And how should one revise one's attitude in response to experience?

5.2   PROSPECTS

The entities that agents take attitudes to, that they believe to be the case, wish were the case, try to bring about, and so on, will be termed 'prospects'.

Prospects include not only possible states of affairs – such as that it will snow in March or that interest rates will rise or that new oil reserves will be discovered – but also consequences or outcomes of actions – such as that I will lose my job or enjoy a holiday by the sea – and the actions themselves – such as that I will refuse to teach logic or that the jury will find the defendant guilty. They also include conditional possibilities – such as that the government will fall if interest rates rise, that crops will be damaged if it snows in March and that I would lose my job if I were to refuse to teach logic. (Prospects are typically termed 'propositions' by philosophers but, at least initially, I will avoid this term, because of the philosophical baggage that comes with it.)

Mainstream decision theory divides prospects into states, which serve as objects of belief, and consequences, which serve as objects of desire. But this distinction, though pragmatically useful, is not fundamental. What counts as a state or a consequence is context-dependent: rain tomorrow is a state of affairs for a couple planning a picnic, a consequence for a shaman performing a rain dance. So, like Jeffrey, I regard all prospects as potential objects of both belief and desire (and, indeed, of other, more complex attitudes). On the other hand, I depart from Jeffrey in explicitly recognising a distinction between unconditional or factual prospects and conditional ones. Or, to put it differently, Jeffrey's theory addresses only attitudes to factual prospects, while in ours attitudes to conditional prospects will play a central role. This feature is crucial, for it is what will allow us to unpack the role played by conditionals in the kinds of hypothetical reasoning that agents engage in when determining what course of action to follow.

As the above examples reveal, prospects are typically identified by natural-language declarative sentences. This is not to say that beliefs and desires are directed at sentences. They are not: it is the features of the world that these sentences pick out – their meanings or contents – that we take attitudes to, not the sentences themselves. But language is important in the development of complex attitudes and essential to their representation and communication. Furthermore, prospects inherit the structure of the propositions expressed by the sentences that we use to identify them. For instance, given any prospect – say that it will snow in March – we can identify (using negation) the complementary prospect, namely that it will not snow in March. Given a second prospect – say that crops will fail – we can speak (using the sentential connectives 'and', 'or' and 'if...then') of the joint prospect of it snowing in March and the crops failing; the disjoint prospect of it snowing in March or the crops failing; and the conditional prospect of the crops failing if it snows in March. And any others that our linguistic and cognitive resources will permit.

Although language is our guide, on some matters it does not speak clearly. In ordinary language the 'same' conditional sentence can have both an indicative and subjunctive form, and there is a good deal of debate as to whether this grammatical difference marks a fundamental difference in semantic content. Certainly, there should be no denying that indicative and subjunctive versions of the same sentence can be evaluated quite differently, as is displayed in Ernest Adams' famous example of the difference between 'If Oswald didn't kill Kennedy, then someone else did' (which is very probably true) and 'If Oswald hadn't killed Kennedy, then someone else would have' (which is quite probably false). A good theory of conditionals should be able to explain these differences. Equally, it should be able to explain the many similarities in the behaviour of the two kinds of conditional.

Complex prospects stand in the same logical relationships to one another as the sentences that identify them: logical relationships that impose significant constraints on the corresponding relationships between the attitudes we take to them. For instance, if one prospect entails another, as the prospect that it will rain on Tuesday entails that it will rain before Friday, then the latter must be believed to no lesser degree than the former. So the assumptions that we make about these logical relationships will affect the account of rationality that we offer. For the theory presented here the basic requirement is that the logic of prospects is that of propositional logic supplemented with some rules for conditionals. This reflects our interest in hypothetical reasoning and decision making under uncertainty. To address other aspects of rational valuation it would make sense to adopt richer logics: for instance, temporal logic for the study of inter-temporal evaluation and predicate logic for multi-criteria evaluation.

5.2.1 Vocabulary

Let us introduce some of the formal vocabulary that will be used throughout. I will use italic Latin capitals to denote factual prospects and lower-case Greek letters to denote arbitrary prospects, both conditional and unconditional. No attempt will be made at this point to distinguish prospects from the sentences that describe them. For any prospects α and β, we denote the prospect that not α' by $\neg\alpha$, the prospect that both α and β by $\alpha \wedge \beta$ (usually shortened to $\alpha\beta$) and the prospect that either α or β by $\alpha \vee \beta$. The necessary and impossible prospects will be denoted by $\top$ and $\bot$, respectively. If α and β are the same prospect (whatever the sentence picking them out) we will write $\alpha = \beta$.

I will denote the (indicative) conditional prospect that if α is the case then β is by $\alpha \mapsto \beta$, but not introduce distinct operations for the other kinds of conditionals. Instead, any conditional prospect that if α is or were the case then β is or would be will be denoted $\alpha \to \beta$. The symbol $\to$ thus denotes a conditional operator *variable* while $\mapsto$ denotes one of its possible values. A conditional with factual antecedent and consequent will be called a simple conditional. If $\alpha\beta = \bot$ then the conditionals $\alpha \to \gamma$ and $\beta \to \delta$ will be said to be orthogonal to one another. A set of prospects $\mathcal{A} = \{\alpha_1, \alpha_2, ..., \alpha_n\}$ is called an *n*-fold partition just in case, for $\alpha_i, \alpha_j \in \mathcal{A}$, $\alpha_i \wedge \alpha_j = \bot$ and $\alpha_1 \vee \alpha_2 \vee ... \vee \alpha_n = \top$. If $\{\alpha_i\}_{i=1}^{n}$ is such a partition then the prospect $(\alpha_1 \to \beta_1)(\alpha_2 \to \beta_2)...(\alpha_n \to \beta_n)$ will be called an (*n*-fold) partitioning conditional and will be denoted by $\bigwedge(\alpha_i \to \beta_i)$. Partitioning conditionals play a prominent role in our deliberations about what to do and what might happen, but they are poorly understood. One of our tasks will be to explain some features of this role.

5.2.2 Boolean Algebras*

Let X be a set of prospects closed under a unary (negation) operation $\neg$, a binary (conjunction) operation $\wedge$, and a binary (disjunction) operation $\vee$. Then the structure $\Omega = \langle X, \wedge, \vee, \neg, \top, \bot \rangle$ is called a *Boolean algebra* of prospects just in case the following laws hold:

Normality	$\alpha \vee \neg\alpha = \top$	$\alpha \wedge \neg\alpha = \bot$
Commutativity	$\alpha \vee \beta = \beta \vee \alpha$	$\alpha \wedge \beta = \beta \wedge \alpha$
Associativity	$\alpha \vee (\beta \vee \gamma) = (\alpha \vee \beta) \vee \gamma$	$\alpha \wedge (\beta \wedge \gamma) = (\alpha \wedge \beta) \wedge \gamma$
Idempotence	$\alpha \vee \alpha = \alpha$	$\alpha \wedge \alpha = \alpha$
Absorption	$\alpha = \alpha \vee (\alpha \wedge \beta)$	$\alpha = \alpha \wedge (\alpha \vee \beta)$
Distributivity	$\alpha \wedge (\beta \vee \gamma) = (\alpha \wedge \beta) \vee (\alpha \wedge \gamma)$	$\alpha \vee (\beta \wedge \gamma) = (\alpha \vee \beta) \wedge (\alpha \vee \gamma)$

Now, from the Boolean algebra so specified, we can define an implication relation $\models$ on X by

$$\alpha \models \beta \Leftrightarrow \alpha \vee \beta = \beta \Leftrightarrow \alpha \wedge \beta = \alpha$$

in which case the Boolean laws ensure that the logic induced by $\models$ is classical propositional logic.[1] It is also possible to begin by specifying the properties of $\models$ and then defining the Boolean operations in terms of it in such a way as to ensure satisfaction of the Boolean laws. Indeed, hereafter I will take a Boolean algebra of prospects to be just a structure $\langle X, \models \rangle$ consisting of a set X of objects together with the right kind of implication relation, with X' denoting the set $X - \{\bot\}$ and Ω' the corresponding algebra with the element $\bot$ removed.

An *atom* of a Boolean algebra $\Omega = \langle X, \models \rangle$ is a non-contradictory element that is implied by no other element in the algebra – i.e. an $\alpha \in X$ such that $\alpha \neq \bot$ and if $\beta \models \alpha$ then $\alpha = \beta$. An *atomless* Boolean algebra is one containing no atoms; a canonical example is the set of all subsets of the unit interval $[0,1]$. When X is finite, then it contains atoms, but we will often need to consider infinite sets that don't. A *complete* Boolean algebra is one in which every set S of elements has an upper and a lower bound relative to $\models$ – i.e. there exist elements α and β such that, for any $\gamma \in S$, $\gamma \models \alpha$ and $\beta \models \gamma$.

A Boolean subalgebra of Ω is simply a Boolean algebra $\langle Z, \models_Z \rangle$ with $Z \subseteq X$ and $\models_Z$ the restriction of $\models$ to Z (the subscript will often be suppressed). There are two salient ways of forming subalgebras: by coarsening the original algebra and by restricting it. An algebra is *coarsened* by a partition π of it by removing all elements in the algebra that imply any element of π. An algebra is *restricted* to a subset Z by removing all the elements not implying $\bigcup\{\varrho_i \in Z\}$. For example, consider the Boolean algebra based on the set of prospects $\{\alpha, \beta\}$. Then the subalgebra formed by restricting it to prospects implying α will be the set $\{\alpha, \alpha\beta, \alpha\neg\beta, \bot\}$, while the subalgebra formed by coarsening it by (removal of) $\{\beta, \neg\beta\}$ will be the set $\{\top, \alpha, \neg\alpha, \bot\}$. More formally, for any $\alpha \in X$, define

---

[1] See Davey & Priestley (2002) for details.

the *principal ideal* generated by α as the set $A := \{\beta \in X : \beta \models \alpha\}$. Then $\Omega_A = \langle A, \models \rangle$ is a Boolean subalgebra of Ω with unit α and zero $\perp$, called a *restricting* subalgebra of Ω. On the other hand, if $\Omega_Z = \langle Z, \models \rangle$ is a subalgebra of Ω with the same unit as Ω, it is called a *coarsening* subalgebra of Ω.

5.2.3 Conditional Algebras*

Let $\Omega = \langle X, \models_X \rangle$ be a Boolean algebra and A a subset of X. Let Y be defined by, for all $\alpha \in A$ and $\beta \in X$, $\alpha \to \beta \in Y$. Then $\Gamma = \langle \Omega, A, Y, \models, \to \rangle$ will be called a *conditional algebra* based on Ω iff (i) Γ is a Boolean algebra, (ii) Ω is a Boolean subalgebra of Γ and (iii) for any $\alpha, \beta, \gamma \in A$ such that $\alpha \neq \perp$:

Conditional Coherence $\alpha \wedge \beta \models \alpha \wedge \gamma \Leftrightarrow \alpha \to \beta \models \alpha \to \gamma$

Conditional Boundedness $\alpha \to \top = \top$

Conditional Complementation $\neg(\alpha \to \beta) = \alpha \to \neg\beta$

Conditional Normality $\top \to \alpha = \alpha$

Conditional Distributivity

1. $\alpha \to (\beta \wedge \gamma) = (\alpha \to \beta) \wedge (\alpha \to \gamma)$
2. $\alpha \to (\beta \vee \gamma) = (\alpha \to \beta) \vee (\alpha \to \gamma)$

The only controversial property here is Conditional Complementation, which implies the law of Conditional Excluded Middle: that $(\alpha \to \beta) \vee (\alpha \to \neg\beta)$ is a logical truth. We will return to this issue in Chapter 8, but for the moment I will take it to be an essential property of a conditional algebra. Some other properties that are often regarded as fundamental will, on the other hand, be treated here as features of only a special, if salient, class of such algebras. In particular, a conditional algebra $\Gamma = \langle \Omega, A, Y, \models, \to \rangle$ will be said to be *regular* iff, for all $\alpha, \beta, \gamma \in A$:

Centring $\alpha \wedge \beta \models \alpha \to \beta$

Modus Ponens $\alpha \wedge (\alpha \to \beta) \models \beta$

When the set Y of prospects contains only simple conditionals – conditionals with factual antecedent and consequent – the conditional algebra Γ will be called a *simple* conditional algebra. Simple conditional algebras have a very important property, namely that they can be thought of as the union of a set of Boolean algebras of sets of conditionals with identical antecedent. This follows from the following theorem.

Theorem 5.1. *(Bradley, 2007c) Suppose that $\Gamma = \langle Y, \models \rangle$ is a simple conditional algebra based on $\Omega = \langle X, \models \rangle$. Let $Y_\alpha := \{\alpha \to \beta \in Y : \beta \in X\}$. Then $\Gamma_\alpha = \langle Y_\alpha, \models \rangle$ is a coarsening Boolean subalgebra of Γ. Furthermore, if Ω is atomless, then so too is Γ_α.*

We will not need to commit ourselves to much regarding conditionals that are not simple. But one property of conditionals with consequents that are themselves conditionals (the so-called 'right-nested conditionals') will turn out to be of interest, namely:

Import-Export $\alpha \to (\beta \to \gamma) = (\alpha \wedge \beta) \to \gamma$

The Import-Export property allows the reduction of right-nested conditionals to simple ones. As we shall see, it is the defining characteristic of indicative conditionals. For this reason, we call any conditional algebra that satisfies import-export an *indicative* algebra.

5.3 ATTITUDES

We can take a wide variety of attitudes to prospects: we hope to avoid illness, we are dismayed by a selfish action, we doubt that the project will be completed on time. But the focus here will be on just belief and desire, which I take to be representative of two broad classes of attitudes: the cognitive ones, which include believing, accepting, doubting and supposing that something is true; and the evaluative ones, which include desiring, wishing, needing and preferring that something be true.

In both common usage and scientific theory, attitudes such as belief and desire are sometimes treated as all-or-nothing affairs and sometimes as states that comes in degrees or, at least, gradations. In the former, categorical mode we speak of believing that it will rain today and of wanting to visit Berlin. In the latter, gradational mode we speak of it being more likely to rain tomorrow than today and of wanting to go to Istanbul more than to Berlin. The categorical mode is typically foremost in discourse and reasoning, the gradational mode in explanations and justifications of action when a balancing of relative strengths of belief and desire is required. It is not surprising, therefore, that different explanatory concerns have motivated different ways of modelling these attitudes.

In the more formal literature, on the other hand, three types of model of the attitudes predominate.

1. In *categorical* models, an attitude is simply represented by its content – the prospect that is the object of the attitude – or by a sentence that picks out this prospect. Correspondingly, the attitudinal state of the agent is represented by a set of propositions or sentences.
2. In *numerical* models, an attitude is represented by a prospect–number pair, with the prospect again representing the object of the attitude and the number the force or strength with which it is held. Correspondingly, an attitudinal state is represented by a real-valued function on the set of all possible contents.

3. In *relational* models, an attitude is represented by a comparison between two prospects expressing the fact that one prospect is more credible, desirable, etc., than the other. Correspondingly, an attitudinal state is represented by a comparative relation on the set of all relevant prospects, representing how each stands in relation to others from the point of view of credibility, desirability, etc.

Decision theory makes little use of the categorical model of the attitudes because the division of prospects into those that are believed or desired and those that are not is simply too crude to permit the formulation of sophisticated decision principles. Instead, agents are typically modelled as having precise numerical degrees of belief (probabilities) and degrees of desire (desirabilities or utilities) for all relevant prospects. This is a practice that we will follow in many parts of the book. But there are a number of reasons for not being satisfied with working with this numerical model alone.

Firstly, categorical notions of belief appear in important theories of belief revision and aggregation, which we will draw from later on. Secondly, and more importantly, the quantitative models of decision theory are in a sense *too* rich, for agents are not typically in a position to make precise numerical cognitive and evaluative judgements. Nor does rationality require that they do so. I may judge something to be more likely than something else without being able to say to what degree. Sometimes I may not even be able to say that. For instance, I might think it more likely that it will rain than that it will snow tomorrow, but be unable to say whether snow or hail is more likely. Given the lack of information I hold about tomorrow's weather and my lack of expertise in judging weather patterns, it would surely not be required of me that I should form definite views about these questions. Indeed, it is arguable that it is wrong for me to do so.

For this reason, I will adopt the relational view of attitudes and take the basic facts of interest to be that an agent believes α more than β or prefers α to β, rather than that they believe or desire that α or even that they believe or desire to degree x that α. There are a couple of advantages to taking this approach. The first is the generality of the relational model, since both categorical and numerical models are derivable as special cases of relational models. When the comparative belief relation is very crude, for instance, so that all prospects belong to one of two equivalence classes (the 'believed' class and the 'not believed' class), then the agent's relational belief state is fully captured by the set of prospects that she believes. On the other hand, if the set of prospects is sufficiently rich and the belief relation satisfies certain structural properties then the relation will be representable by a numerical function, thereby generating a quantitative model (more on this later).

Secondly, although comparative relations impose some kind of order or ranking on prospects, the order doesn't have to be complete: there can be prospects that haven't been or can't be compared from the point of view of the relational attitude. Since only complete relations determine quantitative models of the attitude they represent, moving from the quantitative to the relational increases the scope of our discussions.

Thirdly, adopting a relational view allows a characterisation of the different attitudes in terms of the rationality properties of the comparative relation between prospects that they satisfy. The distinguishing characteristic of cognitive relations, for instance, is that they are *monotonic*: they always rank logically weaker prospects higher than logically stronger ones.[2] If one prospect implies another then the former cannot be more credible/acceptable/plausible than the latter. Suppose, for instance, I am waiting for a package to arrive containing documents that I need in order to write a report, but don't know which day it will arrive on. Then I should regard it as more credible that it will arrive on either Monday or Tuesday than that it will arrive on Monday.

This is not true of the evaluative relations: it is not the case that logically weaker prospects are always more highly valued. On the contrary, some evaluative relations always rank logically weaker prospects *between* the mutually exclusive stronger ones that imply them. I will call such evaluative relations the *averaging* relations. Preference is a canonical example. Suppose the aforementioned report must be done by the end of the week and the sooner it is done the better. It is clear that I would prefer that the package arrived on Monday than on Tuesday, and on Tuesday than on Wednesday, and so on. I would therefore prefer that it arrived on Monday rather than either on Monday or on Tuesday. The reason is that the latter prospect (of it arriving on either day) is consistent with it arriving on Tuesday, the less preferred day, whereas the former prospect is not. Or, to put it slightly differently, the former prospect gives me the package on Monday for sure, whereas the latter gives me an only chance of having it on Monday.

Not all evaluative attitudes are averaging relations, however. A case in point is the 'more important that' relation. Even though I prefer my package to come on Monday than either Monday or Tuesday, it is more important to me that the package arrives on either Monday or Tuesday than that it arrives on Monday and, more generally, that it arrives before the end of the week than that it arrives on any particular day of the week. This is because there is more at stake in my getting it before the end of the week than my getting it on any particular day. Because there is more at stake, there is more of a risk associated with not getting the package by the end of the week than not getting it on any particular day. The risk associated with an action is thus neither simply the probability nor simply the undesirability of its possible consequences, but the importance to the agent of these negative consequences not arising.[3]

---

[2] More exactly, this is true of the 'positive' cognitive attitudes. The 'negative' attitudes, such as doubt, always rank logically weaker prospects lower.

[3] I should really say that *sometimes* we mean this when we speak of risk. As Hansson (1996b) points out, there are many different (and conflicting) usages of the term.

Many cognitive relations, and belief in particular, are *separable* with respect to exclusive disjunction. If I regard it as more likely that the package will arrive on Monday than that it will arrive on Tuesday, then I should regard it more likely that it will arrive on Monday or Wednesday than on Tuesday or Wednesday; and vice versa. This is because the relative likelihood of it arriving on Monday rather than Tuesday does not depend on the likelihood of it arriving on Wednesday (or any other day). In other words, the comparison of Monday and Tuesday can be separated from the question of the likelihood of arrival on Wednesday. The same is true of some evaluative relations as well; some, but not all, are separable with respect to exclusive disjunction. The importance relation is one such case. If it is more important that the package arrives on Monday than on Tuesday, then it should be more important that it arrives on either Monday or Wednesday than on either Tuesday or Wednesday.

5.3.1 Comparative Relations

In our formal treatment of relational attitudes, our starting point will be attitudes that are weak relations on the set of prospects, such as the 'at least as credible as' and the 'at least as desirable as' relations. Strict relations, such as 'the more credible than' and 'the more desirable than' relations, and equivalence relations, such as the 'equally credible as' and 'equally desirable as' relations, will be defined in terms of them.

Formally, let R be a binary relation on a set of objects X, with $\alpha R \gamma$ meaning that α is R-related to γ. Binary relations are typically classified in terms of their satisfaction (or otherwise) of the following properties. For all $\alpha, \gamma, \delta \in X$:

Reflexivity $\alpha R \alpha$

Transitivity If $\alpha R \gamma$ and $\gamma R \delta$ then $\alpha R \delta$

Symmetry $\alpha R \gamma \Leftrightarrow \gamma R \alpha$

Completeness $\alpha R \gamma$ or $\gamma R \alpha$

The relation R is said to be a *weak pre-order* iff it is reflexive and transitive and a *weak order* iff it is also complete. Given any weak pre-order R, we can define an associated strict pre-order P and equivalence relation I, respectively the asymmetric and symmetric parts of R, by

$$\alpha P \gamma \text{ iff } \alpha R \gamma \text{ but not } \gamma R \alpha$$

$$\alpha I \gamma \text{ iff } \alpha R \gamma \text{ and } \gamma R \alpha$$

Our main interest in this book is in weak pre-orders that represent an agent's attitudes to some set of prospects. In this context the following additional properties are important.

TABLE 5.1. *Attitude Types*

	Attitude type		
Properties	Separable cognitive	Averaging	Separable evaluative
Transitivity	X	X	X
Monotonicity	X		
∨-Betweenness		X	
∨-Separability	X		X

Non-Triviality There exist $\alpha \in X$ such that $\alpha P \top$

Boundedness $\top R \alpha$ and $\alpha R \bot$

Monotonicity $\alpha \models \gamma$ then $\gamma R \alpha$

∨-Betweenness If $\alpha \gamma = \bot$, then $\alpha R (\alpha \vee \gamma) \Leftrightarrow (\alpha \vee \gamma) R \gamma$

∨-Separability: If $\alpha \delta = \bot = \gamma \delta$, then $\alpha R \gamma \Leftrightarrow (\alpha \vee \delta) R (\gamma \vee \delta)$

It is clear that these properties of pre-orders are not independent. Firstly, Monotonicity implies Boundedness. Secondly, given Non-Triviality, a weak pre-order cannot be symmetric, nor can it satisfy both Boundedness and ∨-Betweenness (since ∨-Betweenness requires that, for all α, either $\alpha R \top$ or $\neg \alpha R \top$). Hence ∨-Betweenness implies non-monotonicity. Thirdly, if a weak pre-order is ∨-separable and bounded then it is monotonic. (*Proof:* Suppose $\alpha \models \beta$. Then $\neg \beta = \neg \alpha \wedge \neg \beta$. So, by Boundedness, $(\beta \vee (\neg \alpha \wedge \neg \beta)) R (\alpha \vee (\neg \alpha \wedge \neg \beta))$ and, by ∨-Separability, $\beta R \alpha$.) Clearly, monotonicity is inconsistent with ∨-Betweenness. So no bounded weak pre-order can satisfy both ∨-Separability and ∨-Betweenness.

This gives us three salient classes of relations: (1) the bounded, monotonic, ∨-separable relations (called the separable cognitive relations); (2) the non-bounded, ∨-separable relations (called separable evaluative relations) and (3) the non-bounded, ∨-between relations (called averaging relations). Their essential properties are summarised in Table 5.1.

Hereafter we will focus on just two particular weak comparative relations that, respectively, represent the agent's relational beliefs and relational desires or preferences over prospects.[4]

(1) The first, the agent's weak *credibility* or 'at least as credible as' relation, will be denoted by $\unrhd$ and assumed to be a separable cognitive relation – i.e. a non-trivial, transitive, monotonic and ∨-separable pre-order on prospects. Its symmetric and asymmetric parts will be denoted by $\bowtie$ and $\rhd$.

---

4 I assume that relational desires and preferences are the same thing.

(2) The second, the agent's weak *preference* or 'at least as preferred as' relation, will be denoted by $\succsim$ and will be assumed to be an averaging relation – i.e. a non-trivial, transitive and $\vee$-between pre-order. Its symmetric and asymmetric parts will be denoted by $\approx$ and $\succ$, respectively.

5.3.2 Numerical Representation of Comparative Relations*

Let us say that a weak comparative relation R on a set X is numerically represented by a real valued function F just in case for all $\alpha, \gamma \in X$, $\alpha R \gamma \Leftrightarrow F(\alpha) \geq F(\gamma)$.

A question that is of considerable importance both methodologically and normatively in the theory of rational agency is that of the conditions under which a numerical representation exists for a comparative relation. For, when it does, we can substitute comparative talk of one prospect being at least as credible or desirable as another with simpler talk of the credibility or desirability of each of the prospects. And we can justify claims about the latter in terms of properties of the former.

The first observation to be made in this regard is that for any relation to be numerically representable, in the sense of being measurable by a single real-valued function, it is necessary that the relation be complete. Weak orders are numerically representable in a wide range of cases, though far from uniquely. The following theorem, due originally to Georg Cantor, makes this claim precise.[5]

Theorem 5.2. *Let R be a weak order on a set X. If X is countable then there exists a function F that is a numerical representation of R. Furthermore, F' is another such representation iff F' is a positive monotone transformation of F – i.e. there exists a strictly increasing function $f : \mathbb{R} \to \mathbb{R}$ such that $F' = f \circ F$.*

In case X is not countable, numerical representability is not assured for weak orders unless X has an 'order-dense' subset – one containing elements lying (in the weak order) between any two prospects in X. When the preference order is lexicographic, for instance, this condition will be violated (see Kreps, 1988, for details). An atomless Boolean algebra is uncountable, but a weak order on it admits of numerical representation so long as the algebra is complete and the weak order continuous, where the continuity of a weak order R on a Boolean algebra $\langle X, \models \rangle$ is defined as follows:

Continuity Let $\{\alpha_1, \alpha_2, ...\}$ be a countable set such that $\alpha_1 \models \alpha_2 \models ...$; suppose that $\beta R \alpha_i$ and $\alpha_i R \gamma$, for all i; then $\beta R \bigcup \{\alpha_i\}$ and $\bigcup \{\alpha_i\} R \gamma$

Most of the interesting results come from imposing further conditions on such weak orders, and especially those representing rationality constraints,

5 See Krantz *et al*., 1971, chap. 2.1) for a proof.

and investigating the implications for the numerical representation. For the moment, however, let us return briefly to the issue of completeness, even though incomplete attitudinal relations will not be studied until the second half of the book. Completeness is not an essential property of either relational belief or desire, nor is it a requirement of rationality. So the fact that it is necessary for numerical representability may seem unfortunate. But there is an alternative to numerical representability which gives us much of what we want. Let us say that a weak relation R on a set X is represented by a set of real-valued functions Φ just in case for all $\alpha, \gamma \in X$,

$$\alpha R \gamma \Leftrightarrow \forall F \in \Phi, F(\alpha) \geq F(\gamma)$$

Then, remarkably:

Theorem 5.3. *(Evren & Ok, 2011) Let R be a weak pre-order on a set X. Then there exists a set Φ of real-valued functions that represents R.*

5.4 RATIONAL BELIEF AND DESIRE

We have made a number of different distinctions: between cognitive and evaluative attitudes, between attitudes to factual and conditional prospects and between unconditional and conditional attitudes. For an agent to be consistent, these various types of attitude need to cohere in a specific way: her desires must reflect her beliefs, her beliefs regarding conditionals must reflect her conditional beliefs and so on. When they cohere completely then we can say that she is in a state of rational equilibrium. We are probably never in such a state because it takes time to think through the consequences of our judgements and because we are constantly having to revise them as we get new information about the world. Nonetheless it serves as an ideal reference point for our deliberations.

In this section I will offer an account of what consistency requires of an agent with precise degrees of belief and desire, both conditional and unconditional, for various kinds of prospects. This might seem strange in the light of my insistence that agents often don't have precise attitudes and are not rationally required to do so. But, even though our attitudes are imprecise, and reasonably so, they are still indirectly constrained by the rationality conditions on precise ones. For one very basic property that these imprecise attitudes must have is that it be possible to make them precise without landing up in an inconsistent state of mind. And we need a theory of rational precise attitudes to tell us what consistency means here.

A second motivation is that looking at precise attitudes helps to tease out the implications of the various properties of relational attitudes that have been proposed, by showing what they entail in the richest of mental environments, namely those in which the agent has attitudes to all prospects. In fact, this relationship between the two of them is central. For the fact that the rationality properties of relational belief and desire have certain implications for the precise numerical belief and desire serves to justify, or give foundation to, claims about properties of the latter. The argument has the following form: (1) rationality requires a property of a relational attitude, (2) the relational attitude having this property implies that any numerical representation of it has the corresponding property, so (3) the corresponding property is a requirement of rationality on the corresponding numerical attitude.

5.4.1 Numerical Degrees of Belief and Desire

Let us start by giving an exact statement of the properties that degrees of belief and desire should have. Suppose that the degree to which an agent believes that any prospect is true is measured by a real-valued function P. Likewise, suppose that V is a real-valued function on prospects that measures the degree to which she desires that they be true. Then these functions are, respectively, a probability and a desirability on the set of prospects iff they satisfy, for all prospects α and β:

Axioms of Probability

> P0 (*Non-Negativity*) $P(\alpha) \geq 0$
> P1 (*Normality*) $P(\top) = 1$
> P2 (*Additivity*) If $\alpha\beta = \bot$, then:

$$P(\alpha \vee \beta) = P(\alpha) + P(\beta)$$

Axioms of Desirability

> V1 (*Normality*) $V(\top) = 0$
> V2 (*Desirability*) If $\alpha\beta = \bot$, and $P(\alpha \vee \beta) \neq 0$, then:

$$V(\alpha \vee \beta) = \frac{V(\alpha) \cdot P(\alpha) + V(\beta) \cdot P(\beta)}{P(\alpha \vee \beta)}$$

Note that the desirability axioms (V2 in particular) regulate the relationship between the functions P and V, so that when we say something is a desirability function we implicitly refer also to an associated probability function. For this reason it is sometimes more useful to define V in terms of the probability measure P and a normalised signed measure I, which, with an eye to our earlier discussions of evaluative attitudes, I will term an importance measure.[6] To say that I is a normalised signed measure is to say that it is an additive real-valued function on a Boolean algebra (i.e. if $\alpha\beta = \bot$, then $I(\alpha \vee \beta) = I(\alpha) + I(\beta)$) such that $I(\top) = 0$. Note that this implies that $I(\bot) = 0$, since, by the additivity

---

[6] Ethan Bolker (1966) terms it a 'total utility'.

property, $I(\bot) = I(\bot \vee \bot) = I(\bot) + I(\bot)$. Given measures P and I, the function V can then be defined by, for all prospects α,

$$V(\alpha) := \frac{I(\alpha)}{P(\alpha)} \text{ if } P(\alpha) > 0$$

This definition leaves undefined the desirabilities of any prospects of probability zero. This is unfortunate, since for some applications it would be convenient to have them defined.[7] Conceptually, moreover, it seems to me that desirability is the primary notion and importance the derivative one. So, for the most part, we will work with V as our primary evaluative measure.

For the purpose of understanding the connection between desirability, importance and probability an analogy may be helpful. Think of prospects as territories, P as a measure of the proportion of the total area $\top$ occupied by these territories, and V as a measure of their average height relative to that of $\top$. Then what the axiom of desirability expresses is the fact that the average height (relative to $\top$) of a territory is a weighted average of the heights of the sub-territories making it up, where the weights on sub-territories are the proportion of the territory that they occupy. Because V measures average height, the greater the proportion of the total area occupied by a particular territory, the less its average height will deviate from that of the reference point $\top$. At the limit, since $\top$ is the whole area, its height doesn't deviate at all from the average; hence the normalisation $V(\top) = 0$.

These facts hold in virtue of the relationships between the concepts of area, volume and average height. In similar fashion, if we take V to measure the average value of truth of a prospect relative to that of the necessary prospect $\top$, and P to measure its probability of truth, then these measures will respect the relevant axioms simply in virtue of the meaning of the concepts of average value and probability. In particular, because V measures average value relative to $\top$, the more probable the proposition, the less its desirability will diverge from that of $\top$ (and the more its negation will).

5.4.2 Probabilism

We are now in a position to state our first rationality thesis, a claim that has been defended by a wide range of scholars working in different fields, including Ramsey (1990/1926), Bruno de Finetti (1937), Leonard Savage (1974/1954) and Jeffrey (1990/1983). It will be of little surprise to find it here as well.

Thesis 5.1 (*Probabilism*) Rational degrees of belief are probabilities

The literature is replete with different kinds of arguments in support of Probabilism and, collectively, they make for a powerful case in its favour. Some, like the famous Dutch Book arguments, are essentially pragmatic;

---

[7] Especially in game theory, where reasoning about off-equilibrium events is essential.

others appeal to the epistemic goals of belief, such as accuracy (see, for instance, Joyce, 1998, and Leitgeb & Pettigrew, 2010a,b); still others appeal to conditions of rational relational belief or preference that imply it. I will spell out a version of the latter kind of argument later on. For the moment, however, I want to explain the sense in which the probability axioms can be said to characterise rational degrees of belief. The important point to make in this regard is that the probability axioms do not express unadulterated rationality conditions. Rather, they simultaneously establish a convention for the measurement of degrees of belief and impose the constraints on such measures required by rationality in the light of the scaling convention.

Rationality requires at least two things of belief:

1. That nothing be believed to a lesser degree than the impossible prospect and to a greater degree than the necessary prospect. (This is the Boundedness property.)
2. That, when both β and γ are inconsistent with α, then $\alpha \vee \beta$ is believed to a greater degree than $\alpha \vee \gamma$ iff β is believed to a greater degree than γ. (This is the $\vee$-Separability property.)

P0, which requires degrees of belief to be positive, and P1, which requires that the necessary prospect has probability one, establish a measurement convention: that degrees of belief take values in the interval from 0 to 1. Given this measurement convention, the first rationality condition implies that the necessary prospect has probability one and the impossible prospect probability zero, while the second implies, in the presence of some technical conditions (see below), that the probability of any two disjoint prospects be the sum of the probabilities. It is the latter property that is the surprising one, of course, and establishing it is rightly regarded as one of the main achievements of Bayesian scholarship.

Let us make this claim more precise, for it is the foundation of the assertion that rationality requires (given the scaling conventions) that numerical degrees of belief be probabilities. Let X be a set of prospects and $\unrhd$ be a rational credibility order on it – i.e. a non-trivial, monotonic and $\vee$-separable weak order on X. Now we need to make two structural assumptions: first, that the set of prospects X together with the implication relation $\models$ forms a complete atomless Boolean algebra; and, second, that $\unrhd$ is continuous. (Continuity is needed because a complete atomless algebra of prospects allows for a continuum of prospects and we need to ensure that the credibility relation 'fits' with $\models$.) Given these conditions, the existence of a probability representation of rational relational belief is assured.

Theorem 5.4. *(Villegas's theorem) (Villegas, 1964) Let $\Omega = \langle X, \models \rangle$ be a complete, atomless Boolean algebra of prospects. Let $\unrhd$ be a continuous rational credibility order on X. Then there exists a unique (countably additive)* *probability function P on X such that*

$$\alpha \unrhd \beta \Leftrightarrow P(\alpha) \geq P(\beta)$$

5.4.3 Desirabilism

In contrast to Probabilism, our second rationality thesis is peculiar to the version of decision theory floated by Richard Jeffrey. It is:

Thesis 5.2 (*Desirabilism*) Rational degrees of desire are desirabilities

One way to think of a desirability measure is as an extension of the 'usual' utility measure on maximally specific prospects to the full Boolean algebra of prospects based on it.[8] Specifically, given any such utility on a domain D, the desirability of any element E in $\wp(D)$ can be defined as the conditional expectation of utility, given the truth of E. Although this construction makes it clear that *any* interpretation of utility will induce an associated interpretation of desirability, Richard Jeffrey is often said to have defended a specific one, namely the 'news value' conception of benefit. It is true that news value is a type of value that unambiguously satisfies the desirability axioms. Consider getting the news that a trip to the beach is planned and suppose that one enjoys the beach in sunny weather but hates it in the rain. Then, whether this is good news or not will depend on how likely it is that it is going to be sunny or rainy. If you like, what the news means for you, what its implications are, depends on your beliefs. If it's going to rain, then the news means a day of being wet and cold; if it's going to be sunny, then the news means an enjoyable day swimming. In the absence of certainty about the weather, one's attitude to the prospect will lie somewhere between one's attitude to these two prospects, but closer to the one that is more probable. This explains why news value should respect the axiom of desirability. It also gives a rationale for the axiom of normality, for news that is certain is no news at all and hence cannot be good or bad.

Nonetheless, considerable caution should be exercised in giving Desirabilism this interpretation. In particular, it should not be inferred that Jeffrey's claim is that we value something *because* of its news value. News value tracks desirability but does not constitute it. Moreover, it does not always track it accurately. Sometimes getting the news that X tells us more than just that X is the case because of the conditions under which we get the news. To give an extreme example: if I believe that I am isolated, then I cannot receive any news without learning that this is not the case. This 'extra' content is no part of the desirability of X.

---

[8] Jeffrey's theory does not require, however, that there be such maximally specific propositions.

Our main interest is in desirability as a certain kind of grounds for acting in conditions of uncertainty. In this respect, it perhaps more helpful to fix one's intuitions using the concept of willingness to pay than that of news value. For if one imagines that all action is a matter of paying to have prospects made true, then the desirabilities of these prospects will measure (when appropriately scaled) the price that one is willing to pay for them. It is clear that one should not be willing to pay anything to make a tautology true and quite plausible that one should price the prospect of either X or Y by the sum of the probability-discounted prices of the each. So this interpretation is both formally adequate and exhibits the required relationship between desirability and action.

Desirabilism combines a scaling convention for degrees of desire with a substantial claim about rational desire. The scaling convention is established by V1, which normalises the desirability measure with respect to the necessary prospect. Note that there is no desirability axiom corresponding to $P0$, a feature which deserves some comment. In the case of belief we have two natural scaling points, corresponding to the cases of full belief (when the prospect is regarded as certainly true) and full disbelief or empty belief (when the prospect is regarded as certainly false). Moreover, there exists a prospect that must, of rational necessity, be fully believed – the necessary prospect – and a prospect that must of rational necessity be fully disbelieved – the contradictory prospect. The choice of the unit measure for the former and zero measure for the latter is rather natural, if entirely conventional. In the case of desire we have a natural scaling point corresponding to the situation of 'empty desire', where a prospect is neither desired nor undesired. Moreover, there exists a prospect – the necessary one – that is of this kind: it should not be desired by anyone because its truth is certain. Here again the choice of the zero measure for the prospect that is neither desirable nor undesirable is rather natural, if entirely conventional. On the other hand, there is no prospect that should be fully desired by all rational agents in the way that the necessary prospect should be fully believed by all. Hence no candidate for the unit measure. (Indeed, I don't think there is anything like fully desiring a prospect, for this implies an upper bound on our desires. We could cook up a prospect that must, in virtue of its meaning, fit the bill, such as the prospect of a perfect world, or one in which all my desires are satisfied. But semantic manoeuvres of this kind don't really help. For one thing, some argument must be given for why we should think such prospects exist. And, secondly, if they do, there is no reason to think that they have the same meaning for every agent.)

The rationality part of Desirabilism stems from the ∨-Betweenness requirement on preference relations, given that degrees of desire should cohere with relational desires or preferences. For, as we argued before, it follows from the fact that the prospect that $\alpha \vee \beta$ implies, at best, that the more desirable of the two, α or β, is the case and, at worst, that the less desirable is, that it should not be regarded as more desirable than either α or β, or as less desirable than either. Then, given the scaling convention established by V1, ∨-Betweenness implies that degrees of desire must satisfy V2 (again with the help of some technical conditions). This implication, like that of the additivity of probability, is far from obvious, and establishing it is equally significant.

Representation Theorems for Desirabilism*

To make precise the relationship between rational preference and our two rationality theses – Probabilism and Desirabilism – I will draw on two representation theorems, respectively due to Ethan Bolker and James Joyce.

Impartiality Suppose $\alpha \approx \beta$ and that, for some $\gamma \not\approx \alpha$, such that $\alpha\gamma = \perp = \beta\gamma$, it is the case that $\alpha \vee \gamma \approx \beta \vee \gamma$; then, for all such γ, $\alpha \vee \gamma \approx \beta \vee \gamma$

Impartiality plays a central role in Bolker's representation theorem in ensuring that the test he proposes for the equiprobability of co-ranked prospects is consistent. The idea is the following. Take two co-ranked prospects α and β along with some third prospect γ that is inconsistent with both and co-ranked with neither and then observe whether $\alpha \vee \gamma$ and $\beta \vee \gamma$ are ranked together. If they are, then we can conclude that α and β are equiprobable. Suppose, contrary to this inference, that the probability of α is greater than that of β. Then it would be less likely that α given that $\alpha \vee \gamma$ than it would be that β. And so $\alpha \vee \gamma$ would be either a less or a more attractive prospect than $\beta \vee \gamma$ depending on whether $\gamma \succ \alpha, \beta$ or $\alpha, \beta \succ \gamma$. But if it is established that the probability of α and β are the same then it should be the case, for all γ inconsistent with both α and β, that $\alpha \vee \gamma \approx \beta \vee \gamma$ – i.e. Impartiality must hold.

Let $\Omega = \langle X, \models \rangle$ be a Boolean algebra of prospects and $\succsim$ a preference relation on X' – i.e. a non-trivial ∨-between pre-order of X'. Let $\mathfrak{B} = \langle \Omega, \succsim \rangle$ be called a *Bolker structure* iff Ω is complete and atomless and $\succsim$ is complete, continuous and impartial. Then:

Theorem 5.5. *(Bolker's Theorem) (Ethan Bolker, 1966) Let $\mathfrak{B} = \langle \Omega, \succsim \rangle$ be a Bolker structure. Then there exists a pair of desirability and probability functions $\langle P, V \rangle$, respectively on X and X', such that*

$$\alpha \succsim \beta \Leftrightarrow V(\alpha) \geq V(\beta)$$

Furthermore $\langle V^*, P^* \rangle$ is another such pair of functions iff V' is a fractional linear transformation of V – i.e. there exists $a > 0$ and c such that, for all $\alpha \in X'$, $cV(\alpha) > -1$ and

$$P^*(\alpha) = P(\alpha) \cdot (cV(\alpha) + 1)$$

$$V^*(\alpha) = \frac{aV(\alpha)}{cV(\alpha) + 1}$$

A notable feature of Bolker's representation theorem is that his axioms do not sufficiently constrain preference so as to uniquely determine a representation of the agent's degrees of belief. This has some conceptually unattractive implications. For instance, two probability functions P and P' may both cohere with an agent's preferences and yet, for any α and β that are not equally preferred, it can be the case that $P(\alpha) = P(\beta)$ but $P'(\alpha) \neq P'(\beta)$.[9] So, within the Jeffrey–Bolker framework, whether or not an agent believes one thing to the same degree as another is not completely determined by, or revealed in, her preference rankings of prospects. (What is determined, as the discussion of Impartiality shows, is the equiprobability of *co-ranked* propositions and, more generally, ratios of probabilities of co-ranked propositions.)

There is a strong Behaviourist strain in decision theory that views facts about preference, as revealed in intersubjectively observable choice behaviour, as the determinant of the empirical and/or semantic content of ascriptions of belief and desire to agents and which sees decision-theoretic representation theorems as formal demonstrations of what is or is not empirically significant in such mentalistic talk. Behaviourists will infer from Bolker's uniqueness results that there is no empirical foundation for claims such as 'The agent believes α to the same degree as β' and hence no (scientific) sense to them either. Pragmatists too might draw this conclusion, even if they reject the methodological primacy that Behaviourism accords to observable behaviour. For, no matter how preferences are determined, the fact remains that they do not serve, in the Jeffrey–Bolker framework, to determine the agent's relational beliefs. Hence, if differences in properties of belief must be rooted in differences in properties of preferences, Bolker's theorem can be read as showing that there is no fact of the matter as to whether an agent has one particular degree of belief or another, or, indeed, whether she has one set of relational beliefs rather than another. (Of course, *some* belief facts are determined; they are just much less specific than normally assumed.)

This is not the only way of reading Bolker's theorem. One could argue, along the lines of Bradley (1998), that it shows that not all relevant features of preference are captured by Jeffrey's framework, either because there are further rationality conditions that should be imposed or because there are preferences for other types of prospects, or both. Alternatively, one could argue, along the lines of James Joyce (1999), that Bolker's theorem shows that the claim of methodological primacy of preference over belief is false and that foundations for ascriptions of numerical degrees of belief should be sought in relational beliefs rather than preferences. I will return to the former possibility later on. For the moment let's look at Joyce's alternative representation theorem for Desirabilism.

---

[9] **Proof:** Suppose that $P(\alpha) = P(\beta)$, but $V(\alpha) \neq V(\beta)$. Then, by Bolker's theorem, $P'(\alpha) = P(\alpha).(cV(\alpha) + 1)$ and $P'(\beta) = P(\alpha).(cV(\beta) + 1)$. Hence $P'(\alpha) \neq P'(\beta)$.

Joyce takes as his primitives a pair of credibility and preference relations defined on the same Boolean algebra of prospects and argues that the preferences of a rational agent should be consistent with her relational beliefs. To capture this, let us say that a preference relation $\succsim$ *coheres* with a credibility relation $\unrhd$ if only if they jointly satisfy:

Coherence Suppose $\alpha \approx \beta$; then $\alpha \bowtie \beta$ iff, for every γ such that $\alpha\gamma = \perp = \beta\gamma$, it is the case that $\alpha \vee \gamma \approx \beta \vee \gamma$

Note that if an agent's preferences and relational beliefs cohere in this sense then her preferences must satisfy Impartiality. So Joyce's way of proceeding helps us see that the rationality requirement expressed by Impartiality is in fact a constraint on the relationship between an agent's beliefs and desires, not on her preferences alone. This additional constraint turns out to be exactly what is required to establish the existence of a unique numerical representation of the agent's preferences.

As before, let $\Omega = \langle X, \models \rangle$ be a Boolean algebra of prospects and let $\succsim$ and $\unrhd$, respectively, be a preference relation (a non-trivial $\vee$-between pre-order) on X' and a credibility relation (a non-trivial, monotonic and $\vee$-separable pre-order) on X. Let $\mathfrak{J} = \langle \Omega, \unrhd, \succsim \rangle$ be called a *Joyce structure* iff Ω is complete and atomless, $\succsim$ and $\unrhd$ are complete and continuous and $\succsim$ coheres with $\unrhd$. A *Jeffrey representation* of a Joyce structure is a pair of probability and desirability functions $\langle P, V \rangle$, respectively on X and X', that represents $\unrhd$ and $\succsim$ in the sense that

$$\alpha \unrhd \beta \Leftrightarrow P(\alpha) \geq P(\beta)$$
$$\alpha \succsim \beta \Leftrightarrow V(\alpha) \geq V(\beta)$$

Now:

Theorem 5.6. *(Joyce's theorem) (James Joyce, 1999) Every Joyce structure $\mathfrak{J} = \langle \Omega, \unrhd, \succsim \rangle$ has a Jeffrey representation $\langle P, V \rangle$ that is unique up to linear transformation of V, i.e. $\langle V^*, P^* \rangle$ is another such representation of $\mathfrak{J}$ iff $P^* = P$ and $V^* = aV$ for some $a > 0$.*

Note that, under the conditions of Joyce's theorem, rational degrees of belief are uniquely determined by relational beliefs and preferences. This makes his representation theorem especially attractive as a foundation for a theory of ideal rationality, and I shall draw on it repeatedly in subsequent sections.

5.5 ASPIRATIONS OF OBJECTIVITY

The axioms of probability and desirability impose consistency constraints on agents. Although rationality requires no more of us than this, most of us want to do better. We aspire to some form of objectivity: to believe only what is true and perhaps to desire what is best. Such aspirations will be encoded in dispositions to adopt attitudes that conform to what we take ourselves to know or have learnt, or in acceptance of principles expressing such dispositions. For example, the want-to-be-objective agent might adopt as maxims:

(**Poss**) If α is not possible then $P(\alpha) = 0$

(**Truth**) If α is true then $P(\alpha) = 1$

(**Chance**) If the chance of α is x then $P(\alpha) = x$

(**Goodness**) If it is better that α than that β then $V(\alpha) > V(\beta)$

Maxims such as Poss or Truth express external or mind-to-world requirements on the agent. So they are not matters of consistency, and the agent who fails to believe α to degree one, when α is in fact true, commits no sin of irrationality if she does not know that α is true. On the other hand, there are internal surrogates of these requirements that *are* plausible candidates for consistency conditions on conditional attitudes – for instance, the requirement of full conditional belief in α, given that α is true. Such conditions will be at the heart of the next chapter.

Our candidate maxims express commonly accepted reasons for believing or valuing prospects to some degree, and objectivity is found in responsiveness to such reasons. Objectivity may also be sought, however, in the appropriate response to *absence* of reasons; in particular, of reasons for distinguishing cases. The best-known principle of this kind is the Principle of Insufficient Reason, or, as it is often called, the Principle of Indifference.

Principle of Indifference If you have no reason to evaluate two alternatives differently then you should assign them the same value

The Principle of Indifference has been especially influential in probability theory, serving in a central role in the classical theory of probability as well as contemporary logical and objective Bayesian approaches. John Maynard Keynes expressed it thus:

> [I]f there is no known reason for predicating of our subject one rather than another of several alternatives, then relatively to such knowledge the assertions of each of these alternatives have an equal probability.   (Keynes, 1973/1921, 52–53)

The Principle of Indifference powerfully constrains probability assignments. Indeed, as attested by the numerous paradoxes it engenders, it seems to overconstrain them. The basic problem is that its prescriptions are not invariant under redescriptions of the possibilities. Suppose that we are considering what probability to assign to the United Kingdom winning the

Eurovision Song Contest and are in complete ignorance about the matter. Then we might apply the principle by assigning equal probability to each country entering. Or we might assign equal probability to the winner coming from each region of Europe, or of being male or female, and so on. Each application will yield a different probability for a winner from the United Kingdom.[10]

To avoid these paradoxes, we need some way of picking the correct partition. Keynes argued that the Principle of Indifference should be applied only to sets of indivisible alternatives. In the Eurovision contest, the indivisible alternatives are presumably the entrants or possible winners. Since regions field multiple entrants, they are not indivisible and so cannot figure in an application of the principle. But this solution, if it is one, is achieved by relativising the probability assignment to a formulation of the problem. If we had looked at the problem slightly differently, by taking possible winners to be the set of all people eligible to enter the competition, the indivisible alternatives would have been the citizens of the participant countries, or at least those meeting the entry requirements. As the United Kingdom is quite populous, application of the Principle of Indifference would have yielded a higher probability for a British winner than on the original formulation. Of course, we might not be able to apply the principle at all if we don't know how many citizens are eligible to enter, but this hardly vindicates application to official entrants.

This objection applies with equal force to solutions that are based on choosing a formal language in which the problem is to be framed. Then atomic propositions will play the role of indivisible prospects and the Principle of Indifference will yield a unique prescription. But the choice of language remains the preserve of the agent, who must make a subjective judgement as to the appropriateness of the candidate languages.

A closely related, and perhaps even more compelling, principle is what might be called the Principle of Symmetry. The basic idea is that solutions to probability assignment problems should be invariant under transformations of the problem that preserve all relevant features of it. Here is van Fraassen (1989)'s formulation of the idea:

Principle of Symmetry If two problems are essentially the same then they should have the same solution.

So stated, the principle is close to a platitude. All the action will lie in spelling out the conditions under which two problems are essentially the same. One very precise way of doing so is through the notion of a symmetry-preserving transformation: two problems are essentially the same if there is a mapping from one to the other that preserves its structure. For instance, suppose that there are three entrants to the Eurovision contest – Ann, Bob and Carol – and the problem we want to solve is that of determining the

---

[10] The correct answer, of course, is 'Less than 1 per cent'.

probability of Ann winning. If the three are not distinguishable in any relevant respect then the solution to this problem should be the same as that to the problem of determining the probability of Bob winning, since this can be obtained from the original problem by a simple mapping f such that f(Ann) is {Bob}, f(Bob) is {Ann}, and f(Carol) is {Carol}. So the Principle of Symmetry yields the same answer in this case as the Principle of Indifference (namely one-third).

As John Norton (2008) demonstrates, the Principle of Symmetry is also subject to paradoxes. The problem is, essentially, that when there are many symmetries then too many different transformations are possible and this leads to an over-constraining of the solution. For instance, suppose that we add Dudley to our list of entrants. We might think that this should allow for a different assignment of probability to Ann winning. But there is a simple transformation from our new problem into the old: the 'coarsening' transformation g, such that g(Ann) is {Ann}, g(Bob) is {Bob}, g(Carol) is {Carol} and g(Dudley) is {Carol}. So the Principle of Symmetry requires the same probability assignment in this case. Requiring the same solution to be invariant under each such transformation is therefore too demanding.

Once again, we could save the principle from paradox by making the partition of prospects an essential feature of the problem. But now we have clearly returned to the realm of subjective judgement. Only in textbooks and exam papers do problems present themselves with an official description attached. In the real world, agents must decide for themselves what the essential features are of the situation they face. This is not to deny the importance of the Principle of Symmetry to both science and philosophy. Rather, it is to argue that it is better regarded as a tool that an agent can avail herself of in organising her thinking, and which will yield different solutions in accordance with how (subjectively) the agent determines the essential features of the problems she faces.

> These two principles express platitudes of evidence whose acceptance seems irresistible. They follow directly from the simple idea that we must have reasons for our beliefs. So if no reasons distinguish among outcomes, we must assign equal belief to them; or if two descriptions of the outcomes are exactly the same in every noncosmetic aspect, then we must distribute beliefs alike in each (John Norton, 2008, p. 52).

Exactly so. But what reasons there are for partiality, and what parts of descriptions are noncosmetic, must be settled by subjective judgement.

6

Conditional Attitudes

We not only take attitudes to prospects, we also take conditional attitudes to them: attitudes on the hypothesis, supposition or assumption that something is true. On the assumption that I will not be tired tonight, I would rather go out than stay in. On the supposition that interest rates will rise, I believe that housing prices will fall. And so on. Conditional attitudes, like their unconditional counterparts, come in grades, a feature that is crucial to the role they play in decision making. I might find the prospect of going to the beach desirable, conditional on it being a sunny day, but not at all so, conditional on it being rainy. Whether I should go to the beach depends not only on the relative probability of sun and rain but also on the relative strengths of these conditional desires. One aim of this chapter is to identify the different kinds of suppositions involved in forming graded conditional attitudes, to propose rationality conditions that are characteristic of them and then to examine the implications of these conditions. A second is to explore the role that conditional attitudes play in hypothetical reasoning. Throughout I will assume the conceptual framework developed in the previous chapter.

First, two points should be made about what conditional attitudes are not. A common view is that conditional beliefs and desires given some β are dispositions to believe and desire conditional on β being true or on coming to believe that β. But this view is not quite right. While we should expect there to be a close relation between our conditional attitudes and the attitudes we take when we learn that the condition in question holds, sometimes this relation is disturbed by the conditions under which the learning occurs. I would prefer, were I confronted by a bully, that I would act bravely rather than run away. If I were confronted by a bully, however, I would (I predict) prefer to run. I believe that, were I to drink a lot, my driving ability would be impaired. Were I to drink too much, however, I would believe my driving to be better than ever. So an agent's conditional attitude to any prospect, given that α, is not the attitude she will or would have to it in the event that the condition is or were realised or found to be true. Rather, it is her *current* attitude to the prospect on the supposition that α is or were true.

Conditional attitudes to prospects are closely related to unconditional attitudes to conditional prospects. For instance, as we will see later, rationality requires that one should believe that β, on the supposition that α, just to the degree that one believes that, if α is or were the case, then β is or would be. But they aren't the same thing. A conditional attitude to β, given that α, is an attitude to the prospect of β from a certain mental standpoint, namely that achieved by supposing that α. On the other hand, in the unconditional attitude to the prospect that if α then β, α is a constituent of the *prospect* that is the object of the attitude, not of the attitude itself.

Not only are conditional attitudes and attitudes to conditionals not the same thing, they often don't even take the same value. Suppose a coin will be tossed and that I may or may not be given £100. I certainly favour getting the £100, but am indifferent as to whether it is accompanied by the coin landing heads, or it landing tails. So the degree to which I favour or desire that I get £100, on the supposition that the coin lands heads just equals, the degree to which I desire getting £100. On the other hand, my attitude to the prospect that if the coin lands heads I get £100 is not the same as my attitude to getting it (unconditionally). In fact, it is distinctly less favourable, for the former prospect is not that of getting the money for sure but only of getting it in the event of the coin landing a particular way. So it cannot be that my conditional attitude to getting the money, supposing that the coin lands heads, is the same as my attitude to the prospect that if it lands heads I get the money.

This point has a very important implication. Whilst it makes sense to compare attitudes to prospects within the scope of a supposition, it is far from clear what would be involved in comparisons across different suppositions. For example, one may ask, 'Supposing that you are not feeling tired tonight, would you prefer to go out or stay in?', but not, on pain of creating confusion, 'Do you prefer to go out, on the supposition that you are not tired, or to stay in, on the supposition that you are?'. Decision theorists often ignore this difficulty and assume cross-suppositional comparisons without explaining how it is possible to do so without serious ambiguity. We take this up in Section 6.5.

6.1 SUPPOSITIONS

The ability to take conditional attitudes is fundamental to prospective agency, for it allows us to deliberate about possible future conditions, and in particular about those that are caused or influenced by our own actions. As James Joyce (1999) observes, however, there is more than one way of supposing that something is true. In general, if we suppose that α is true, then we should form a set of (suppositional) beliefs that includes the belief that α and diverges as little as possible from our actual beliefs. But there are many different standards for minimal divergence. We might suppose that as a matter of fact α is true, such as when I suppose, to help with my financial planning, that I won't have enough money at the end of the month to pay the rent. Suppositions of this kind should respect to as great a degree as possible current unconditional beliefs; I should not, for instance, adopt the belief that I will secure a large inheritance to cover the rent. Things are quite different when we suppose or imagine that, contrary to the facts, α is true. A supposition of this kind may well be best accommodated by giving up some of one's beliefs not contradicted by α, to allow retention of well-entrenched ideas about the way that the world works. For example, when supposing that it rained yesterday, in order to think about what I would have done had this been the case, I might have to give up my belief that I went for a walk in the mountains that day, even if I did in fact do so (and have sore feet to prove it).

A related distinction – between evidential and interventional suppositions – has played an important role in the development of causal decision theory. When you make a supposition as part of evidential reasoning, you reason as if you have received evidence that implies the truth of the supposition. In contrast, when you suppose something interventionally you imagine that there has been some intervention in the course of events which makes the supposition true. In this latter case, unlike the former, you do not revise your degrees of belief in any of the causes of the condition supposed true because you do not treat your supposition as positive evidence for them.

Consider, for instance, the situation modelled by the graph in Figure 6.1, in which the arrows represent relations of causal influence between variables.[1] According to the graph, the presence or otherwise of a certain gene is a cause of both lung cancer and of smoking, while both of these are causes of breathing difficulties. These causal relations will induce probabilistic correlations between the variables relevant to evidential reasoning. In particular, evidence that the agent will in fact smoke makes its more probable that he has the gene in question, which in turn makes it more likely that he will get lung cancer. On the other hand, if we suppose that there is an intervention from outside the causal system represented by the graph (for instance, a freely made choice) which makes it true that the agent will smoke, then his smoking no longer provides evidence for the presence of the gene. So interventional supposition of the agent smoking should not lead to revision of the degree to which we believe that the agent will get cancer.

How are these two sets of distinctions – factual versus counterfactual and evidential versus interventional – related? Evidential supposition and matter-of-fact supposition are just the same thing, I think. But the relation between counterfactual and interventional supposition is less straightforward. What counterfactual and interventional supposition have in common is that

---

[1] The use of graphs to model causal relations is now well entrenched. See, for instance, Spirtes, Glymour & Scheines (2000) and Pearl (2009).

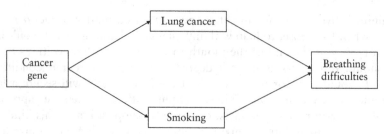

FIGURE 6.1. Causal Graph for Breathing Difficulties

they lead to revision of beliefs about the facts in a way which retains
entrenched beliefs about causal connections. They do so in slightly different
ways, however. When we engage in interventional supposition we do not
presuppose that the hypothesised prospect is as a matter of fact false. When
I think about what would happen if I were to cancel my classes and take the
day off, for instance, I need not be sure that I won't do it. On the contrary, the
very reason why I engage in the supposition is to help me to decide whether
I should do it. In doing so I suspend belief about whether the supposition is
true, rather than presume that it is false. So interventional supposition is not a
form of counterfactual supposition.

Equally, not all cases of counterfactual supposition seem to involve
interventional reasoning (though some clearly do). When we make an
interventional supposition we don't revise our probabilities for the causes of
the things supposed true. In entertaining the supposition that I smoke (by
choice), I must not, as we saw, revise my probability for having the gene.
On the other hand, when I suppose, say contrary to the facts, that I didn't get
lung cancer, then I don't need to imagine that someone intervened to prevent
it. Consequently, I may revise my degrees of belief for the possible causes
of the lung cancer in a way that would be inappropriate for interventional
supposition. For example, I might infer that I must not have had the gene,
even though having it is a causal antecedent, not consequence, of having lung
cancer.

6.2 SUPPOSITIONAL PROBABILITY AND DESIRABILITY

Let us now consider what rationality properties conditional beliefs and desires
should satisfy, both in themselves and in relation to unconditional beliefs and
desires. As the manner in which something can be supposed true can take
more than one form, there are relatively few rationality properties that hold
with complete generality. But since conditional belief is a form of belief, and
conditional desire a form of desire, it is reasonable to expect them at least to
satisfy the usual rationality constraints on these types of attitude. Furthermore,
an attitude conditional on something being true should reflect the fact that the

supposition renders that something certain from the judgemental perspective being adopted.

To make these claims more precise we need to characterise supposition more formally. A well-developed theory of suppositional belief exists for a categorical model of belief states – the so called AGM theory[2] – but here we need such a theory for relational and quantitative models of both belief and desire. There are two subtly different ways of doing so: by describing the characteristics of the belief and desire states that result from supposition (as in Joyce, 1999) or by characterising the operation of supposition that produces these states (as in Gärdenfors, 1988). I will follow the second path here, construing supposition as a function that takes as its arguments pairs of probability and desirability functions on prospects, together with the prospect supposed true, and returns a pair of real-valued functions on prospects; intuitively, measures of the degree to which prospects are believed and desired, on the supposition that the prospect is or were the case. An arbitrary supposition function will be denoted by the symbol $\circledast$ and the value that it assigns to a pair of probability and desirability functions, P and V, and prospect α, will be denoted by P_α^* and V_α^* – i.e. $\circledast(P, V, \alpha) = (P_\alpha^*, V_\alpha^*)$. In the special case where the mode of supposition is evidential, the supposition operation will be denoted by the symbol $\oplus$ and its outputs by P_α^+ and V_α^+.

Let us call the induced functions P_α^* and V_α^* suppositional probability and desirability functions. Then our two basic rationality claims regarding conditional attitudes can be expressed by:

Thesis 6.1: (*Suppositional Belief*) Rational conditional degrees of belief are suppositional probabilities

Thesis 6.2: (*Suppositional Desire*) Rational conditional degrees of desire are suppositional desirabilities

To give these theses content we need to specify the properties of the supposition operation. We do so as follows. For any pair of probability and desirability functions, P and V, and prospect $\alpha \neq \perp$:

Axioms of Supposition – Probability

P^*1 (*Probability*) P_α^* is a probability function
P^*2 (*Certainty*) $P_\alpha^*(\alpha) = P_\alpha^*(\top)$
P^*3 (*Anchoring*) $P_\top^*(\alpha) = P(\alpha)$

Axioms of Supposition – Desirability

V^*1 (*Desirability*) V_α^* is a desirability function

---

[2] See Gärdenfors (1988) for an overview.

V^*2 *(Certainty)* $V_\alpha^*(\alpha) = V_\alpha^*(\top)$
V^*3 *(Anchoring)* $V_\top^*(\alpha) = V(\alpha)$

Axioms P^*1 and P^*2 and their duals, axioms V^*1 and $V^*2,$ require suppositional degrees of belief and desire, given a particular prospect, to be probability and desirability functions that treat the prospect that is supposed true as a certainty and therefore on a par with the necessary prospect. Note the scaling convention that is thereby introduced, for it implies immediately that $P_\alpha^*(\alpha) = 1$ and $V_\alpha^*(\alpha) = 0$. These conventions are the natural ones to adopt given those adopted for probability and desirability, namely that full belief should get the unit measure and empty belief and desire the zero measure. For under the supposition that α one should fully believe that α, but not desire it at all. Finally, the Anchoring axioms P^*3 and V^*3 require the supposition of the necessary prospect to have no effect on the probability and desirability of other prospects. This is natural given that the truth of $\top$ is always already a given in judgement.

Together, $P^*1 - P^*3$ and $V^*1 - V^*3$ make up the weakest axiom system for rational supposition, but there are additional properties of interest that will be considered as we go along. In particular, it seems to me that it is intrinsic to the role that supposition plays in hypothetical reasoning that it satisfies

Supposition Averaging   If $\alpha\beta = \perp$, then:

$$P_\alpha^*(\gamma) \geq P_\beta^*(\gamma) \Leftrightarrow P_\alpha^*(\gamma) \geq P_{\alpha\vee\beta}^*(\gamma) \geq P_\beta^*(\gamma)$$

Supposition Averaging expresses the idea that the probability one assigns to any prospect can be achieved by weighing its probability under one supposition against its probability under the contrary supposition, so that its 'overall' probability should not be greater than its probability under both suppositions, or less than its probability under both. As is proved as Theorem A.3 in the Appendix, it implies:

Regularity $P_\alpha^*(\beta) \geq P(\alpha\beta)$

Regularity in turn implies that if $P(\alpha) = 1$ then $P_\alpha^* = P$ and hence the Anchoring condition P^*3 (in virtue of the fact that $P(\top) = 1$). *Proof*: $P(\alpha) = 1$ then $P(\alpha\beta) = P(\beta)$. Hence, by Regularity, $P_\alpha^*(\beta) \geq P(\beta)$ and $P_\alpha^*(\neg\beta) = 1 - P_\alpha^*(\beta) \geq P(\neg\beta) = 1 - P(\beta)$. So $P_\alpha^*(\beta) = P(\beta)$.

James Joyce (1999) argues that Regularity is an essential characteristic of suppositional probability, expressing as it does the idea that supposition is uncertainty-reducing. For when evaluating β under the supposition that α there is no doubt as to α or not, in contrast to evaluating it without any supposition in play. But there are reasons to doubt that Regularity holds with complete generality. Suppose you are not sure whether it will rain tomorrow or not, but you are sure that if it doesn't rain then the match will be played ($\neg rain \rightarrow play$). Nevertheless, under the supposition that, as a matter of fact, the match won't be played, you are sure that (even) if it doesn't rain then the match will *not* be played. So, contrary to Regularity, $P(\neg play \wedge (\neg rain \rightarrow play)) > P^+_{\neg play}(\neg rain \rightarrow play) = 0$.

Examples such as this are hardly decisive, but do suggest that Regularity, and hence Supposition Averaging, should be restricted to factual prospects. In the next chapter I will discuss conditional prospects more carefully, showing why this restriction is necessary and suggesting how to extend such principles to conditionals. For the moment, however, let's simply set them aside.

6.3 EVIDENTIAL SUPPOSITION

6.3.1 Evidential Conditional Belief

When we reason by assuming that some condition is as a matter of fact true, then we engage in what I have termed evidential supposition. Suppositional reasoning of this kind is subject to further rationality constraints not binding on other forms of supposition. For instance, if I regard it as more probable than not that it will rain tomorrow and that the match will be cancelled than that it will rain tomorrow and the match will be played, then I should regard it more likely, on the supposition that, as a matter of fact, it will rain tomorrow, that the match will be cancelled than that it will be played. And vice versa. More generally:

Conditional Belief If $P(A) > 0$ then:

$$P^+_A(B) \geq P^+_A(C) \Leftrightarrow P(AB) \geq P(AC)$$

Two observations. Firstly, Conditional Belief does not generally hold for interventional or counterfactual supposition, as our example with smoking and cancer genes serves to illustrate. Suppose I consider it unlikely that I have the cancer gene and hence unlikely that I will either take up smoking or get lung cancer. Nonetheless, in virtue of the correlation between the smoking and lung cancer, I regard it more likely that I smoke and get cancer than that I smoke and don't. But on the other hand, given the presumed causal set-up, I believe that it is *less* likely that were I to smoke I would get cancer than that I would not.

Secondly, note that the scope of Conditional Belief is restricted to factual prospects. This is because it is unclear how one should compare a conditional whose antecedent is inconsistent with what is supposed true with one whose antecedent is not – e.g. how to compare, under the supposition that it will rain, the prospect that if it rains the match will be called off with the prospect that if it doesn't rain then it will be played.

With this qualification, we can agree with James Joyce (1999) that Conditional Belief is a characteristic principle of evidential supposition.

Although it is essentially a relational rationality constraint on belief, it serves, in conjunction with the chosen scaling of our measures of degrees of belief and some background technical conditions, to determine our next rationality thesis:[3]

Thesis 6.3: (*Conditional Evidential Belief*) Rational degrees of belief on the evidential supposition that A are conditional probabilities given A – i.e. for all factual prospects X

$$P_A^+(X) = P(X|A)$$

To make this precise, recall that a conditional probability function $P(\cdot|\cdot)$ is a two-place, real-valued function on a Boolean algebra $\Omega = \langle X, \models \rangle$, such that for α and β in X:

Conditional Probability If $P(\alpha) \neq 0$, then:

$$P(\beta|\alpha) = \frac{P(\beta\alpha)}{P(\alpha)}$$

The thesis of Conditional Evidential Belief is widely accepted: luminary defendants include Frank Ramsey (1990/1926), Bruno de Finetti (1937), Leonard Savage (1974/1954), and Jeffrey (1990/65). Indeed, many Bayesians adopt the much stronger view that conditional degrees of belief are rationally required to be conditional probabilities, without restriction to evidential supposition. As with Probabilism, a range of different kinds of arguments have been put forward to support it, including the Dutch Book arguments of Bruno de Finetti (1937) based on a notion of conditional bets, and arguments based on considerations of either rational relational belief (e.g. Suppes & Zanotti, 1976) or rational preference (e.g. Leonard Savage, 1974/1954). A version of these latter arguments will be given in due course. But all them smuggle in 'evidentialist' assumptions of one kind or another and so really support no more than our more restricted thesis. For instance, the procedure for settling conditional bets used in Dutch Book arguments is clearly appropriate only for evidential supposition, since these bets are called off if the condition isn't in fact true.

6.3.2 Evidential Conditional Desire

Suppositional reasoning about preference or desirability displays many similar properties to belief. If I prefer that it rains and the match is called off to that it rains and it be played then I should prefer that, if it rains, the match be called

---

3 A proof that Conditional Belief implies Thesis 5 when the algebra of prospects is atomless is given by Joyce (1999).

off rather than played. And vice versa. For in both cases I compare a state of the world in which it rains and the match is played to one in which it rains and the match is called off. Hence, for all factual prospects B and C consistent with A:

Conditional Desire   If $P(A) > 0$ then:

$$V_A^+(B) \geq V_A^+(C) \Leftrightarrow V(AB) \geq V(AC)$$

Conditional Desire is the characteristic (essentially relational) constraint on evidential supposition involving factual prospects. Given our scaling of measures of degrees of desire, and under some background technical conditions, it implies:

Thesis 6.4:   (*Conditional Evidential Desire*) Rational degrees of desire on the evidential supposition that A are conditional desirabilities given A – i.e. for all factual prospects X

$$V_A^+(X) = V(X|A)$$

Formally, a conditional desirability function $V(\cdot|\cdot)$ is a two-place, real-valued function on a Boolean algebra $\langle X, \models \rangle$, such that for α and β in X':

Conditional Desirability   If $P(\alpha) \neq 0$ and $\alpha\beta \neq \perp$ then:

$$V(\beta|\alpha) = V(\alpha\beta) - V(\alpha)$$

Thesis 6.4, is the natural counterpart, for desire, of Thesis 6.3. In fact, in the presence of Desirabilism the former implies the latter (for a proof, see theorem 2 in Bradley, 1999). But, while the ratio definition of conditional probability is well known, that of conditional desirability is likely to be less so, and a few words of explanation and justification of it are in order. Intuitively, what you learn by getting the news that α and that β is more than just the sum of what you learn from the news items taken separately, since there are things you can infer from them both being true. What you learn is the sum of what you learn from getting the news that α and of getting the news that β, given that you already know that α. News value follows news content in this regard. For example, the value of the news that a picnic has been arranged with some friends and that it is going to be sunny is greater than the sum of the news value of each taken separately, because sun makes the picnic better and the picnic makes the sun better. What it does equal is the value of the news that it is going to be sunny plus the value of the news of the picnic, given that we already know it will be sunny. In general, therefore, $V(\alpha\beta) = V(\alpha) + V(\beta|\alpha)$, which is just a rearrangement of our definition of conditional desirability.

Why the mathematical sum? Essentially, because of the way we use desirability to represent preference. Suppose that the utility of money was linear. Then if we entertain the fiction that truths can be bought or sold at will (suppose offerings to gods are effective) we could define the desirability of a prospect as the fair price for its truth and the conditional desirability of a prospect, given some α, as the fair price for its truth on the supposition that α is true. Then if your conditional desirability for β given α did not satisfy our expression, you would be vulnerable to a money pump. Note that, since the fair price for $\top$ is zero, it follows that $V(\top) = 0$. Now suppose that $V(\alpha\beta) > V(\beta|\alpha) + V(\alpha)$ – i.e. that you are prepared to pay more for the truth of $\alpha\beta$ than the sum of what you are prepared to pay for α and for β, on the supposition that α is true. In this case someone could buy the truth of α from you for $V(\alpha)$ and then the truth of β for $V(\beta|\alpha)$. Finally, they could sell the truth of α and β back to you for $V(\alpha\beta)$ at a profit! (A similar argument will establish that you face sure loss if $V(\alpha\beta) < V(\beta|\alpha) + V(\alpha)$).

A final note. The fact that conditional probabilities and desirabilities are not defined for probability zero conditions can be inconvenient. The standard way of getting around this problem is to use Popper–Renyi measures (see Popper, 1959, and Rényi, 1955). In this framework, what this amounts to is replacing Theses 6.3 and 6.4 with the claim that rational evidential conditional degrees of belief in, and desire for, factual prospects are suppositional probabilities and desirabilities satisfying, for all α, β and γ:

P^+4 (*Multiplication*) $P_\alpha^+(\beta\gamma) = P_{\alpha\beta}^+(\gamma) \cdot P_\alpha^+(\beta)$
V^+4 (*Addition*) $V_\alpha^+(\beta\gamma) = V_{\alpha\beta}^+(\gamma) + V_\alpha^+(\beta)$

From which the standard formulations follow.

6.4 INDEPENDENCE

The relationship between my attitude to a prospect and my attitude to it under various suppositions gives an indication of the degree to which my attitude to the prospect depends on what I suppose to be true. A special case of this is when my attitude to some prospect β does not depend at all on whether I suppose that some other prospect α is true or not. In this case we say that my attitude to β is independent of the truth of α. Formally:

Independence β is independent of the supposition that α, relative to ⊛:

      1. *Probabilistically* iff $P_\alpha^*(\beta) = P(\beta)$
      2. *Desirabilistically* iff $V_\alpha^*(\beta) = V(\beta)$

Some points to note about this definition. First, suppositional probabilistic and desirabilistic independence are related. In particular, if both β and $\neg\beta$ are desirabilistically independent of α, then β and $\neg\beta$ are probabilistically independent of α. (This is proved as Theorem A.4 in the Appendix. The converse does not hold, however.)

Second, whether or not a prospect β is either probabilistically or desirabilistically independent of the supposition that α can depend on the mode in which α is supposed true. Consider again Figure 6.1. On its depiction of causal relations, getting lung cancer is causally but not evidentially independent of the supposition of smoking. Likewise, the probability of lung cancer on the interventional supposition of smoking equals the unconditional probability of lung cancer, while the conditional probability of lung cancer, given that someone smokes, is higher than it.

Finally, in the special case of evidential supposition involving factual prospects, our definitions of independence imply that X is, respectively, probabilistically and desirabilistically independent of the evidential supposition that A iff

$$P_A^+(X) = P(X|A) = P(X)$$

$$V_A^+(X) = V(X|A) = V(X)$$

Together with the definitions of conditional probability and desirability, this implies the standard definitions of probabilistic and desirabilistic independence:

Evidential Independence X is evidentially independent of A:

> 1. Probabilistically iff $P(AX) = P(A) \cdot P(X)$
> 2. Desirabilistically iff $V(AX) = V(A) + V(X)$

These definitions make no reference to the notion of a supposition: so construed, probabilistic and desirabilistic independence are simply symmetric relations between prospects characterising a feature of the agent's unconditional attitudes.[4] Our derivation of them suggests, however, that the standard notions are essentially evidential ones. This is manifest in the fact that X can be probabilistically independent of A without it being causally – for example, when the causal influence of A on X is offset by the causal influence of another variable which is probabilistically correlated with A.

6.5 FOUNDATIONS *

Let us now turn to the question of whether our quantitative theory of rational conditional attitudes has a foundation in a corresponding relational theory. There is an extensive literature on the topic, which typically attacks it by introducing a new class of object, namely prospects of the form 'α given that β' – frequently written $\alpha|\beta$. Comparative relations are then defined over this class, allowing for expressions of the form $\alpha|\beta \succsim \gamma|\delta$. This approach certainly

---

4   See Bradley (1999) for a proof of the claim that these relations are symmetric.

delivers the goods in some sense, for there exist numerous characterisations of both conditional probability and conditional expected utility theory in terms of representations of the comparisons of this kind.[5] But in fact I think this approach sheds little light on these quantitative concepts.

The question is what interpretation to give to expressions such as $\alpha|\beta \succsim \gamma|\delta$. The usual answer, namely that it is a comparison (in terms of credibility, desirability, etc.) between α given that β is the case, and γ given that δ is, offers little clarification. For what exactly is being compared: the prospects α and γ? The givens β and δ? Both? Some relationship that holds between both α and β and γ and δ? To draw on an analogy, if we are asked to say which of two buildings is taller, we have no difficulty in understanding what we are being asked to do. But if we are asked to say which is taller, the first building, given that you adopt one vantage point, or the second, given that you adopt another vantage point, we struggle to answer. Perhaps the question is: is the difference in height between the first building and the first vantage point greater than between the second building and the second vantage point? This question is comprehensible, but its answer presupposes the existence of a quantitative framework whose foundations are precisely what is at issue.

My conclusion is that the usual approach is deeply flawed. When we talk of evaluating some prospect α, given that β, we refer to β not as a constituent of the prospect being evaluated but as a characteristic of the standpoint from which the evaluation is being made. So we can meaningfully speak of α in comparison to γ, given that β, but not of α given that β in comparison to γ given that δ. This being the case, we should take as our starting point not attitudinal relations over mysterious conditional objects but conditional attitudinal relations over ordinary objects, where a conditional attitudinal relation, under some supposition, is simply a relation representing comparisons made from the standpoint achieved by performing the supposition in question.

6.5.1 Representation of Conditional Attitudes

To characterise conditional attitudinal relations more formally, it is convenient to do so via the notion of a supposition function on relations. Informally, a supposition function maps an attitudinal relation and the proposition being supposed true to a corresponding attitudinal relation under the supposition concerned. More formally, let $\Omega = \langle X, \models \rangle$, be a Boolean algebra and $\mathcal{R}$ a class of binary relations defined on X. For any $\alpha \in X'$ (i.e. $X - \{\bot\}$), let $\Omega_A = \langle A, \models_A \rangle$ be the corresponding Boolean subalgebra of prospects in $A := \{\gamma \in X : \gamma \models \alpha\}$ and $\mathcal{R}_A$ be the restrictions of the binary relations in $\mathcal{R}$ to the domain A. A supposition function $\circledast$ on $\mathcal{R}$ is then a mapping from every $R \in \mathcal{R}$ and $\alpha \in X$ to a relation $R_\alpha^* \in \mathcal{R}_A$, such that, in particular, $\circledast(R, \top) = R$. The relation

---

5   See, for instance, Fishburn (1973), Joyce (1999), Luce & Krantz (1971) and Suppes (2002).

$R_\alpha^* = \circledast(R,\alpha)$ is called the conditional relation, under the supposition that α, corresponding to R. A real-valued function Φ numerically represents $\circledast$ on the relation R just in case, for all $\alpha \in X'$ and $\beta, \gamma \in A$:

$$\Phi(\beta,\alpha) \geq \Phi(\gamma,\alpha) \Leftrightarrow \beta R_\alpha^* \gamma$$

Our interest is in the properties of a supposition function on relational beliefs and preferences that are necessary and sufficient for the existence of numerical suppositional probabilities and desirabilities measuring the agent's degrees of conditional belief and desire. It turns out that the crucial requirement is that the agent's relational beliefs and preferences under a supposition cohere whenever her unconditional relational beliefs and preferences do – i.e. that supposition does not undermine the coherence between her attitudes. Formally, let $\mathcal{R}^{\unrhd}$ and $\mathcal{R}^{\succsim}$, respectively, be a class of complete and continuous credibility on Boolean algebra Ω and a class of complete and continuous preference relations on Ω' and let $\circledast$ be a supposition function on both. Then:

Coherent Supposition If $\unrhd \in \mathcal{R}^{\unrhd}$ and $\succsim \in \mathcal{R}^{\succsim}$ cohere then, for all $\alpha \neq \bot$, the corresponding relations $\unrhd_\alpha^*$ and $\succsim_\alpha^*$ cohere

To make the representation claim more precise, let us call the two-place functions, $P(.,.)$ and $V(.,.)$ defined by for all $\alpha \in X'$, $P(.,\alpha) = P_\alpha^*$ and $V(.,\alpha) = V_\alpha^*$, respectively a suppositional probability and a suppositional desirability function. Then, as is proved as Theorem A.5 in the Appendix:

Representation of Conditional Attitudes Let $\Omega = \langle X, \models \rangle$ be an atomless Boolean algebra of prospects and, for $\unrhd \in \mathcal{R}^{\unrhd}$ and $\succsim \in \mathcal{R}^{\succsim}$, let $J = \langle \Omega, \unrhd, \succsim \rangle$ be a Joyce structure. Let $\circledast$ be a supposition function on both $\mathcal{R}^{\unrhd}$ and $\mathcal{R}^{\succsim}$ that satisfies Coherent Supposition. Then:

1. (Existence) *There exists a suppositional probability $P(\cdot,\cdot)$ on $X \times X'$ that numerically represents $\circledast$ on $\unrhd$ and a suppositional desirability function $V(\cdot,\cdot)$ on $X' \times X'$ that numerically represents $\circledast$ on $\succsim$.*
2. (Uniqueness) *$P(\cdot,\cdot)$ is unique. $V'(\cdot,\cdot)$ is another such suppositional desirability iff, for all $\alpha \in X'$, there exists $a > 0$ such that, for all $\beta \in A'$, $V'(\beta,\alpha) = aV(\beta,\alpha)$.*

This theorem establishes that, given some technical assumptions, if both an agent's credibility and preference relations and her corresponding conditional credibility and preference relations are coherent with each other, then her conditional degrees of belief and desire must be suppositional probabilities and desirabilities. Technical assumptions aside, the theorem depends only on the general rationality conditions on relational belief (transitivity and

∨-Separability) and preference (transitivity and ∨-Betweenness), the consistency conditions built into the characterisation of the supposition function, and the requirement of coherence between an agent's preferences and beliefs, both conditional and unconditional. So we can conclude that rationality requires that agents' conditional degrees of belief and desire satisfy axioms $P^*1 - P^*3$ and $V^*1 - V^*3$. This result, though not particularly surprising given the weakness of these axioms, is nonetheless significant in that it gives relational foundations to a minimal quantitative theory of rational conditional belief and desire without appeal to mysterious conditional prospects.

6.5.2 Evidential Supposition

It is interesting to consider further conditions on supposition functions on attitudinal relations, especially with an eye to distinguishing evidential and other forms of supposition. The following two are particularly salient.

Dominance Let $\{\alpha_i\}$ be any partition of the set of prospects X; if $\beta R^*_{\alpha_i} \gamma$, for all
    α_i such that $\alpha_i \beta \neq \bot \neq \alpha_i \gamma$, then $\beta R \gamma$

Suppositional Rigidity $\beta R^*_\alpha \gamma \Leftrightarrow (\alpha\beta) R(\alpha\gamma)$

Dominance says that if β is ranked higher than γ on every supposition then it must be ranked higher overall. Suppositional Rigidity, on the other hand, requires that the order of the prospects that imply α be undisturbed by the supposition that α is true. Applied to belief, it appears as the relational correlate of the Conditional Belief principle; applied to preference, as the relational correlate of the Conditional Desire principle.

Although apparently quite different these two conditions are not independent. Suppositional Rigidity is implied by Dominance and Dominance is implied by Suppositional Rigidity together with ∨-Separability (a formal statement and proof of these claims is given in the Appendix as Lemma A.7). The result suggests that credibility relations will satisfy both conditions or neither, which is somewhat surprising as it turns out that Suppositional Rigidity is *the* characteristic property of evidential supposition, in the sense that conditional credibility and preference relations that satisfy it determine the existence of conditional probability and desirability measures of the agent's suppositional beliefs and desires. More formally, as is established in the Appendix as a corollary to Theorem A.5:

Representation of Evidential Conditional Attitudes *As before, let ⊛ be a supposition function for both ⊵ and ≿ that satisfies Coherent Supposition. Then:*

1. *If the $\unrhd^*_\alpha$ satisfy Suppositional Rigidity relative to ⊵ then there exists a unique conditional probability function $P(\cdot|\cdot)$ that represents ⊛ on ⊵.*

   2. *If the* $\succsim_\alpha^*$ *satisfy Suppositional Rigidity with respect to* $\succsim$ *then there*
      exists a conditional desirability function $V(\cdot|\cdot)$, *unique up to positive*
      linear transformation, that represents ⊛ *on* $\succsim$.

Although rather similar on the surface to the theorem of the previous
section, the claim being made here is a good deal stronger. Firstly, the formal
properties imposed on the representation of the agent's conditional degrees of
belief and desire are much richer and, secondly, the permitted transformations
of it are much reduced. In particular, whereas a suppositional desirability
representation of the agent's conditional preferences is unique only up to
the choice of a scaling point for each supposition, a conditional desirability
representation of them is unique up to linear transformation – i.e. the scaling
for each supposition must be the same. In this sense a conditional desirability
representation allows one to say that, for instance, the desirability of β given
that α is greater than the desirability of γ given that δ – a type of comparison
that a suppositional desirability representation does not permit.

6.5.3 Independence and Separability

Despite the importance of notions of independence to decision and probability
models, their foundations are not well established. There are two different
paths one can explore in this regard. The first is to take judgements of
independence as primitives, of the same status as preference and relational
belief judgements, and then use the former to constrain the latter. In particular,
it would be natural to require that the relational order between any prospects
that are independent of some prospect α be undisturbed by the supposition
that α is or were true. More formally, let us denote the independence of β
from α, relative to the relation R, by $\beta\|^R\alpha$. Then it is required that:

Supposition Independence If $\beta\|^R\alpha$ and $\gamma\|^R\alpha$ then $\beta R_\alpha^* \gamma \Leftrightarrow \beta R\gamma$

   A second, more 'fundamentalist' approach is to try and derive independence
from properties of relational belief and preference. This can be done only
partially, but the exercise is revealing nonetheless. The crucial notion
here is the separability of prospects (or sets of prospects) under a binary
relation. Intuitively, a compound prospect is separable if its components
make independent contributions to its overall value. Disjunctions of mutually
exclusive prospects are separable under the preference relation, for instance
– this is what we termed 'v-Separability' – and later on we will consider
whether orthogonal conditionals are as well. But conjunctions of prospects
are not typically separable under either preference or relational belief, because
their elements may either complement each other (e.g. having bread and
having butter) or supplement each other (e.g. having white bread and having
brown bread). Only if two propositions are independent of each other will the
conjunction of them be separable.

This last observation suggests turning Supposition Independence around so that we can use the separability of conjunctions of prospects to derive their independence. Formally, let $\Omega = \langle X, \models \rangle$ be Boolean algebra and R be a binary relation defined on it. Let $\mathcal{A}$ and $\mathcal{B}$ be any two subsets of X such that $\forall \alpha \in \mathcal{A}, \beta \in \mathcal{B}$ and $\alpha \wedge \beta \neq \bot$ and let $\mathcal{A} \otimes \mathcal{B} := \{\alpha \wedge \beta \in X : \alpha \in \mathcal{A}, \beta \in \mathcal{B}\}$. Then:

Definition 6.1. *The binary relation R is said to be **separable** on the sets $\mathcal{A}$ and $\mathcal{B}$ iff, for all $\alpha \in \mathcal{A}$ and $\forall \beta, \gamma \in \mathcal{B}$:*

$$(\alpha \wedge \beta) R (\alpha \wedge \gamma) \Leftrightarrow \beta R \gamma$$

In this case $\mathcal{A} \otimes \mathcal{B}$ is said to be a separable subset of X under the relation R.

Now if the agent's relational beliefs and preferences are separable over two sets of prospects then her conditional degrees of belief and desire for the elements of one are independent of the truth of the elements of the other. More formally, as we prove as Theorem A.11 in the Appendix:

Representation of Separable Relations *Suppose $\mathcal{J} = \langle \Omega, \unrhd, \succsim \rangle$ is a Joyce structure and $\langle P, V \rangle$ a corresponding Jeffrey representation of it. Then:*

1. *If $\mathcal{A} \otimes \mathcal{B}$ is separable under $\unrhd$ then $\forall \alpha \in \mathcal{A}$ and $\forall \beta \in \mathcal{B}$:*

$$P(\alpha \beta) = P(\alpha) \cdot P(\beta)$$

2. *If $\mathcal{A} \otimes \mathcal{B}$ is separable under $\succsim$ then $\forall \alpha \in \mathcal{A}$ there exists $a > 0$ such that $\forall \beta \in \mathcal{B}$:*

$$P(\alpha \beta) = P(\alpha) \cdot P(\beta)$$
$$V(\alpha \beta) = V(\alpha) + a V(\beta)$$

6.6 HYPOTHETICAL REASONING

6.6.1 Dominance Arguments

We often use judgements about what is or would be true under a variety of circumstances in deliberations aimed at establishing what is true, probable or desirable. In thinking about whether a friend will be at home tomorrow I might reason: 'If the weather is good he will be doing his gardening, but if the weather is bad he won't want to go out of doors; so either way he is likely to be at home.' Or, when deciding whether to invite someone to your party, you reason: 'If she is in a good mood, then she won't be upset if she is not invited, but if she is in a bad mood she will be poor company; so either way it is better not to invite her.'

In these cases hypothetical reasoning takes the specific form of a dominance argument: we reason from the judgement that one prospect is more probable/desirable than another given each possible contingency to the judgement that the former prospect is more probable/desirable overall. Dominance arguments are much loved by decision theorists and are often treated as if they were universally applicable. Savage writes, for instance:

> It is all too seldom that decisions can be arrived at on the basis of the [Dominance] principle, but ... I know of no other extralogical principle governing decisions that finds such ready acceptance (Leonard Savage, 1974/1954, p. 21).

In fact, however, dominance arguments of the kind that Savage refers to are only valid under very special circumstances. To see why, let us examine a general version of the dominance principle that Savage is referring to. Let $A = \{\alpha_i\}$ be a partition of the space of prospects. Then consider:

Preference Dominance If $\forall \alpha_i \in A$, $\beta \succsim^*_{\alpha_i} \gamma$, then $\beta \succsim \gamma$

Preference Dominance says that if one prospect is at least preferable as another, on supposition of the truth of each of the members of some partition of prospects, then it is as least as preferable as the second unconditionally. Compelling though this principle may seem at first, it is not in fact rational to adhere to it, for so doing severely constrains what suppositional probabilities one can adopt. Firstly, it forces one to treat the α_i as probabilistically independent of all other prospects. For, if they were not, then it could be the case that one prospect preference dominates another but is less desirable because the former being true makes it much more likely that some particular supposition is true. For instance, in our opening example, it could well be the case that the potential invitee's mood is affected by whether she is invited or not, making it better to invite her – contrary to the conclusion of the dominance argument. Secondly, in combination with some mild structural assumptions, it forces one to regard all prospects logically independent of the α_i as probabilistically independent of them. (Facts of this kind are established, with slightly different structural assumptions, in Broome, 1990, and Bradley, 2005a, theorem 28, within the framework of Desirabilism; and in Mongin, 1995, within Savage's framework.)

Are dominance arguments involving beliefs more secure? Consider:

Belief Dominance If, for all $\alpha_i \in A$, $P^*_{\alpha_i}(\beta) \geq P^*_{\alpha_i}(\gamma)$ then $P(\beta) \geq P(\gamma)$

Belief Dominance clearly holds when the mode of supposition is evidential – i.e. when $P^*_{\alpha_i}(.) = P(.| \alpha_i)$. For by the Law of Total Probability – a consequence of the definition of conditional probability – we have

$$(LTP)\ P(\beta) = \sum_i P(\beta|\alpha_i).P(\alpha_i)$$

From which it follows immediately that, if $P(\beta|\alpha_i) \geq P(\gamma|\alpha_i)$ for all $\alpha_i \in A$, then $P(\beta) \geq P(\gamma)$. And so it will come as no surprise to learn that Belief Dominance is a logical consequence of Conditional Belief, our main principle of evidential suppositional belief, even though the two principles have rather different sorts of justification.

What *is* surprising is that the converse is true: Belief Dominance implies Conditional Belief. To see this, let α be some member of the partition $\{\alpha_i\}$. Then, for all $\alpha_i \neq \alpha$, $P^*_{\alpha_i}(\alpha\beta) = 0 = P^*_{\alpha_i}(\alpha\gamma)$ in virtue of both $\alpha\beta$ and $\alpha\gamma$ being incompatible with the supposition that α_i. So it follows by Belief Dominance that $P^*_\alpha(\beta) \geq P^*_\alpha(\gamma) \Leftrightarrow P(\alpha\beta) \geq P(\alpha\gamma)$ in accordance with Conditional Belief. Since we already know that Conditional Belief does not apply to non-evidential supposition, we must conclude that Belief Dominance is also not a generally valid rationality constraint on the relation between conditional and unconditional beliefs. Or, to put it slightly differently, arguments exploiting Belief Dominance are valid only if the mode of reasoning is evidential.

6.6.2 Averaging Arguments

In hypothetical reasoning we often reach judgements about how probable or desirable some prospect is by weighing up how probable or desirable it is under the supposition of various possible contingencies, and then giving weight to these suppositional judgements in accordance with how probable the contingencies are. In evidential reasoning we can turn to the Law of Total Probability to tell us how unconditional beliefs should be informed by conditionals beliefs. But are there more general principles of this kind that would apply to a wider variety of types of reasoning?

One plausible candidate is a principle that is implied by Belief Dominance: that a rational agent's degrees of belief should be a weighted average of her suppositional degrees of belief. Formally, relative to a partition $\{\alpha_i\}$, there exist positive weights k_i that sum to one and such that, for all prospects β:

Constant Averaging $P(\beta) = \sum_i k_i \cdot P^*_{\alpha_i}(\beta)$

Constant Averaging has the appearance of a natural generalisation of the Law of Total Probability, giving formal expression to the idea that one's degree of belief in a prospect should be a weighted average of one's degrees of belief in it under various suppositions, without assuming that the mode of supposition in question is evidential. But in fact the appearance of greater generality is an illusion. For it follows from Constant Averaging that $P^*_{\alpha_i}(\beta) = P(\beta|\alpha_i)$ and hence that reasoning involving this principle *is* evidential in nature. Indeed, it turns out Belief Dominance, Constant Averaging and the claim that suppositional probabilities are conditional probabilities are all equivalent in the presence of the basic axioms of suppositional probability and thus all serve as ways of characterising evidential reasoning. A formal statement of this claim and its proof are given in the Appendix as Theorem A.9.

Constant Averaging, it would seem, is too strong to serve as a general constraint on hypothetical reasoning aimed at reaching conclusions about what is likely to be true. But the principle of Suppositional Averaging that was tentatively proposed earlier on suggests a natural weakening of it, namely that, for all prospects β, there exist non-zero weights k_i^β on the suppositional probabilities that sum to one, such that:

Variable Averaging $P(\beta) = \sum_i k_i^\beta \cdot P_{\alpha_i}^*(\beta)$

Variable Averaging is much weaker than Constant Averaging in that the former allows the weights on the suppositional probabilities to vary with the prospect being evaluated. Hence Variable Averaging is a more plausible candidate for the kind of general constraint on hypothetical reasoning that we are seeking.

If Variable Averaging is indeed a general principle of hypothetical reasoning, then it is natural to ask whether there are conditions weaker than those considered thus far which, in the presence of the basic axioms of suppositional probability and desirability together with Variable Averaging, suffice to characterise purely evidential reasoning. It transpires that the crucial property of supposition in this regard is that it be order-independent, in the sense that the outcome of supposing that α is the case and then that β is should be the same as supposing that β is the case and then α is. Formally, this is equivalent to adopting as alternative axioms of evidential supposition the following:

P^+4' (*Commutativity*) $(P_\alpha^+)_\beta^+ = P_{\alpha\beta}^+$
V^+4' (*Commutativity*) $(V_\alpha^+)_\beta^+ = V_{\alpha\beta}^+$

Then:

Commutative Supposition Suppose that a supposition operation on ordered pairs of probability functions and prospects satisfies $P*1$, $P*2$ and Variable Averaging. Then it satisfies Constant Averaging iff it satisfies P^+4'.

The left-to-right implication follows immediately from the fact that satisfaction of Constant Averaging implies that suppositional probabilities are conditional probabilities and that

$$\frac{P(\gamma\beta|\alpha)}{P(\beta|\alpha)} = \frac{P(\gamma\alpha|\beta)}{P(\alpha|\beta)}$$

The right-to-left implication, on the other hand, is anything but immediate. It is proved in the Appendix as Theorem A.10. Note that a corollary of this is that, in the presence of $P*1 - P*3$, the Multiplicativity axiom P^+4 is equivalent to P^+4' and Supposition Averaging.

Commutativity seems like a natural requirement on suppositions. But, in fact, counterfactual supposition is not order-independent. If I suppose that I can afford to go out tonight, and then suppose, contrary to the facts, that my wallet has been stolen, then I might well give up my belief that I can afford to go out in order to accommodate this latter supposition. On the other hand, if I first suppose that my wallet has been stolen and only then that I can afford to go out, I must, in virtue of the Certainty condition, find myself in a state in which I do believe that I can afford to go out (perhaps giving up the belief that my wallet has been stolen in order to accommodate it). So it seems that Commutativity has strong credentials as the defining characteristic of evidential supposition.

6.6.3 Summary

Evidential probabilistic hypothetical reasoning is distinguished by satisfaction of a number of different principles – Conditional Evidential Belief, Belief Dominance and Constant Averaging – that are equivalent in the presence of the basic axioms of supposition $P*1 - P*3$. The fact that these principles are, on the face of it, both rather different from one another and individually quite compelling perhaps explains why the study of probabilistic supposition and hypothetical reasoning has come to be identified so strongly with the theory of conditional probability. But, as we have seen, these principles do not govern all forms of supposition.

In this chapter I have argued the following.

- Suppositional degrees of belief and desire must satisfy $P*1 - P*3$ and $V*1 - V*3$.
- Suppositional degrees of belief in factual prospects must respect Supposition Averaging (equivalently Variable Averaging).
- Rational evidential supposition is order-independent – i.e. satisfies P^+4' and V^+4'.

Jointly these claims imply that rational degrees of belief in factual prospects under evidential supposition are conditional probabilities.

7

Conditionals and the Ramsey Test

7.1 THE RAMSEY TEST FOR BELIEF

> If two people are arguing 'If p will q?' and are both in doubt as to p, they are adding p hypothetically to their stock of knowledge and arguing on that basis about q; so that in a sense 'If p, q' and 'If p, ¬q' are contradictories. (Ramsey, 1990/1929, p. 155)

This remark of Frank Ramsey appears only as a footnote to his paper 'General Propositions and Causality', but it has sufficed to lend his name to a hypothesis that has figured prominently in contemporary debate in both the semantics and pragmatics of conditionals.[1] This interest in the Ramsey Test hypothesis, as it is usually called, is fuelled by widespread dissatisfaction with the material conditional as a rendition of the semantic content of ordinary language conditionals. Discontent is focused on two points: the fact that the material conditional interpretation appears to support fallacious reasoning and the fact that reasonable belief in conditionals appears to diverge from that demanded by the material conditional interpretation of them.

On the material conditional construal of conditionals, $\alpha \rightarrow \beta$ is logically equivalent to $\neg\alpha \vee \alpha\beta$. Hence, for example, $\neg(\alpha \rightarrow \beta)$ implies that α. But the inference from 'It is not the case that if it snows tomorrow then the government will fall' to 'It will snow tomorrow' is clearly not valid, because denying that the weather will have an impact on the government's fortunes does not commit one to any particular meteorological prognosis. Likewise, disbelieving that the government will fall if it snows does not mean believing that it will snow (and, in summer, should not).

---

[1] The literature on both probabilistic and non-probabilistic versions of the Ramsey Test hypothesis is now very large. See, for instance, Gärdenfors (1988) and Edgington (1995).

Nor does the material conditional do any better as an interpretation of counterfactuals – indeed, if anything, it does worse. On a material conditional interpretation, the claim expressed by the sentence 'If George Bush had been concerned to protect the environment, then he would have lowered the tax on fuel' should be highly credible, because of the improbability of its antecedent (this follows from the fact that this view implies that $P(\alpha \to \beta) = P(\neg\alpha) + P(\alpha\beta)$). But intuitively the claim is implausible because environmentalists typically believe that fuel taxes should be raised.

The Ramsey Test hypothesis does not directly offer an alternative account of the semantic content of conditionals. Instead, it makes a claim about the nature of rational belief in conditionals, a claim that should constrain any proposal for such a semantics. The basic idea is simple and compelling. To judge whether it is credible that if P then Q, first suppose that P is true. Then adjust your beliefs no more than is necessary to accommodate this supposition. Finally, observe whether your new beliefs entail that Q. If so, you should adopt the belief that if P then Q.

This procedure certainly seems to give the right answer in our two examples. If I suppose that it will snow tomorrow and find that this does not lead to the belief that government will fall, then the Ramsey Test does not commit me to any belief concerning tomorrow's weather. Equally, the claim that George Bush would have lowered the tax on fuel, if he had been concerned to protect the environment, does not pass the Ramsey Test, because the supposition that he was concerned to protect the environment leads, if anything, to the belief that he would have raised fuel taxes. But to ascertain whether it is viable as a general constraint on belief attitudes we need to make it more precise.

Versions of the Ramsey Test hypothesis can be formulated for the different models of belief discussed before. In a qualitative model of categorical belief it can be rendered as the requirement that $\alpha \to \beta$ belongs to the agent's set of beliefs iff β belongs to her belief set obtained by supposing that α.[2] In the relational framework it is the requirement that the conditionals $\alpha \to \beta$ and $\alpha \to \gamma$ be ordered by a credibility relation $\unrhd$ in accordance with the ordering of β and γ by the conditional relation $\unrhd^*_\alpha$ – i.e. $\alpha \to \beta \unrhd \alpha \to \gamma$ iff $\beta \unrhd^*_\alpha \gamma$ (more on this later). Finally, the most natural way of applying the general idea of the Ramsey Test to a framework in which belief comes in degrees is to tie a rational agent's degrees of belief in a conditional to her degrees of belief in its consequent on the supposition of the truth of its antecedent. Formally:

Thesis 7.1: (*Ramsey Test for Belief*) Rational degrees of belief in conditionals equal degrees of belief in their consequent on the supposition of their antecedent; i.e.

$$P(\alpha \to \beta) = P^*_\alpha(\beta)$$

---

[2]   This is how it is formulated in Gärdenfors (1988).

We observed earlier that there is more than one kind of ordinary language conditional. One advantage of the Ramsey Test hypothesis is that it allows us to link this observation to the fact that there are different kinds of suppositions or ways of supposing something true. Indeed, the Ramsey Test is best viewed as a test schema with different types of belief revision associated with different modes of supposition being suitable for testing the credibility of different kinds of conditionals: evidential supposition for indicative conditionals and interventional and/or counterfactual supposition for subjunctive and/or counterfactual conditionals. This claim is made explicit, in the case of indicative conditionals at least, by our next rationality thesis.

Thesis 7.2: $P(\alpha \mapsto \beta) = P_{\alpha}^{+}(\beta)$

Thesis 7.2 and Thesis 6.3 (Conditional Evidential Belief) have an important joint implication. Let A and B be any factual prospects. Then, by the former, $P(A \mapsto B) = P_{A}^{+}(B)$ and, by the latter, $P_{A}^{+}(B) = P(B|A)$ – i.e. evidential supposition involving factual prospects is achieved by ordinary Bayesian conditioning. Hence they jointly imply that, for any factual prospects A and B:

$$(Adams' \ Thesis) \ P(A \mapsto B) = P(B|A)$$

Adams' Thesis is widely recognised both to capture our intuitions about rational belief in conditionals and to provide the best explanation for the empirical evidence concerning the role played by conditionals in the inferences that people make.[3] But it has proved very difficult to accommodate it within standard semantic and probability theory, a problem that we return to below.

A final remark. Note that it follows from the Ramsey Test for belief that both $P(\alpha \to (\beta \to \gamma)) = P_{\alpha}^{*}(\beta \to \gamma) = (P_{\alpha}^{*})_{\beta}^{*}(\gamma)$ and $P(\alpha\beta \to \gamma) = P_{\alpha\beta}^{*}(\gamma)$. Hence, by the Import-Export property, $(P_{\alpha}^{*})_{\beta}^{*}(\gamma) = P_{\alpha\beta}^{*}(\gamma)$. So these two conditions jointly imply the commutativity of supposition, a property that we discovered to be characteristic of *evidential* supposition. Since the Ramsey Test for belief holds independently of the mode of supposition, this vindicates our earlier claim that the Import-Export property characterises indicative conditionals: the type of conditional which is evaluated using evidential supposition.

7.2 THE RAMSEY TEST FOR DESIRE

So far we have said nothing about the desirability of conditional prospects, but it is natural to ask whether an equivalent test applies in this case. I believe it does. To determine whether it is desirable that $\alpha \to \beta$, one can suppose that α is the case and then assess from the standpoint induced by this supposition whether it is desirable that β or not. Indeed, without saying anything about how desirabilities under a supposition should be co-scaled with

---

3 See, for instance, Adams (1975), Stalnaker (1981b), McGee (1989) and Over & Evans (2003).

(non-suppositional) desirabilities, one can compare the relative desirability of the prospect that $\alpha \to \beta$ and that $\alpha \to \gamma$ by considering the relative desirability of the prospect that β or that γ under the supposition that α is the case. This suggests:

Thesis 7.3 (*Ramsey Test for Desire*) Rational degrees of desire in conditionals are proportional to the conditional desirabilities of their consequents on the supposition of their antecedent. More exactly, for every prospect $\alpha \neq \perp$ there exists a real number $k_\alpha > 0$, such that, for all prospects β,

$$V(\alpha \to \beta) = k_\alpha \cdot V_\alpha^*(\beta)$$

The Ramsey Test for Desire is the natural counterpart of the Ramsey Test for Belief. In fact, in the presence of Probabilism and Desirabilism, it implies the latter (see Theorem A.12 in the Appendix). Like its credal equivalent, it should be read as a test schema relating the desirabilities of different kinds of conditionals to the desirabilities of their consequents under the corresponding mode of supposition of their antecedents: the desirabilities of indicative conditionals to desirabilities under the matter of fact supposition of their antecedent and so on.

The Ramsey Test for Desire only partially constrains the desirability that should be assigned to a conditional. But some progress can be made if we focus on simple indicative conditionals, whose desirabilities are evaluated using evidential supposition. Recall our earlier example when a coin will be tossed and I may or may not be given £100 once it has been determined which way it has landed. It is desirable that I win the £100, on the supposition that the coin landed heads. It is also desirable that if the coin lands heads I get £100. But less so, for in this case getting the money is contingent on the coin landing heads, whereas in the former case it was given that the coin landed heads. So the desirability of the getting £100 needs to be discounted, in the latter case, by the improbability of the antecedent condition being met. My conjecture is that this is a general feature of the desirability of simple indicative conditionals: they are determined by the conditional desirability of their consequents, given the truth of their antecedents, just to the extent that their antecedents are probable. Formally:

Thesis 7.4: $V(\alpha \mapsto \beta) = V_\alpha^+(\beta) \cdot P(\alpha)$

Thesis 7.4 and Thesis 6.4 (Conditional Evidential Desire) also have an important joint implication. Let A and B be any factual prospects. Then, by the former, $V(A \mapsto B) = V_A^+(B) \cdot P(A)$ and, by the latter, $V_A^+(B) = V(B|A)$. Hence they jointly imply that

$$(Bradley's\ Thesis)\ V(A \mapsto B) = V(B|A) \cdot P(A)$$

A final technical note. In the presence of the axioms of probability and desirability, Thesis 7.4 implies Thesis 7.2 and Bradley's Thesis implies Adams'

Thesis (this is proved as part of Theorem A.12 in the Appendix). The latter are not equivalent, however. For the other direction an additional assumption is required, namely that

(Restricted Actualism) $V(\neg A \wedge (A \mapsto B)) = V(\neg A)$

Restricted Actualism is implied by Bradley's Thesis (proved as Corollary A.17) and, jointly with Adams' Thesis, implies it (proved as theorem 10 in Bradley & Stefánsson, 2015).

This assumption is extensively discussed in the next chapter. For now we turn to question of the specific form taken by the Ramsey Test hypothesis when applied to counterfactual or interventional conditionals. To answer it, we must first give some consideration to chances and the attitudes that agents might rationally take to them, a topic of importance to decision making quite independently of its relation to conditionals.

7.3 CHANCES AND COUNTERFACTUALS

Some of the uncertainty that an agent faces is objective or structural, or at least may appear as such to her. When this is the case, it is reasonable for her to make her decisions in the light of her beliefs and preferences regarding such objective uncertainty; and, indeed, in the light of preferences regarding it as well. Let us suppose that such structural uncertainty is measured by a chance function on a sub-domain of the Boolean algebra of factual prospects consisting of those prospects that she considers to be governed by objectively uncertain factors. What the true chances are is something that the agent will typically be (subjectively) uncertain about and so she will need to take account of different hypotheses about what these chances are. Maximally specific chance hypotheses will be complete specifications of the degree of objective uncertainty associated with each prospect in the relevant domain and will consequently imply more coarse-grained propositions regarding the chances of particular events.

To capture this more formally, let $\mathcal{Z} = \{Z, \models\}$ be a Boolean subalgebra of the background set of factual prospects; intuitively, Z contains those prospects to which it is meaningful to ascribe chances. Let $\Pi = \{ch\}$ be the set of all probability functions on $\mathcal{Z}$ and let $\Delta = \wp(\Pi)$ be the set of all subsets of Π. The elements of Δ serve here as what will be called 'chance prospects'. In particular, for any $X \in Z$, and $x \in [0,1]$, let $Ch(X) = x$ denote the chance prospect defined by $\{ch \in \Delta : ch(X) = x\}$. Intuitively, $Ch(X) = x$ is the proposition that the chance of X is x (and the chance of $\neg X$ is $1 - x$). A maximally specific chance hypothesis is simply the conjunction of a consistent and exhaustive set of propositions regarding the chances (at some point in time) of the factual prospects. In particular, Ch will denote the hypothesis that says that the chances are as given by the probability function ch and Ch_A the hypothesis that says that the chances conditional chances given that A are as given by the probability function $ch(\cdot|A)$. The focus of our interest is the product set $Z \times \Delta$ whose elements are combinations of factual and chance prospects. For instance, the prospect $(Y, Ch(X) = x)$ is the element of this set that is true when it is the case both that Y and that the chance of X is x. Hereafter, for simplicity, I will write Y for (Y, Δ) and Ch for (Z, Ch).

Our question is: what attitudes are agents rationally permitted to hold with respect to these prospects? Different theories of chance will imply different constraints and a full evaluation of them would take us too far afield. Some principles, however, have a very natural home within a broadly Bayesian theory; in particular, variants of what David Lewis dubbed the Principal Principle. Roughly, the principle says that a rational agent should set her degrees of belief in accordance with their expected chances or, equivalently, that the probability of any prospect conditional on the truth of some chance hypothesis is just whatever chances that hypothesis accords it. A simple version of the principle can be stated more formally as follows. Let P be a subjective probability function that is defined on a Boolean algebra containing the product set $Z \times \Delta$ and that measures the agent's degrees of belief in both factual and chance prospects. Then:

Principal Principle For all $X \in Z$ and $Ch \in \Delta$, if $P(X) \in (0, 1)$ then:

$$P(X|Ch) = ch(X)$$

Lewis' (less simple) version of the principle makes essential reference to a notion of admissibility, something for which no commonly agreed account has been given. I hope to bypass this problem by formulating a more general version of the principle as follows. Let Ch_A be any maximally specific hypothesis as to the conditional chances of the factual prospects in Z, given the truth of prospect $A \in Z$. Then:

Belief-Chance Principle For all $X, A \in Z$ and any $Ch_A \in \Delta$, if $P(X) \in (0, 1)$ and $ch \in Ch_A$ then:

$$P(X|A, Ch_A) = ch(X|A)$$

The Belief-Chance Principle says that the degree of belief a rational agent should have in a factual prospect X, conditional on any factual prospect A and corresponding maximally specific conditional chance hypothesis Ch_A, equals the conditional chance of X given that A according to that hypothesis.[4] The simple version of the Principal Principle is the special case obtained when $A = Z$.

These principles will figure in discussions in later chapters, but our concern now is to use them to shed some light on interventional and counterfactual supposition. Suppose we are considering various hypotheses about the chances

---

4   This principle is very close to the New Principal Principle proposed by Hall (2004).

of developing lung cancer conditional on smoking 20 a day for the next ten years. The Belief-Chance Principle tells us that our degree of belief in developing lung cancer, on the supposition of our smoking 20 a day for ten years and one such hypothesis about the chances, should equal the conditional chances of lung cancer, given the smoking, according to that chance hypothesis.

Now, how would the supposition that one will smoke affect things? Clearly, not at all if the mode of supposition is evidential, since the fact of smoking is already being conditioned on. But equally not if the mode of supposition is interventional. For entertaining the possibility of smoking does not alter the conditional chance, given smoking, that the chance hypothesis under consideration confers on lung cancer (it is built into the hypothesis). And the latter uniquely determines what degrees of belief for lung cancer it is rational to adopt. So it follows that:

Rigidity of Chances For all $X, A \in Z$ and any $Ch_A \in \Delta$:

$$P_A^*(X|A, Ch_A) = P(X|A, Ch_A)$$

Now, from Rigidity of Chances and the Belief-Chance Principle, it follows that

$$P_A^*(X|Ch_A) = P_A^*(X|A, Ch_A) = ch(X|A)$$

And hence, by the law of total probability, that

$$P_A^*(X) = \sum P_A^*(X|Ch_A) \cdot P_A^*(Ch_A)$$
$$= \sum ch(X|A) \cdot P_A^*(Ch_A) \tag{7.1}$$

In other words, the probability of prospect X under the interventional supposition that A equals the expected conditional chance of X, given that A, calculated relative to the probabilities of the conditional chance hypotheses, under the interventional supposition that A.

So far we have succeeded only in relating the suppositional probabilities of factual prospects to the suppositional probabilities of conditional chance hypotheses. To eliminate all reference to non-evidential probabilities one further assumption is required, namely that the maximally specific hypotheses regarding the conditional chances, given A, are probabilistically independent of A – i.e. that:

Chance Independence $P_A^*(Ch_A) = P(Ch_A)$

When Chance Independence holds it follows from Equation 7.1 that

$$P_A^*(X) = \sum ch(X|A) \cdot P(Ch_A)$$

In other words, the probability of X under the interventional supposition that A is just the expected conditional chance of X, given the truth of A.

The last step is to apply this result to counterfactual conditionals. This is very straightforward. By the Ramsey Test Hypothesis, $P(A \to X) = P^*_A(X)$. So it follows, in line with the proposal of Skyrms (1981), that the probabilities of counterfactuals equal the expected conditional chances of their consequents given the truth of their antecedents. Formally:

Skyrms' Thesis $P(A \to X) = \sum ch(X|A) \cdot P(Ch_A)$

The status of Skyrms' Thesis depends on that of Chance Independence. The latter is an attractive principle, but I don't think that it holds with complete generality. It will fail when the probabilities of some chance hypotheses are not independent of the supposition that certain kinds of interventions will be performed. This might happen if these interventions are probabilistically correlated with prospects in the domain of the chance function. Suppose, for example, that the causal influence of smoking on lung cancer was mediated by a gene; in particular, that the chances of lung cancer given smoking were much higher with the gene than without it. Suppose also that anyone with this gene was more likely to smoke. Then smoking and having the gene would be probabilistically correlated and the probability that the chances of lung cancer given smoking were high would be greater under the supposition that one would smoke (than without the supposition).

7.4 FOUNDATIONS

Stripped of its numerical casing, the Ramsey Test hypotheses express a rationality requirement on relational beliefs and preferences that is best captured by the following property of correspondence between attitudes to conditionals and conditional attitudes to their consequents:

Ramsey Property $(\alpha \to \beta)R(\alpha \to \gamma)$ iff $\beta R^*_\alpha \gamma$

Two questions now present themselves. Firstly, under what conditions will credibility and preference relations having the Ramsey Property imply the existence of corresponding probability and desirability measures respecting the Ramsey Test hypotheses for numerical belief in, and desire for, conditionals? And, secondly, what additional properties must credibility and preference relations have in order that these measures satisfy the Adams and Bradley Theses?

The first question can be addressed immediately by drawing on a representation result which is formally stated and proved in the Appendix (as Theorem A.14), but whose content and implications are informally described here:

Representation of Attitudes to Simple Conditionals Suppose that an agent's credibility and preference relations are defined on a simple conditional algebra of prospects (as defined in Section 5.2.3). Then:

1. If her relational beliefs have the Ramsey Property, then her degrees of desire satisfy the Ramsey Test for belief.
2. If her preferences cohere with her relational beliefs and also have the Ramsey Property, then her degrees of desire satisfy the Ramsey Test for desire.

To address the second question we need to consider what additional properties relational belief and desire must satisfy when taking indicative conditionals as their objects. Recall that indicative conditionals are evaluated by supposing that their antecedents are true as a matter of fact and that such evidential supposition is characterised by satisfaction of Suppositional Rigidity. But, in the presence of the Ramsey Test hypothesis, Suppositional Rigidity implies:

Indicative Property $(\alpha \mapsto \beta)R(\alpha \mapsto \gamma)$ iff $(\alpha \wedge \beta)R(\alpha \wedge \gamma)$

Quite independently of this derivation, it is very plausible that relational belief and preference for indicative conditionals should have the Indicative Property. The indicative conditionals $\alpha \mapsto \beta$ and $\alpha \mapsto \gamma$ respectively, promise β and γ in the event that α is the case, but are equally silent on what the case is if α is not. Since both are silent on the $\neg\alpha$ case, the comparison between the two should depend only on what they promise when α is true – i.e. on the comparison between β together with α and γ together with α.

On the other hand, counterfactual conditionals can fail to have the Indicative Property. For instance, if developing lung cancer is correlated with, but not caused by, smoking, then it could be less credible that if I were to smoke then I would develop lung cancer than that I would not, but more credible that I smoke and develop lung cancer than that I smoke and don't develop it. So I think this property is a reasonable contender to be a characteristic property of indicative conditionals.

Now, the crucial fact about the Indicative Property is that, within a regular logic of conditionals, it implies the separability of indicative conditionals from their antecedents under the relation R (and vice versa). For, by Modus Ponens and Centring, $\alpha \wedge \beta = \alpha \wedge (\alpha \mapsto \beta)$, and so any relation with the Indicative Property must be such that

$$(\alpha \mapsto \beta)R(\alpha \mapsto \gamma) \Leftrightarrow (\alpha \wedge (\alpha \mapsto \beta))R(\alpha \wedge (\alpha \mapsto \gamma))$$

This fact allows us to exploit our earlier representation result for separable relations to derive properties of numerical belief in, and desire for, partitioning conditionals, without assuming anything about rational conditional attitudes.

Recall that separable preferences have an additive representation and relational beliefs a multiplicative one. Hence, in particular, separable preferences and relational beliefs for indicative conditional respectively satisfy, for some $a > 0$:

$$V(\alpha \wedge (\alpha \mapsto \beta)) = V(\alpha) + a V(\alpha \mapsto \beta)$$

$$P(\alpha \wedge (\alpha \mapsto \beta)) = P(\alpha) \cdot P(\alpha \mapsto \beta)$$

So, by Modus Ponens, $V(\alpha\beta) = V(\alpha) + a V(\alpha \mapsto \beta)$ and $P(\alpha\beta) = P(\alpha) \cdot P(\alpha \mapsto \beta)$.

When restricted to factual prospects, the latter result is just Adams' Thesis, while the former takes us close to Bradley's Thesis, but not all the way. One more important property of preference, independent of any conditions we have assumed thus far, is required for this. What is needed is that an indicative partitioning indicative conditional lies, in the preference ranking, between the factual prospects consistent with it. The basic thought is this. Any prospect may be realised in a number of different possible ways, but generally one will not be certain as to which of these is the actual one. But at worst it will be realised in the least preferred of the possible ways it can be true and at best by the most preferred of the ways. So a prospect can never be better than its best realisation or worse than its worst one. Now consider in particular the partitioning indicative conditional $(\alpha \mapsto \beta)(\neg\alpha \mapsto \gamma)$. Its truth is consistent with only two prospects: that it is the case that α and β and that it is the case that $\neg\alpha$ and γ. So it should lie between the prospects $\alpha\beta$ and $\neg\alpha\gamma$ in the preference ranking. Formally:

$\mapsto$**-Betweenness** $\alpha\beta \succsim \neg\alpha\gamma \Leftrightarrow \alpha\beta \succsim (\alpha \mapsto \beta)(\neg\alpha \mapsto \gamma) \succsim \neg\alpha\gamma$

We are now able to answer our second question, which draws on Theorem A.16, a representation result which is formally stated and proved in the Appendix, but whose content and implications are informally described here.

Representation of Attitudes to Indicative Conditionals Suppose that an agent's credibility and preference relations are defined on regular conditional algebra of prospects. Then:

1. If her relational beliefs have the Indicative Property, then her degrees of belief conform to Adams' Thesis.
2. If her preferences also have the Indicative Property, cohere with her relational beliefs and satisfy $\mapsto$-Betweenness, then her degrees of desire satisfy Bradley's Thesis.

7.5   FACING THE TRIVIALITY RESULTS

Although there are many reasons, both empirical and conceptual, for endorsing Adams' Thesis, a series of triviality results due to David Lewis

(1976) and others (e.g. Hajek, 1989, Edgington, 1995, Döring, 1994 and Bradley, 1998) show that it is impossible to reconcile a general version of it with standard semantic and probability theory. The basic problem is simple to demonstrate. Note firstly that Adams' Thesis implies that:

Conditional Certainty If $P(A) > 0$, then:
   1. If $P(B) = 1$ then $P(A \mapsto B) = 1$
   2. If $P(B) = 0$ then $P(A \mapsto B) = 0$

Now let us assume that Adams' Thesis is preserved under conditionalisation. (If it were not, then his thesis would be incompatible with a rational agent generally revising her beliefs in this way; on the face of it, a very undesirable consequence.) It then follows from Conditional Certainty that $P(A \mapsto B|B) = 1$ and $P(A \mapsto B|\neg B) = 0$, because $P(B|B) = 1$ and $P(B|\neg B) = 0$. Then, by the law of total probability:

$$P(A \mapsto B) = P(A \mapsto B|B) \cdot P(B) + P(A \mapsto B|\neg B) \cdot P(\neg B)$$
$$= 1.P(B) + 0.P(\neg B)$$
$$= P(B)$$

But then, by Adams' Thesis, $P(B|A) = P(B)$. But this cannot generally be the case. For instance, if $1 > P(A) > 0$ then $P(A|A) = 1 \neq P(A)$. So it seems that Adams' Thesis must be false. And this in turn implies that one of the background assumptions used to derive the result doesn't apply here or that either Thesis 7 – the Ramsey Test hypothesis – or Thesis 6.3 – the claim that rational degrees of belief on the evidential supposition of some condition are conditional probabilities of truth given the condition – is false.

In fact, the problem is rather more subtle. Note firstly that Thesis 6.3 makes a claim whose scope is restricted to factual prospects. This carries over to the version of Adams' Thesis that we derived from it and the Ramsey Test hypothesis for belief. Now, to get the triviality result, we argued that Adams' Thesis should be preserved under conditionalisation. But there is an ambiguity in this requirement. We could take it to mean that

$$\text{(i) } P(A \mapsto B|C) = \frac{P((A \mapsto B)C)}{P(C)}$$

or that

$$\text{(ii) } P(A \mapsto B|C) = \frac{P(AB|C)}{P(A|C)} = P(B|AC)$$

These will not generally be the same. Consider, for instance, the case where C is $\neg B$. Then the two will be the same only if

$$\frac{P((A \mapsto B)\neg B)}{P(\neg B)} = P(A \mapsto B|\neg B) = 0$$

But this is guaranteed to be the case only if $\neg B$ is inconsistent with $A \mapsto B$ (which, on the face of it, it is not – think of the case when A is false).

The triviality result given above drew on both interpretations. The first interpretation was required in order to apply the law of total probability and the second in order to apply Conditional Certainty to derive that $P(A \mapsto B|B) = 1$ and $P(A \mapsto B|\neg B) = 0$. Once we clear up the ambiguity between the two, this triviality argument fails. The right way to do this is to retain the standard definition of conditional probability as employed in (i) and hence reject (ii). Rejecting (ii) means dropping the requirement that Adams' Thesis is preserved under conditionalisation. This is perfectly justified, however, since it is not Adams' Thesis that is the primary rationality requirement but Thesis 7.4: Adams' Thesis is just a consequence of it and Thesis 6.3, for the special case of factual prospects. Thesis 7.4, on the other hand, *must* be preserved under conditionalisation. And from this it follows, together with the Commutativity axiom for evidential supposition, that

$$\text{(iii)} \ P_C^+(A \mapsto B) = (P_C^+)_A^+(B) = P_{AC}^+(B) = P(B|AC)$$

The upshot, from (i) and (iii), is that $P_C^+(A \mapsto B) \neq P(A \mapsto B|C)$, contrary to (ii). The explanation of the Lewis-style triviality results is thus to be found in the conflation of these two quantities. But, while they are equal for factual prospects, they are not so in general. Indeed, as (iii) shows, the Ramsey Test for belief implies that they are different for conditional prospects.

Unfortunately, this does not mean that we are home and dry. Although carefully distinguishing probability under evidential supposition from conditional probability deals with the triviality results that, like Lewis', are based on considerations of belief revision, there are other similar results that make different assumptions. All of them take for granted, as Lewis does, that conditionals have bivalent truth conditions, or, more generally, that they have a Boolean semantics, and that rational degrees of belief in conditionals are probabilities. But Lewis' assumption that degrees of belief must be revised by conditionalisation can be replaced with the assumption that every probability function on prospects represents a permissible belief state. Given these three assumptions, far weaker claims than Adams' Thesis lead to triviality (more on this in the next chapter).

I cannot hope to exhaustively survey all the responses to these triviality results to be found in the literature, but it will be helpful to at least set out my general stance on them. There are, broadly, three main classes of them. Authors such as Lewis (1976), Jackson (1979) and Douven (2007) argue that these results show that Adams' Thesis is false as a claim about rational degrees of belief and that the evidence that we are disposed to assert conditionals to a degree equal to the conditional probability of their consequent given their antecedent should be explained by pragmatic principles of one kind or another, not by the semantic content of indicative conditionals (which they take to be that of the material conditional). But these alternative accounts are rather unsatisfactory on the whole because they don't extend in a natural way to sentences containing conditionals. For example, the sentence 'If I try to climb Mt. Everest, then I will succeed' is, on these accounts, very probably true (because I won't attempt the ascent) but not necessarily assertible. But then, why is the sentence 'Most likely, if I try to climb Mt. Everest, then I will succeed' also not assertible?

At the other extreme, non-factualists such as Edgington (1995) and Gibbard & Harper (1981) argue that the triviality results show that conditionals don't make factual claims and hence do not have (standard) truth conditions.[5] The thought is that when we say such things as 'If interest rates rises, housing prices will fall' we are expressing a conditional belief regarding house price falls, not an unconditional belief in the facts (the proposition) expressed by the conditional sentence just voiced. Indeed, this line of thinking goes, the sentence in question does not express a proposition at all. It merely serves as a vehicle for the articulation of a judgement which the speaker regards as justified for some reason, no doubt partially in virtue of the facts, but not entirely so.

This non-factualist view has sometimes been dismissed on the grounds that, when we express ourselves by means of conditional sentences, we are not engaged in an exercise of self-description. It is of the world that we speak, not ourselves. But this objection misses the strength of the non-factualist's position. The non-factualist is not claiming that in uttering a conditional sentence we are merely reporting our conditional belief, any more than when uttering a non-conditional sentence we merely report our unconditional beliefs. Our utterance implies a belief, or conditional belief, but this is not the object of the judgement we make. Rather, the non-factualist's point is that the judgement is not a factual one; in particular, it is not generally something that is either true or false. No doubt, the facts are relevant to the quality of the judgement expressed, but they do not determine it.

Nonetheless, non-factualism does face some difficulties. Firstly, as I have already argued, our attitudes to conditionals don't always take the same values as the corresponding conditional attitudes, so it is not clear that the non-factualist actually *has* an explanation of the Ramsey Test. Secondly, non-factualism makes it something of a mystery that we argue over the claims expressed by conditional sentences in much the same way as we argue over factual claims (i.e. by arguing over what is the case, not over what we believe to be the case). And, thirdly, without some account of semantic value, it is difficult to explain how we compound conditional sentences with other conditional and factual sentences using the usual sentential connectives and how we can make inferences with conditionals that eventuate in sentences that make factual claims. Together these considerations make non-factualism unsatisfactory as a response to the triviality results.

5 Adams himself should be included in this group.

The third common type of response is to drop the assumption that conditionals have bivalent truth conditions. Bradley (2002), Milne (1997) and McDermott (1996), for instance, all argue for a three-valued semantics based on the values of 'truth', 'falsity' and 'neither', while Stalnaker & Jeffrey (1994) allow conditionals to take any real number between 0 and 1 as their semantic value. I won't try to review all these theories here, but will simply note that none of them has gained wide acceptance, partly because of problems with the predictions they make regarding rational attitudes to compounded conditional sentences (in particular, partitioning sentences) and partly because of the difficulty in explaining what the probabilities occurring in Adams' Thesis are, if they are not probabilities of truth.

My own diagnosis is that it is the third of the assumptions used to derive triviality that needs to be rejected, namely that any probability function on the Boolean algebra of prospects is a permissible belief state. Indeed, Adams' Thesis in effect tells us that it cannot be true. The challenge is to explain how this restriction arises – i.e. what is it about the semantic content of conditionals that explains the restrictions on the probabilities that we can assign to combinations of conditional and unconditional prospects. This will be my task in the first part of the next chapter. Before we do so, however, there is a last putative rationality thesis to consider.

7.6 PARTITIONING CONDITIONALS

Partitioning conditionals of the form 'If A is the case then X, but if A is not, then Y' play a central role in the kind of hypothetical reasoning described in Section 6.6. In this section we consider two closely related theses about how the probabilities and desirabilities of the constituents of such partitioning conditionals depend on each other. Informally, what they respectively say is that the probability (or desirability) of a conditional describing what is the case if some condition is true is independent of the truth of any conditional describing what is the case if the condition is false. More precisely:

Thesis 7.5 (*Belief Independence*) Orthogonal indicative conditionals are probabilistically independent – i.e. if $\alpha \wedge \alpha' = \bot$, then, for all β and γ,

$$P(\alpha \mapsto \beta | \alpha' \mapsto \gamma) = P(\alpha \mapsto \beta)$$

Thesis 7.6: (*Value Independence*) Orthogonal indicative conditionals are desirabilistically independent – i.e. if $\alpha \wedge \alpha' = \bot$, then, for all β and γ,

$$V(\alpha \mapsto \beta | \alpha' \mapsto \gamma) = V(\alpha \mapsto \beta)$$

Belief and Value Independence respectively say that how probable or desirable it is that β is the case if α is does not depend on what is the case if α is not. The motivating intuition behind both seems to be this. Either α

is true, in which case the fact that something or other is true if α is false is irrelevant (because α isn't false). Or else α is false, in which case the fact that β is true if α is is itself of no relevance to the decision maker (because α isn't true).

To claim that what is the case if α is false is irrelevant when α is in fact true is *not* to claim that what *would* have been the case if α were false is irrelevant to the probability or desirability of what *is* the case if α is true. For what would have been the case had some condition been true can cause us to regret what is actually the case, or carry some information about what is the case that is relevant to both its probability and its desirability. For instance, suppose that I must choose between two boxes each containing a monetary prize, which I may keep if and only if it is larger than the one in the box that I didn't pick. I pick the box A. Then the desirability of the prospect that, if I pick box A then I win £100 is clearly not independent of the prospect that if I had picked the other box, I would have won £200. It would seem thus that neither independence condition carries over to orthogonal counterfactual conditionals.

This is important, because the Value Independence thesis has *very* powerful consequences. Note, first, that since $\neg(\alpha' \mapsto \gamma) = \alpha' \mapsto \neg\gamma$ it follows from Value Independence that $\alpha \mapsto \beta$ is desirabilistically independent of both $\alpha' \mapsto \gamma$ and $\neg(\alpha' \mapsto \gamma)$. So Belief Independence follows from it by Theorem A.4 in the Appendix. More generally, as is proved in the Appendix as part of Theorem A.19:

Independence of Conditionals Let (P, V) be a pair of probability and desirability functions on a conditional algebra of prospects satisfying Value Independence. Let $\{\alpha_i\}_{i=1}^{n}$ be an n-fold partition of the set of prospects. Then:

1. *(Multiplicativity)* $P(\bigwedge_i(\alpha_i \mapsto \beta_i)) = \prod_i P(\alpha_i \mapsto \beta_i)$
2. *(Additivity)* $V(\bigwedge_i(\alpha_i \mapsto \beta_i)) = \sum_i V(\alpha_i \mapsto \beta_i)$

These implications are controversial. Consider an example due to Mark Lance (1991). A werewolf visits the neighbourhood around half the nights of each year and attacks and kills anyone out of their house. So there is a 50 per cent probability that if you leave your house at night through the back door you will be killed (by the werewolf). So too if you leave through the front door. What, then, is the probability that if you leave your house through the back door you will be killed, *given* that if you leave through the front door you will be killed? Now, it is clear that if the conditional whose truth is being given was a counterfactual then the probability would be 100 per cent since the fact that you would be killed by the werewolf if you left through the frontdoor, implies that the werewolf is, in fact, in the neighbourhood (supposing that the werewolf's presence is causally independent of your leaving through one door rather than another). So Multiplicativity, as expected, does not hold for counterfactual conditionals.

Lance took his example to refute Multiplicativity for indicative conditionals as well, but here the situation is less clear. Suppose that you are in fact going to leave by the back door. Then the claim that you will be eaten if you leave by the front door could be construed as completely uninformative, since the asserted content of this claim depends on the truth of a condition that is in fact false. As it's uninformative, supposing it true should not affect how probable other prospects are. But without a proper understanding of the content of conditionals it is difficult to adjudicate these claims, so I will defer further discussion until the next chapter.

In view of how far-reaching these consequences are, it is important to ask whether the Value Independence thesis has any foundation in properties of relational attitudes. Not surprisingly, the answer is in the affirmative, with the essential property being that preferences over orthogonal indicative conditionals are separable. Formally:

$\mapsto$-**Separability** $(\alpha \mapsto \beta)(\neg\alpha \mapsto \gamma) \succsim (\alpha \mapsto \delta)(\neg\alpha \mapsto \gamma) \Leftrightarrow \alpha \mapsto \beta \succsim \alpha \mapsto \delta$

$\mapsto$-Separability says that the relation between two partitioning conditionals that differ in what they claim to be case in the event that α, but that agree in what they claim in the event that $\neg\alpha$, should depend only on the relation between the prospects on which they differ – i.e. on what is the case if α. The $\mapsto$-Separability condition belongs to a family of separability conditions found in decision theory and which includes the von Neumann and Morgenstern Independence axiom and Savage's Sure-Thing Principle, the main role of which is to ensure that prospects that are realised in different states of the world are evaluated independently of each other. We have already rehearsed some of the arguments for and against separability in the context of Savage's theory, and will return to them in the chapter after the next, so I won't repeat them here.

Our final representation result, stated more formally in the Appendix as Corollary A.18, establishes the implications of such a separability condition within our framework.

Representation of Partitioning Conditionals Suppose that the agent's relational beliefs and preferences are defined on a simple conditional algebra
of prospects and that her preferences cohere with her relational beliefs
and satisfy $\mapsto$-Separability. Then:

1. Her degrees of belief accord with Belief Independence.
2. If her preferences satisfy $\mapsto$-Betweeness, then her degrees of desire
   accord with Value Independence.

TABLE 7.1. *Axioms and Rationality Properties*

Rationality thesis	Characteristic axioms
Numerical representability	Completeness Transitivity Continuity
(1) Probabilism	Boundedness of $\unrhd$ ∨-Separability of $\unrhd$
(2) Desirabilism	∨-Betweenness of $\succsim$ Coherence of $\unrhd$ and $\succsim$
(3) Suppositional Belief (4) Suppositional Desire	Coherent Supposition
(7) Ramsey Test for belief (9) Ramsey Test for desire	Ramsey Property
(5) Cond. Evidential Belief (6) Cond. Evidential Desire	Suppositional Rigidity
(8) Adams' Thesis (10) Bradley's Thesis	Indicative Property + $\mapsto$-Betweenness
(11) Belief Independence (12) Value Independence	$\mapsto$ -Separability + $\mapsto$-Betweenness

7.7 SUMMARY

This concludes our account of rational prospective agency, a form of rationality characteristic of agents who are capable not only of reacting to their environment but of engaging in hypothetical reasoning both about possible features of this environment and about the effects of interventions in it. The normative core of the account is contained in the 12 rationality claims that have been defended, primarily by showing how they follow from more fundamental rationality conditions on relational beliefs and preferences. The relations between these relational properties and the corresponding rationality theses are summarised in Table 7.1.

PART III

FACING THE WORLD

8

Multidimensional Possible-World Semantics

8.1 PROSPECTS AND POSSIBLE WORLDS

> Any account of mental representation must explain the parallels between the objects or contents of speech acts and the objects or contents of propositional attitudes.
>
> (Stalnaker, 1984, p. 59)

Agency requires a number of abilities: to recognise and represent salient contingencies or prospects, to form consistent attitudes to them and to act in the light of these attitudes. In the second part of the book I gave an account of the requirements that rationality imposes on the beliefs and desires that an agent forms in negotiating her way around her environment. In doing so I refrained from saying anything specific about the contents of these attitudes, simply bundling them under the category of 'prospects'. This has certain advantages, not least that it allows this account of rationality to be paired with more than one conception of what the objects of belief and desire are. But on the downside it left open the question of whether the proposed rationality constraints are consistent with any account of prospects, a question whose importance is highlighted by the triviality results previously alluded to. So before tackling how decisions should be guided by beliefs and desires we need to address the question of what prospects are.

The philosophical importance of this question lies in the central role that an account of prospects must play in any general theory of thought, reasoning and action. Prospects serve not only as the objects of attitudes such as belief and desire but also as the contents of sentences that agents use to express such beliefs and desires and to reason about the world and their place in it. An agent who says 'It will rain on Thursday' is able to express her belief that it will rain on Thursday by uttering this sentence in virtue of the fact that it has as its content or meaning just the prospect that she believes true. When she reasons from this to the thought 'I will get wet if I don't take an umbrella' the inference she makes is assisted by the semantic relationships between these sentences and others. To explain the link between, on the one hand, agents' capacity to reason and communicate and, on the other, their ability to form the attitudes that serve as their basis for deciding what to do, we must be able to identify the common content of their thought and speech. To say what prospects are is to give a theory of such content.

8.1.1 Possible-Worlds Models

My starting point will be the possible-worlds account of content; specifically in the form developed by Robert Stalnaker (1984), though its main features are common to many working within this framework. The central claim of this account is that prospects or propositions are *sets of possible worlds*, where a possible world is a way that things might be or might have been. As such, a possible world is a characteristic of our 'reality', a characteristic that makes modal claims true or false. (More on this in due course.)

In philosophical logic, possible worlds are often taken as primitives to be used to explain notions such as consistency and possibility. Here, on the other hand, we take for granted a background conception of what is possible and use possible worlds to represent those possibilities that are maximally specific with regard to all that is *relevant*, something that depends on the purpose of the representation. When a decision maker draws up a model of the decision problem she faces, she will take as the possible worlds those states picked out by descriptions that are maximally specific with regard to all that matters, in her view, to the decision at hand. On the other hand, when we use a model to describe a decision made by someone else we must use descriptions that are maximally specific with regard to all that we think *they* view as important (even if we do not). When we want to employ it normatively, say to criticise someone else's decisions, then we must include descriptions that we regard as relevant even if they do not. And so on.

This kind of usage of possible worlds is appropriate to 'small-world' modelling, when we are interested only in a limited range of possibilities. In these models, possible worlds are not metaphysically atomic in any sense and what is a possible world in one model will be a coarse-grained prospect in another, more refined, model. Nonetheless, it is useful at times to speak in terms of a 'grand-world' model in which the possible worlds are metaphysical atoms. Different agents may recognise different possibilities and the same agent may recognise different ones at different times (as she acquires or loses cognitive capabilities, for instance). Working with a grand-world model allows us to represent such differences in terms of different partitions of a single underlying space and to treat an agent's attitudes to coarser-grained prospects as if they were derived from attitudes to atomic possible worlds. But the fiction involved here should be kept in mind. To fully describe such worlds one must give a complete specification of all the metaphysically possible ways things might be, something that is clearly beyond the capabilities of resource-limited agents. To take an agent to have attitudes towards a grand world is like taking a physical object to be located at a precise point in space. Just as real objects occupy spatial regions, so too real agents take attitudes only to the coarser-grained propositions or events. It is these attitudes – to the prospects that the agent is capable of recognising and representing – that we are really interested in.

Let us now see how the possible-worlds account of content provides a platform for a general theory of thought and agency. I will reserve the formal vocabulary of the previous chapters for sentences and add some new vocabulary for their contents. So italic roman capitals will denote factual sentences (these being sentences in which the conditional operator introduced below does not occur) while lower-case Greek letters will denote arbitrary sentences. The content of a sentence α will generally be denoted by $[\alpha]$. When the context makes for clear application of this convention, however, the set of worlds at which the factual sentence A is true will be denoted by the non-italic letter A, and vice versa. The symbols $\neg$, $\wedge$ and $\vee$ will, respectively, denote the sentential operations of negation, conjunction and disjunction. As before the symbol $\rightarrow$ will denote a conditional operator variable and $\mapsto$ the indicative conditional operator.

In what follows we work with a background language $\mathcal{L}$ and a set $W = \{w_1, w_2, ..., w_n\}$ of possible worlds, assumed for simplicity to be finite (nothing of substance depends on this assumption). The Boolean algebra based on the power set of W – i.e. $\wp(S)$, the set of all subsets of W – is denoted by Ω and that based on the power set of any subset A of W by Ω_A. Throughout, I will use $\bar{A}$ to denote $W - A$.

By convention, when p is a probability mass function on any set of worlds S, then P will be the corresponding probability function on the power set of S, and hence such that the measure that P places on any subset of S is the sum of the masses of its world elements, as measured by p; i.e. for any $X \subseteq S$

$$P(X) = \sum_{w \in X} p(w)$$

The basic possible-worlds framework can now be captured by four propositions.

(PW1) **Semantics** The meaning of any sentence α is given by the set of possible worlds at which it is true that α. More precisely, the semantic contents of the $\mathcal{L}$-sentences can be specified by an interpretation: a mapping v from pairs of sentences and possible worlds to the semantic values '1' (for truth) or '0' (for falsity) satisfying the Boolean laws of composition – i.e. such that, if the semantic value assigned to $\mathcal{L}$-sentence α at a world w is denoted by $v_w(\alpha)$, then

$$v_w(\alpha \wedge \beta) = v_w(\alpha) \cdot v_w(\beta)$$
$$v_w(\neg \alpha) = 1 - v_w(\alpha)$$
$$v_w(\alpha \vee \beta) = v_w(\alpha) + v_w(\beta) - v_w(\alpha \wedge \beta)$$

(PW2) **Logic** A sentence β is a semantic consequence of another sentence α, relative to an interpretation v, just in case the truth of α ensures that of β. Formally:

$$\alpha \models_v \beta \Leftrightarrow [\alpha] \subseteq [\beta]$$

Similarly, β is a tautology relative to an interpretation v (denoted $\models_v$ β), just in case it is true in all possible worlds – i.e. in case $[\beta] = W$. Finally, β is a valid consequence of α (denoted $\alpha \models \beta$) iff it is a semantic consequence of α under every interpretation, and β is valid (denoted $\models \beta$) iff it is a tautology under every interpretation.

(PW3) **Pragmatics** The degree to which a rational agent believes a sentence is given by her subjective probability for the sentence being true. More formally, let p be a probability mass function on the set of worlds that measures the probability of each world being the actual one. Then a rational agent's degrees of belief in sentences will equal her expectation of their semantic value – i.e. be given by a probability function Pr on $\mathcal{L}$ such that, for all $\mathcal{L}$-sentences α,

$$\text{Pr}(\alpha) = \sum_{w \in W} p(w) \cdot v_w(\alpha) = P([\alpha])$$

(PW4) **Models** The final claim concerns the relationship between the semantics, pragmatics and logic of a language. Loosely, it is this: what belief attitudes it is rational to take to sentences and what inferences it is correct to make with them is determined by what sentences mean and what beliefs one has about possible worlds, and not the other way round. To make this more precise, let $\Pi = \{p_j\}$ be the set of all probability mass functions on the set of possible worlds W, interpreted as the set of rationally permissible degrees of belief. And let $\mathcal{I} = \{v_i\}$ be the set of all permissible assignments of semantic values to sentences of $\mathcal{L}$. A possible-worlds model (PW-model, for short) of $\mathcal{L}$ is a structure $<W,v,p>$ where W is the background set of worlds, v belongs to $\mathcal{I}$ and p to Π. Such a structure determines both what belief attitudes the speaker can rationally take to $\mathcal{L}$-sentences and what inferences she can rationally make with them. In particular, if Pr and $\models_v$ are, respectively, a probability measure and a semantic consequence relation on $\mathcal{L}$-sentences then we can say that a PW-model $<W,v,p>$ explains the pair (Pr, $\models_v$) just in case Pr and $\models_v$ are related to v and p as required by (PW2) and (PW3). That is, it yields explanations of the form '$\alpha \models_v \beta$ because $[\alpha] \subseteq [\beta]$' and 'Pr(X) = x because $P([X]) = x$'. The final assumption underlying standard applications of possible-worlds models can now be made explicit:

$$\forall v \in \mathcal{I}, p \in \Pi, \; <W,v,p> \text{ is a PW-model of } \mathcal{L}$$

These four propositions provide the core of the possible-worlds framework. But they do not suffice to determine an interpretation of the entire theory of rationality developed in the previous chapters. For this we need an account of conditional propositions consistent with the constraints we imposed on rational attitudes towards them. The possible-worlds framework allows for numerous different specific theories about the truth conditions of conditionals, but the natural place to start is with the one that has become standard, namely that developed by David Lewis (1973) and Robert Stalnaker (1968). Indeed Stalnaker conceived of his semantics as the natural partner to the Ramsey Test hypothesis for belief in that it formalises the idea that conditionals should be evaluated by determining the truth of their consequent under the supposition that their antecedent was true. So there is a deep connection between his work and the rationality claims of this book.

On Lewis and Stalnaker's theory, whether a conditional is true or false at some world w depends on whether its consequent is true or false at all worlds that are 'closest' or 'most similar' to w in some relevant respect. The minimal departure from actuality involved in evaluating conditionals is handled in Stalnaker's semantics by a selection function on possible worlds and in Lewis' by a similarity relation on worlds. Nothing of substance hangs on which way one chooses to do it and we will follow Stalnaker, generalising slightly so as to encompass Lewis' views. Thus:

(PW5) **Conditionals** A conditional $\alpha \to \beta$ is true at world w just in case its consequent is true at all worlds most similar to w at which it is true that α. Formally, let f be a selection function from $W \times \mathcal{L}$ to Ω, mapping a world-sentence pair (w, α) to a set of 'closest' α-worlds. Then

$$[\alpha \to \beta] = \{w : f(w, \alpha) \subseteq [\beta]\}$$

Lewis and Stalnaker disagreed on an important detail, namely whether the selection function f should be constrained to pick a single most similar world for any given world-sentence pair or whether there could be similarity ties. At stake is the validity of the law of conditional excluded middle (i.e. whether $(\alpha \to \beta) \vee (\alpha \to \neg\beta)$ is a logical truth), denied by Lewis and endorsed by Stalnaker. Consider the pair of sentences that Lewis appeals to:

'If Bizet and Verdi had been compatriots, Bizet would have been Italian.'

'If Bizet and Verdi had been compatriots, Verdi would have been French.'

Most people would hesitate to affirm either sentence. According to Lewis, this is because no world in which Bizet and Verdi were both Italian is more similar to the actual world than all the worlds in which they were both French, and vice versa. So, according to proposition PW5, both sentences are false. Not so, says Stalnaker: we hesitate to affirm either because the context typically underdetermines the relevant minimal departure from actuality which the selection function represents. But on every full specification of a standard for minimal departure, one of the two sentences will be true and the other false.

Lewis and Stalnaker also disagreed on the scope of the theory. Lewis restricted PW5 to counterfactual conditionals, holding that the truth conditions for indicative conditional sentences are given by the material conditional – i.e. that $\alpha \mapsto \beta$ is equivalent to $\neg \alpha \vee \beta$. Stalnaker, on the other hand, took PW5 to give the correct truth conditions for both indicative and counterfactual conditionals, with differences between to be captured by properties of the selection function. These properties in turn should depend on the kind of supposition the selection function encodes. We return to both issues below.

8.1.2 Triviality Strikes Again

The possible-worlds framework, as contained in the five propositions PW1–PW5, offers a simple and elegant theory of the relation between language and belief. Unfortunately it is inconsistent with the theory of rational agency developed in the previous chapters. In particular, in a sense that will be made clear, there is no assignment of semantic values to conditionals within this framework that is non-trivially consistent with Adams' Thesis: the claim that the probability of an indicative conditional is the conditional probability of its consequent given its antecedent.

The conflict between Adams' Thesis and the Lewis–Stalnaker semantics for conditionals is easy enough to demonstrate. To do so, let us work with the model illustrated in Figure 8.1. Although extremely simple, it will serve to illustrate many of the issues that will be discussed in this chapter. Suppose that our background language $\mathcal{L}$ contains at least three factual sentences A, B and C and the conditional sentence $A \to B$. Consider a set $W = \{w_1, w_2, w_3, w_4, w_5\}$ of just five possible worlds and the corresponding set Ω of its subsets, including the events $A = \{w_1, w_2, w_3\}$, $B = \{w_1, w_2, w_4\}$ and $C = \{w_1, w_3, w_5\}$, which are, respectively, the sets of worlds at which it is true that A, B and C (i.e. $[A] = A$, $[B] = B$, etc.).

To construct a counterexample to the Adams' Thesis in this model, consider a selection function f such that $f(w_5, A) \notin B$, and probability mass function p on worlds such that $p(w_2) = 0$ and $p(w_5) \neq 0$. By proposition PW5 of the possible-worlds framework, such a probability is as an admissible degree of belief function. Now $P(B|A) = 1$, but $P([A \to B]) < 1$ because $w_5 \notin [A \to B]$. So the probability of $A \to B$ does not equal the conditional probability that A given that B, contrary to Adams' Thesis.

The immediate explanation of this conflict, as Lewis (1976) observed, is that conditionalising on the truth of A redistributes probability from worlds in which it is false that A to those in which it is true in a very different way from that induced by the counterfactual supposition that A is true. So a simple solution to the triviality result presents itself. As conditionalisation is characteristic of evidential supposition, a form of supposition not appropriate for evaluating counterfactuals, the Lewis–Stalnaker semantics should be applied to counterfactuals only. Adams' Thesis, on the other hand, applies only to indicative conditionals.

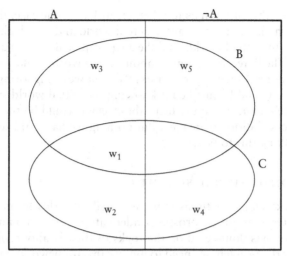

FIGURE 8.1. A Small-World Model

Unfortunately, this will not do. For the conflict between the possible-worlds framework and Adams' Thesis runs far deeper than this example might suggest. There is in fact *no* non-trivial way of assigning sets of worlds to conditional sentences which is consistent with both Adams' Thesis and the first, third and fourth PW propositions. By 'non-trivial assignment', I mean one that allows for the possibility that, for some sentences A and B, $\neg A$ and $A \rightarrow B$ are both true at some world (i.e. it is not case, that for all sentences A and B, $A \rightarrow B$ implies that A). Again we can illustrate why with our toy model.[1] Suppose that an assignment of worlds is made to $A \rightarrow B$ such that $[A \rightarrow B] \nsubseteq A$. Let p be a probability mass function p on worlds such that $p(w_1) = 0 = p(w_3)$, but all other worlds have non-zero probability. Then $P(B|A) = 0$, but $P([A \rightarrow B]) > 0$. So $P([A \rightarrow B]) \neq P(B|A)$.

The problem, I would suggest, lies much deeper than Lewis thought. It is this. In line with the Lewis–Stalnaker theory we introduced a new semantic parameter – the selection function – in order to give a truth-conditional account of conditionals, but we did so without considering whether the possible-worlds framework as contained in propositions PW1 to PW4 required modification. But clearly it will, since the selection function amounts to a new parameter in the interpretation of a language. We can easily modify the Semantic and Logic propositions to take this into account, by making it explicit that the mapping v depends on f. But what should we do about the Pragmatics proposition?

In the standard possible-worlds framework an agent's uncertainty as to whether A is the case is equated with her uncertainty as to whether the actual world is one of the worlds at which it is true that A. But this will not do for

---

[1] A formal proof is given in Bradley (2000).

conditionals, for the question as to whether, if *A* were the case, then *B* would be depends both on what is true at the actual world and on which world is the most similar to the actual world. And the agent can be uncertain about both of these matters. The latter uncertainty – about which world is most similar to the actual one – is not reflected in proposition PW3, however. Ignoring uncertainty about similarity would be justified if knowing the actual world sufficed to fix belief about the counterfacts: the facts about what would be the case if some contrary-to-fact condition were true. But nothing we have said so far gives us a reason for taking this to be true.

8.2 MULTIDIMENSIONAL SEMANTICS

Let's take stock. Our hypothesis that the possible-worlds framework would provide an interpretation of prospects adequate to our account of rational belief and desire has floundered on the rocks of the triviality results. To steer our way out of the problem we need to modify this framework in order to take account of agents' uncertainty regarding the counterfacts: about what would be true if some supposed condition were true. We could do so by introducing a second probability measure on selection functions. But the same thing can be achieved in a more natural way by having possible worlds serve not only as candidates for the way things are or could be, but also as candidates for the ways things are or could be under the supposition that one or another condition holds.

If world w_A could be the case under the supposition that *A*, then we will say that w_A is a possible counteractual *A*-world. If *A* is false, w_A will be said to be *strictly counterfactual*. (Any counteractual *A*-world is strictly counterfactual relative to any possible world in $\bar{A}$, for instance. But counteractual worlds are not always strictly counterfactual, for if *A* is true then w_A may not only be a possible way that things are under that supposition that *A* but in fact the way things actually are.) Now, what I want to suggest is the following. *Possible counteractual worlds make conditional sentences true in the same way that possible actual worlds make factual sentences true.* The conditional 'If *A* then *B*' is made true by any possible counteractual *A*-world w_A at which it is true that *B*. For example, the counteractual world in which Obama is born in Kenya and goes to school in Nairobi makes it true that had Obama been born in Kenya he would have gone to school in Nairobi, while the counteractual world in which he is born in Kenya but goes to school in Mombasa makes it false.

At this point one could object, in the spirit of Lewis, that there may be no fact of the matter as to what would be the case if something that is currently true had not been. Would Verdi and Bizet have been Italian or French had they been compatriots? There seems to be no truth of the matter. This objection rests on a confusion between two questions: that of whether something is true or false and that of whether we can determine which is the case. It is entirely plausible that the facts regarding Bizet and Verdi's life and times (including the similarities that Lewis appeals to in his theory) do not determine the relevant counterfacts regarding the nationality that they would have had, had they been compatriots. But they still would have had some nationality: either Italian or French or something else. So, in speaking of the counteractual A-world to w, I do not require that the facts about world w determine which is the counteractual A-world. I require only that we can speak of different candidates for the position and be able to say that some of them make the sentence 'If A then B' true and others make it false.

To illustrate the implications of the proposed treatment of conditionals, let us continue to work with our simple five-world model exhibited in Figure 8.1, elaborating it to allow for representation of conditional prospects. Relative to the set of possible worlds W, each supposition induces a set of possible counteractual worlds. The supposition that A, for instance, induces the set of counteractual A-worlds $W_A = \{w_1, w_2, w_3\}$, and corresponding set Ω_A, containing conditional events $B_A = \{w_i \in W_A : w_i \in B\} = \{w_1, w_2\}$, $C_A = \{w_1, w_3\}$, and so on. The supposition that not A induces a different set of counteractual worlds – namely $W_{\bar{A}} = \{w_4, w_5\}$ – and corresponding set of conditional events $\Omega_{\bar{A}}$; the supposition that B yet another; and so on. (Note that I have extended the convention of denoting sets of worlds with non-italicised letters, by using B_A to denote the set of A-worlds at which it is true that B in the same way as A was used to denote the set of worlds at which it is true that A. Note also that $B_A = A \cap B$.)

For simplicity, we restrict attention to a single supposition for the moment, namely the supposition that A. The set of elementary possibilities is then given by a subset F of the cross-product of W and W_A, which can be presented in tabular form as in Table 8.1. Each ordered pair $\omega_{ij} = \langle w_i, w_j \rangle$ appearing in the cells of the table represents an elementary possibility: that w_i is the actual world and that w_j is the counteractual A-world. Prospects are just sets of such possibilities and thus subsets of the space F of possible world-pairs. For any X $\subseteq$ W and $Y_A \subseteq W_A$, we denote by (X, Y_A) the prospect that is the subset of F lying in the cross-product of X and Y_A, namely the prospect that X is the case and that Y is or would be on the supposition that A. More precisely:

$$(X, Y_A) = \{\langle w_i, w_j \rangle \in F : w_i \in X, w_j \in Y_A\}$$

In Table 8.1, for instance, the prospect that A and B is given by the set (A $\cap$ B, W_A), while the prospect that if A is or were the case then B is or would be, is given by the set (W, B_A).

Prospects serve both as the contents of sentences and the objects of agents' attitudes. The contents of factual sentences are given by rows of the table. The prospect that A, for instance, is given by the first, second and third rows of the table, while that of B by the first, second and fourth. The contents of conditional sentences, on the other hand, are given by columns of the table. The prospect that $A \to B$, for instance, has as its content the first and second columns of the table and that of $A \to C$ the second and third. The contents of conjunctions, disjunctions and negations of prospects (conditional or otherwise) are given by the intersection, union and complements of their contents.

TABLE 8.1. *Multidimensional Possibility Space*

Worlds	Supposed A-worlds		
	w_1	w_2	w_3
w_1	$\langle w_1, w_1 \rangle$	$\langle w_1, w_2 \rangle$	$\langle w_1, w_3 \rangle$
w_2	$\langle w_2, w_1 \rangle$	$\langle w_2, w_2 \rangle$	$\langle w_2, w_3 \rangle$
w_3	$\langle w_3, w_1 \rangle$	$\langle w_3, w_2 \rangle$	$\langle w_3, w_3 \rangle$
w_4	$\langle w_4, w_1 \rangle$	$\langle w_4, w_2 \rangle$	$\langle w_4, w_3 \rangle$
w_5	$\langle w_5, w_1 \rangle$	$\langle w_5, w_2 \rangle$	$\langle w_5, w_3 \rangle$

When a pair of worlds $\langle w_i, w_j \rangle$ is part of the content of a sentence S we can say that this pair makes it true that S. In this sense, the truth conditions of sentences are given by ordered pairs of worlds. An immediate implication is that we cannot in general speak of a prospect as being true or false at a world *simpliciter*. For instance, while $A \to B$ is true at $\langle w_3, w_1 \rangle$, we cannot say that it is made true (or false) by the facts at w_3 because these facts alone do not determine its truth or falsity independently of the relevant counterfacts – i.e. of whether the counteractual A-world is w_1, w_2 or w_3.

The upshot is that the theory being proposed gives a truth-conditional semantics for conditionals, while at the same time allowing that the truth values of conditionals are not determined by the facts alone. The former property allows for an explanation both of the role that conditionals play in discourse aimed at establishing the truth and of how they compound with other sentences. The latter property explains the difficulty we have in some situations in saying whether a conditional is true or false, a difficulty that Lewis used to motivate his rejection of the law of conditional excluded middle and that non-factualists point to in their rejection of truth-conditional accounts of conditionals.

8.2.1 Centring

The formal distinction between facts and counterfacts does not preclude that they be related in various specific ways, both semantically and pragmatically. Any view about their relation must square with the fact there are questions such as whether the allies would have lost the Second World War if Hitler had captured Moscow, or whether I would have been a philosopher if I had been born in a different family, that seem impossible to settle no matter how much evidence concerning what actually happened we can hope to bring to bear on them. On the other hand there are questions such as whether the sugar would have dissolved if I had added it to my coffee that do seem to be decided by features of the actual world: the chemical properties of the sugar, the temperature of the coffee, how much sugar had already been added, and so on.

TABLE 8.2. *Centred Possibilities*

Worlds	Supposed A-worlds		
	w_1	w_2	w_3
w_1	$\langle w_1, w_1 \rangle$	–	–
w_2	–	$\langle w_2, w_2 \rangle$	–
w_3	–	–	$\langle w_3, w_3 \rangle$
w_4	$\langle w_4, w_1 \rangle$	$\langle w_4, w_2 \rangle$	$\langle w_4, w_3 \rangle$
w_5	$\langle w_5, w_1 \rangle$	$\langle w_5, w_2 \rangle$	$\langle w_5, w_3 \rangle$

There are two extreme views on counterfactuality that fare badly because they have trouble explaining one of these two classes of cases. On the Autonomy view the counterfacts are completely independent of the facts and hence any combination of facts and counterfacts is possible. On the Reductionist view, on the other hand, the counterfacts are completely determined by the facts. The Reductionist view leads back to the orthodoxy and incompatibility with Adams' Thesis. Not so, the Autonomy view. But it has an implausible implication, namely that whether the sugar will dissolve if added to the coffee is independent of whether or not the sugar was in fact added to the coffee and whether or not it dissolved when added.

More promising is some kind of intermediate, non-reductionist view that recognises a variety of possible relations between facts and counterfacts. Some classes of counterfacts might be completely determined by a broad enough specification of the facts pertaining at a possible world (as in the sugar case). Others might be constrained, though not fully determined, by the chances of relevant events. For instance, the fact that a coin is very biased towards heads can be taken as grounds for saying that it would have landed heads if it had been tossed. Still others might be hardly constrained at all. Views about these relations can be accommodated in two ways: by restricting the possible combinations of facts and counterfacts serving as semantic values of sentences and by restricting the joint attitudes that an agent can rationally take to them.

Let's leave consideration of the latter to later sections and focus for now on a widely adopted semantic principle that I will call Centring, in line with the terminology introduced by Lewis (1973). The Centring condition rules that certain combinations of factuality and counteractuality are impossible. In particular, it says that if world w_i is the actual world and A is true at w_i, then w_i must also be the counteractual A-world. For instance, if I added sugar to my coffee and it dissolved then, under the supposition that I added it to my coffee, the sugar dissolved. After all, the thought goes, I did add it, and that's what happened. In our toy model, the effect of restricting the elementary possibilities in this way is to eliminate some cells, leaving us with Table 8.2.

Centring is not the only assumption that we could make about the relation between facts and counterfacts, and different conditions may be appropriate for different modes of supposition. Any condition restricting the two-dimensional space F of elementary possibilities can be perspicuously represented by a table like that of Table 8.2. But, if we want to represent multiple suppositions, there are technical advantages to constructing this space by reintroducing selection functions, now serving to pick out, for each possible world w and supposition A, which counteractual worlds w_A are admissible. Formally, a selection function f in this context is a mapping from $W \times \Omega$ to Ω satisfying, for all $A \subseteq W$:

1. $f(w, A) \subseteq A$
2. $f(w, A) = \varnothing \Leftrightarrow A = \varnothing$
3. $f(w, W) = \{w\}$
4. If $w \in A$ then $w \in f(w, A)$

The first condition simply states that counteractual worlds under the supposition that A must be worlds at which it is true that A. The second says that the set of counteractual worlds is empty only if the supposition is contradictory. The third condition requires that supposition of a tautology (the entire space of possible worlds) is inert, in the sense that supposing it true does not change the admissible facts. The fourth condition requires that any world at which it is true that A must be a possible counteractual A-world. This condition is usually termed Weak Centring, in contrast to the stronger condition informally introduced before and which we can now state as:

Centring: If $w \in A$ then $f(w, A) = \{w\}$

Stated this way, Centring expresses a particular conception of what is possible, according to which what is actually true determines what might have been true under any supposition consistent with the actual truth. This is surely right for *epistemic* possibility: at any world w at which it is true that A it is not epistemically possible that any world other than w be the case on the supposition that A. And epistemic possibility would seem to be what is at issue when we reason evidentially using indicative conditional sentences. On the other hand, it is much more controversial whether Centring governs *causal* possibility and hence whether it is appropriate to counterfactual or interventional reasoning. Both Lewis and Stalnaker assume that it is, perhaps because they take counterfactual and evidential reasoning to coincide when what is being supposed is in fact true. But in the absence of a deterministic relationship between two events it does not seem right to regard the fact of their co-occurrence to imply that the occurrence of one causally necessitates the other. In any case, we do not need to settle the issue here, for it suffices that different modes of supposition may be represented by different constraints on the spaces of world n-tuples for multidimensional possible-worlds models to accommodate a variety of views about the differences between indicative and counterfactual conditionals.

8.2.2 Multidimensional Possible-World Semantics

We now have all the ingredients we require to state our semantic theory more formally. On the account being offered here the meanings of sentences are still given by the conditions in which they are true, but these conditions are represented by n-tuples of possible-worlds rather than single worlds. It follows that to give an interpretation of a language we need to specify three things: the background set of worlds W, the selection function f determining the space F and an assignment v^* of subsets of this space to sentences of the language.

Let $\mathcal{FL}$ be a language of factual sentences that is closed under conjunction, disjunction and negation and let v be a possible-worlds interpretation of it – i.e. a Boolean mapping from pairs of $\mathcal{FL}$-sentences and worlds in W to the truth values $\{0,1\}$. As before, let $v_w(\alpha)$ denote the truth value assigned by v to sentence α and $[\alpha]$ be the set of worlds at which it is true. Let $S = \{A_i\}_{i=1}^{n-1} \subseteq \mathcal{FL}$ be an indexed set of suppositions and F be the set of world n-tuples $\omega = (w, w_1, ..., w_{n-1})$ determined by the selection function f, with w being any member of W and w_i belonging to $f(w, [A_i])$. Finally, we define the simple conditional language $\mathcal{SCL}$ as a set of sentences containing $\mathcal{FL}$ that is closed under conjunction, disjunction and negation and is such that, for all $A_i \in S$ and $B \in \mathcal{FL}$, $A_i \to B \in \mathcal{SCL}$.

The multidimensional possible-worlds framework that I am proposing can now be captured by four propositions.

(MPW1) **Semantics** The semantic contents of $\mathcal{SCL}$-sentences are given by a mapping v^* from pairs of sentences and n-tuples of possible worlds to the truth values $\{0,1\}$ satisfying the Boolean laws of composition and such that, if $v_\omega^*(\alpha)$ denotes the truth value assigned by v^* to sentence α, then, for all $B \in \mathcal{FL}$,

$$v_\omega^*(A_i \to B) = v_{w_i}(B)$$

In other words, the truth value of $A_i \to B$ at n-tuple $\omega = (w, w_1, ..., w_i, ..., w_{n-1})$ is the truth value of B at the possible (counteractual) world w_i.

(MPW2) **Logic** Let $[\alpha]^*$ denote the set of world n-tuples ω at which α is true. Then

$$\alpha \models_{v^*} \beta \Leftrightarrow [\alpha]^* \subseteq [\beta]^*$$

Correspondingly, $\alpha \models \beta$ iff, for all such v^*, $\alpha \models_{v^*} \beta$.

(MPW3) **Pragmatics** Let p be a probability mass function on the set of world n-tuples $\omega = (w, ..., w_{n-1}) \in F$. Then the rational agent's degrees of belief in $\mathcal{SCL}$-sentences are measured by a probability function Pr on $\mathcal{SCL}$ such that, for all $\alpha \in \mathcal{SCL}$,

$$\Pr(\alpha) = \sum_{\omega \in F} p(\omega) \cdot v_\omega^*(\alpha)$$

(**MPW4**) **Models** Let $\Pi = \{p\}$ be the set of all permissible joint probability mass functions on F and $\mathcal{I}^* = \{v^*\}$ be the set of all permissible assignments of semantic values to sentences of $\mathcal{SCL}$. A multidimensional possible worlds model (MPW-model for short) of $\mathcal{SCL}$ is a structure $< F, v^*, p >$ where F is the background set of world n-tuples, v^* belongs to $\mathcal{I}^*$ and p to Π. Then

$$\forall v \in \mathcal{I}^*, p \in \Pi, < W, v^*, p > \text{ is an MPW-model of } \mathcal{SCL}$$

Let's now explore the implications of the third of these propositions for the measurement of uncertainty.

8.3   PROBABILITY

We observed earlier that an agent can be uncertain both about what is actually the case (the facts) and about what is or would be the case if some condition is or were true (the counterfacts). These two forms of uncertainty are (at least partially) independent of each other. Someone might be pretty sure that the match is to be played tomorrow, but quite unsure as to whether it would be played were it to rain. Equally they could be sure that the match will not be played were it to rain, but quite unsure as to whether it will rain or not. Both kinds of uncertainty play an important role in our deliberations about what to do: whether to buy a ticket for the match, for instance, and whether to bring an umbrella.

The first kind of uncertainty will be represented here by a probability mass function p on the set of possible worlds W and corresponding probability function P on Ω. The second kind of uncertainty – her uncertainty about what would be case if A were true – will be represented by a probability mass function p_A on the set of possible counteractual worlds W_A and corresponding probability function P_A^* on Ω_A. Finally, to measure her joint uncertainty regarding both the facts and the counterfacts under the supposition that A, we extend p to a joint density on the set F of pairs of worlds that lie in the table cells, subject to the constraint that p_A is the marginal of the extended density p – i.e. that, for all $w_j \in \Omega_A$,

$$p_A(w_j) = \sum_{w_i : \langle w_i, w_j \rangle \in F} p(\langle w_i, w_j \rangle)$$

where $p(\langle w_i, w_j \rangle)$ is the probability that w_i is the actual world and w_j is/would be the counteractual world on the supposition that A.

The joint density p determines a corresponding extension of P to a joint probability function on Γ that measures the joint probabilities of actuality and counteractuality under the supposition that A. So, just as $P(X)$ measures the probability that X is the case and $P_A^*(Y_A)$ measures the probability that Y is or would be the case if A is or were, $P(X, Y_A)$ measures the joint probability that

X is the case and that Y is or would be the case if A. When there is no risk of ambiguity I will drop 'empty' terms, writing $P(X)$ for $P(X, W_A)$, $P(Y_A)$ for $P(W, Y_A)$ and $P(Y_A|X)$ for $P((W,Y_A)|(X,W_A))$.

The probability function P encodes, within a multidimensional possible-worlds model, the agent's state of belief regarding both the facts and the counterfacts. It is straightforward to establish that it follows from the way that P is constructed that the agent's degrees of belief must satisfy the core rationality properties advanced in the previous chapters. In particular:

1. *(Probabilism)* P is obviously defined in a way which accords with the probability axioms P0, P1 and P2.
2. *(Suppositional Probability)* P_A^* is a suppositional probability function. Satisfaction of P*1 follows from the fact that P_A^* is defined in accordance with the probability axioms and of P*2 from the fact that A = W_A, the domain of P_A^* (which by definition has measure one). Finally, P*3 follows from the fact that supposition by a tautology is inert, a feature imposed by condition 3 on the selection function.
3. *(Regularity)* If Centring holds, then P_A^* satisfies Regularity. For, by Lemma A.20 in the Appendix, Centring implies that, for all $X \in \Omega$, $(A \cap X, W_A) = (A, X_A) \subseteq (W, X_A)$. Hence

$$P_A^*(X_A) = P(W, X_A) \geq P(A, X_A) = P(A \cap X, W_A) = P(A \cap X)$$

4. *(Ramsey Test)* P and P_A^* jointly satisfy the Ramsey Test for Belief. To see this recall that the content of the conditional prospect $A \rightarrow B$ is given by the set (W, B_A), the union of the first and second columns of Table 8.1. But, since P_A^* is the marginal of P on W_A, necessarily $P(W, B_A) = P_A^*(B_A)$. Hence the probability of $A \rightarrow B$ must equal the probability of B on the supposition that A, as required.

Two conclusions can be drawn from this. Firstly, that it is possible for an agent to have rational beliefs regarding prospects with contents specified by the world n-tuples making them true: in particular, degrees of belief that are probabilities of truth and conditional degrees of belief that are suppositional probabilities of truth. Secondly, an agent who has probabilistic degrees of belief and conditional belief not only can but *must* have degrees of belief in conditionals that satisfy the Ramsey Test. It is this second feature that is the more noteworthy of the two, since it is unique to multidimensional possible-worlds models.

8.3.1 Independence

What further conditions are required to characterise evidential supposition within the multidimensional possible-worlds framework? We have already argued that Centring is an essential characteristic of evidential reasoning.

TABLE 8.3. *Centred Probabilities*

Worlds	Supposed A-worlds		
	w_1	w_2	w_3
w_1	$p(w_1)$	–	–
w_2	–	$p(w_2)$	–
w_3	–	–	$p(w_3)$
w_4	$p(\langle w_4, w_1 \rangle)$	$p(\langle w_4, w_2 \rangle)$	$p(\langle w_4, w_3 \rangle)$
w_5	$p(\langle w_5, w_1 \rangle)$	$p(\langle w_5, w_2 \rangle)$	$p(\langle w_5, w_3 \rangle)$

Given it, the agent's total state of uncertainty in our toy model can be summarised as in Table 8.3. As it shows, Centring completely determines the relationship between the facts and the counterfacts under the supposition that A, whenever A is in fact true. But it leaves the relationship between the facts and the *strict* counterfacts completely open. One salient possibility is that the strict counterfacts are probabilistically independent of the facts. Formally:

World Independence $\forall w_i \in W_{\bar{A}}, p(\langle w_i, w_j \rangle) = p(w_i) \times p_A(w_j)$

World Independence is stochastic independence on the sub-domain of F consisting of pairs of facts and strict counterfacts. This property is interesting for a number of reasons. Firstly, it allows for a decomposition of joint uncertainty into its factual and counterfactual components. And, secondly, it implies that uncertainty about what is or would be the case under some supposition is completely reducible to factual uncertainty. To show this takes a number of steps. The first is to observe that World Independence is equivalent to a seemingly more general condition on prospects that requires that any prospect inconsistent with it being the case that A be probabilistically independent of what would be the case if A were true (the equivalence of the two conditions is proved as Theorem A.21 in the Appendix). More formally, it says:

Fact–Counterfact Independence $\forall X \subseteq W_{\bar{A}}, \forall Y_A \subseteq W_A,$

$$P(X, Y_A) = P(X) \cdot P(Y_A)$$

It is not difficult to show that Fact–Counterfact Independence does not hold in general for counterfactual supposition. If I know that a prize is in one of two boxes, pick one of them and discover that there is no prize in it, then I can be sure that if I had picked the other box then I would have got the prize. So what is the case, namely that the prize is not in the box I picked, completely determines what would have been case had I picked the other one. On the other hand, the fact that if, as a matter of fact, I have picked the other box, then I have the prize, does not seem to be informative at all about whether there is a prize in the box I actually picked, because I did *not*, as a matter of fact, pick the other box. So such examples do not settle the status of Fact–Counterfact Independence as a condition on evidential supposition.

Evaluation of Fact–Counterfact Independence is best achieved by considering its implications. For now we focus on one of them, a much weaker independence condition, called Supposition Independence, that turns out to be particularly significant. Supposition Independence says that the probability that A is the case is independent of what is or would be the case if A were true. Formally:

Supposition Independence $\forall Y_A \subseteq W_A, P(A|Y_A) = P(A)$

From Centring and Supposition Independence follow both Thesis 6.1 and Thesis 7.2, the two rationality claims made in previous chapters about the relation between evidential supposition and indicative conditionals.

1. *(Conditional Evidential Belief)* Given Centring, it follows from Lemma A.20 (in the Appendix) that $P(A, Y_A) = P(A \cap Y)$. But, by Supposition Independence, $P(A, Y_A) = P(A) \cdot P(Y_A)$. So $P(A \cap Y) = P(A) \cdot P(Y_A)$. Since we are concerned with evidential supposition, let's denote the marginal of P on Ω_A by P_A^+. Then, since $P_A^+(Y_A) = P(Y_A)$, it follows that, for all $Y_A \subseteq W_A$, if $P(A) > 0$:

$$P_A^+(Y_A) = \frac{P(A \cap Y)}{P(A)} = P(Y|A)$$

Hence, in accordance with Thesis 6.1, probabilities on the evidential supposition that A are conditional probabilities given that A.

2. *(Adams' Thesis)* Recall that the content of the indicative conditional $A \mapsto Y$ is given by the prospect (W, Y_A). But by the above:

$$P(W, Y_A) = P_A^+(Y_A) = P(Y|A)$$

So, in accordance with Adams' Thesis:

$$P([A \mapsto Y]) = P(Y|A)$$

Jointly, these results vindicate the claim that Supposition Independence is the defining characteristic of evidential supposition within Centred multidimensional possible-worlds models. Note also that, since Supposition Independence is implied by World Independence, we have now shown that the latter implies that

$$p(\langle w_i, w_j \rangle) = p(w_i) \times p_A(w_j) = p(w_i) \times p(w_j|A)$$

This completes the demonstration that World Independence plus Centring yields a complete reduction of all uncertainty to uncertainty about the facts.

That such a reduction is possible is, of course, just what would be expected of a form of supposition that concerns only what is, as a matter of fact, true.

We have now secured one of the main aims of introducing the multidimensional possible-worlds models, namely to provide a means of accommodating Adams' Thesis within a truth-conditional semantics for conditionals. This is of no small significance given the fact that it has widely been considered impossible to do this. It is true that we cannot claim to have proved that adherence to Adams' Thesis is a general requirement of rationality, for the derivation of this principle required Supposition Independence. Nonetheless, we can claim to have shown that, insofar as it is reasonable to consider Supposition Independence to be a characteristic principle of evidential supposition, that Adams' Thesis is a valid principle of rational belief in indicative conditionals.

One may nonetheless wonder why the triviality results for Adams' Thesis do not also apply in this framework. Constructing a triviality argument certainly seems simple enough. Let any occurrence of the phrase 'possible-world' in your favourite triviality result be replaced by that of 'n-tuple of possible-worlds' and you appear to get a triviality result for the theory presented here. In fact this is not so, however, for the fourth proposition of the multidimensional possible-worlds framework is less permissive than the corresponding proposition in the orthodox framework. MPW4, unlike PW4, implicitly allows for restrictions on permissible belief measures above and beyond the requirement that they be probabilities. In particular, it allows for restrictions on the relation between the joint probabilities on the space of world n-tuples and the marginal probabilities on possible worlds (such as those contained in the various independence principles canvassed above); restrictions that cannot be formulated without the additional structure contained in multidimensional possible-worlds framework. Crucially, amongst those belief measures ruled out by the independence constraints are just those whose existence is required for the triviality results to go through (see Bradley, 2012, 2011, for more details).

8.4 DESIRABILITY AND COUNTERFACTUAL VALUE

Just as beliefs about counterfactual possibilities play an important role in our reasoning, so too do our evaluative attitudes to them. Consider two rather different examples of psychological attitudes that are directed at counterfactual possibilities: regret and fairness. Someone might be disinclined to forgo an opportunity to go to the theatre even though she thinks the production will be terrible, because she anticipates the regret that she would experience if the production were to turn out to be wonderful (contrary to expectations). And someone might feel differently about not getting a job in cases in which, had she had better qualifications, she would have got it from cases in which she would not have got it no matter how well qualified she was.

Similarly, just as our uncertainty about what is the case can be different from our uncertainty about what would be the case if some or another condition were true, so too our assessment of how desirable something is can differ from our assessment of how desirable its truth is or would be on the supposition of some condition or other. For instance, I might prefer that I be served a cold beer rather than a hot chocolate tonight, but under the supposition that it will be a very cold evening my preference would be reversed.

To reflect this, let's introduce measures of value on both the facts and the counterfacts in the same way that we introduced probability measures on both. The desirability of possible actual worlds will be represented here by a utility function u on W, while the desirability of possible counteractual worlds, under the supposition that A, will be represented by a utility function u_A on W_A.[2] Then to measure the joint desirability of worlds we extend u to the set F of pairs of worlds, so that, for example, $u(\omega_{ij}) = u(\langle w_i, w_j \rangle)$ measures the desirability that w_i is the actual world and w_j the counteractual world on the supposition that A. For convenience, we assume that u_A and u are both zero-normalised, in the sense that

$$\sum_{w_j \in W_A} u_A(w_j) \cdot p_A(w_j) = \sum_{\omega_{ij} \in F} u(\omega_{ij}) \cdot p(\omega_{ij}) = 0$$

From the function u we can determine a corresponding desirability function V on the set of prospects Γ by defining the desirability of any prospect α to be the conditional expectation of utility value given α. Formally, if $P(\alpha) > 0$:

$$V(\alpha) := \sum_{\omega_{ij} \in \alpha} \frac{u(\omega_{ij}) \cdot p(\omega_{ij})}{P(\alpha)}$$

In identical fashion, desirability function V_A^* on Ω_A can be determined from u_A on W_A, with $V_A^*(Y_A)$ measuring the desirability that Y is or would be the case under the supposition that A. As before, I will drop 'empty' terms and write $V(X)$ for $V(X, W_A)$, $V(Y_A)$ for $V(W, Y_A)$, and so on.

The function V serves to encode within our model the agent's desires for the truth of both facts and counterfacts. Intuitively, V measures, for any pair (X, Y_A), the desirability that X is the case and that Y is/would be the case if A is/were. So, informally, we can speak of V as a measure of joint desirability value. But caution is required since V is not a measure in the strict mathematical sense and the relationship between it and the 'marginal' V_A^* that is imposed by the multidimensional possible-worlds structure is less strict than in the case of the probability measures. The zero-normalisation of the desirability functions ensures that $V_A^*(W_A) = V(W, W_A) = 0$, but since the unit of the desirability functions is not specified (in contrast to the case of

---

[2] A utility function is nothing more than a numerical index of some underlying order, understood here to be the agent's preference order over worlds.

probability measures) all that is required is that there exists a constant $k_A > 0$ such that, for all $X \in \Omega$ and $Y_A \in \Omega_A$,

$$V(W, Y_A) = k_A \cdot V_A^*(Y_A)$$

Now, what properties of desirability are entailed by these basic definitions and conditions?

1. *(Desirabilism)* The first thing to note is that V is a desirability function in the formal sense of satisfying the desirability axioms V1 and V2. To see this, let α and β be two disjoint prospects. Then, in virtue of the zero-normalisation of V,

$$V(W, W_A) = \sum_{\omega_{ij} \in F} u(\omega_{ij}) \cdot p(\omega_{ij}) = 0$$

So V satisfies V1, the Normality axiom. And, since $\alpha \cap \beta = \varnothing$,

$$
\begin{aligned}
V(\alpha \cup \beta) &= \sum_{\omega_{ij} \in \alpha \vee \beta} \frac{u(\omega_{ij}) \cdot p(\omega_{ij})}{P(\alpha \cup \beta)} \\
&= \sum_{\omega_{ij} \in \alpha} \frac{u(\omega_{ij}) \cdot p(\omega_{ij})}{P(\alpha \cup \beta)} + \sum_{\omega_{ij} \in \beta} \frac{u(\omega_{ij}) \cdot p(\omega_{ij})}{P(\alpha \cup \beta)} \\
&= \frac{V(\alpha) \cdot P(\alpha) + V(\beta) \cdot P(\beta)}{P(\alpha \cup \beta)}
\end{aligned}
$$

in accordance with V2, the Averaging axiom.

2. *(Suppositional Desirability)* To show that V_A^* is a suppositional desirability function we must establish satisfaction of V*1 and V*2. The first is achieved by showing that V_A^* satisfies the axioms of desirability (the demonstration is the same as the above one for V). V*2 follows from V2, the zero-normalisation of V_A^* with respect to its domain W_A and the fact that $W_A = A$. For then

$$V_A^*(A) = \sum_{w_j \in W_A} u_A(w_j) \cdot p_A(w_j) = 0$$

Finally, V*3, like P*3, follows from condition (3) on the selection function.

3. *(The Ramsey Test for Desire)* Recall that the content of the conditional prospect that $A \to B$ is given by the set (W, B_A). Hence it follows immediately that, for some $k_A > 0$,

$$V([A \to B]) = V(W, B_A) = k_A \cdot V_A^*(B)$$

So the Ramsey Test for desire also holds on our model. Indeed, it too is built into the way conditional prospects are represented.

TABLE 8.4. *Desirabilities of the Counterfacts under World Actualism*

Worlds	Supposed A-worlds		
	w_1	w_2	w_3
w_1	$u(w_1)$	—	—
w_2	—	$u(w_2)$	—
w_3	—	—	$u(w_3)$
w_4	$u(w_4)$	$u(w_4)$	$u(w_4)$
w_5	$u(w_5)$	$u(w_5)$	$u(w_5)$

8.5 VALUE ACTUALISM

Many decision theories implicitly assume in one form or another a doctrine that I will term 'Value Actualism'. The basic intuition behind this doctrine is that *only the actual world matters*, so that the desirability of combinations of facts and counterfacts depends only on the desirability of the facts. As in the case of the independence conditions for probabilities, some caution is required in making this intuition more precise. The Centring condition already determines the relation between the desirability of a possible world w_i in which A is true and any possible counteractual A-world $w_j \in W_A$. For then either w_j is w_i, in which case the desirability of the pair just reduces to that of w_i, or it is not, in which case the desirability is not defined, since this pair of worlds would represent an impossible combination. So the proper scope of Value Actualism is the relation between the facts and the *strict* counterfacts (which is unconstrained by Centring).

8.5.1 World Actualism

One way of expressing Value Actualism more formally is as follows:

World Actualism $\forall w_j \in W_A, \forall w_i \in \bar{A}, u(\langle w_i, w_j \rangle) = u(w_i)$

World Actualism says that the desirability that some world w_i, inconsistent with A, be the case and that some world w_j would be the case if A were depends only on the desirability of w_i. In other words, once it has been established what world is the actual one, then it should be a matter of indifference what would have been the case had some false condition been true. Given Centring and World Actualism, we can represent the agent's evaluative state as in Table 8.4.

The applicability of World Actualism rests on the possibility of giving a complete description of everything that matters. If we are able to do so, then any way in which the counterfacts matter to us in the actual world could be registered in the description we give of that world. It is not that the counterfacts themselves must be written into the descriptions of worlds – this would lead to contradiction when the counterfacts specified in the description of a world differed from those in counteractual worlds – but that any way in which these counterfacts bear on our evaluation of the facts must be specified. For instance, suppose that the desirability of dining at home is sensitive to how good a meal one would have had had one dined out at the local restaurant, because the fact that one would have had a better meal at the restaurant causes one to regret eating at home and the fact that one would have had a worse meal makes one appreciate the home cooked meal all the more. Then these facts – the regret or the appreciation one experiences in the light of the counterfacts – must be built into the description of the actual world if World Actualism is to obtain.

The problem with the condition of World Actualism is therefore that it is a partition-dependent condition: it might hold for one specification of the possible worlds, but not when they are specified more coarsely. So perhaps we should not think of it as a condition that applies to every model of counterfactual possibility but, rather, as a methodological principle; one which requires contingencies to be sufficiently finely individuated for World Actualism to hold within the model. So construed, the principle is one that many decision theorists seem to endorse, at least implicitly. Indeed, the strategy of fine individuation advocated by John Broome (1999) in response to apparent counterexamples to the rationality claims of Bayesian decision theory can be viewed along these lines.

8.5.2 Prospect Actualism

Complete specification of all that matters is something that is very difficult for real agents to do. And so there is interest in giving a partition-independent formulation of Value Actualism. The most natural way of doing so is as follows.

Prospect Actualism $\forall X \subseteq \bar{A}, \forall Y_A \in W_A, V(X, Y_A) = V(X)$

Prospect Actualism says that, if X being the case implies that A is not, then the desirability of the joint prospect that X and that Y, on the supposition that A, depends only on the desirability that Y. Or, to put it slightly differently, once it is given that X (and hence that not A) then it does not matter what is or would be the case if A were. Prospect Actualism is a much stronger constraint on evaluative attitudes than World Actualism (which it implies). But is it reasonable?

Suppose, for instance, you have to choose between two restaurants. You go to restaurant A and are served a very poor meal. An acquaintance goes to the other restaurant and reports that she was served a very good meal. Are things worse overall than they would have been if it had been the case that you had been served a poor meal at the other restaurant as well? The issue is not whether your judgement concerning the meal at restaurant A can depend on what the meal at restaurant B would have been like – surely it should not – but whether the prospect of having a poor meal at restaurant A when you would have had a good one at restaurant B is a worse one than that of having the poor meal at restaurant A when you would also have had a poor one at restaurant B.

The issue boils down in this case to whether the badness associated with the difference between what is the case and what might have been if some other course of action had been pursued is built into the description of the actual state of affairs: for instance, the regret you might feel in not having gone to the other restaurant. In other cases, it depends on the information contained in the description of the counterfactual circumstances. Suppose, for instance, that the acquaintance in our example reports that standards of food hygiene were very poor at the other restaurant. You know they have the same owner, so you infer that standards will also be poor at the restaurant you chose. This affects your view about the desirability of your choice. In other words, the desirability of the prospect of going to restaurant A is not independent of the supposition that had you gone to restaurant B you would have found food hygiene standards to be very poor. So Prospect Actualism will be violated whenever there are either probabilistic or desirabilistic dependencies between the facts and the strict counterfacts. This makes it much less plausible than World Actualism.

Even if Prospect Actualism fails as a general principle, however, it may nonetheless be valid as a principle of evidential supposition. When we make an evidential supposition we suppose that some condition is, as a matter of fact, true. In the cases falling under the scope of Prospect Actualism, that supposition is false, so arguably there can be no informational or desirabilistic import associated with it. In contrast, when the mode of supposition is counterfactual the supposition is not undermined by the fact that the condition being supposed is, as a matter of fact, false. On the contrary, what we are doing is assuming its truth for the sake of deliberation, in the knowledge that it is not actually so. So Prospect Actualism will fail.

8.5.3 Restricted Actualism and Evidential Desirability

Our final actualist principle of interest, Restricted Actualism, says that the joint desirability of it not being the case that A and of Y being the case under the supposition that A depends only on the desirability of A not being the case. Or, to put it slightly differently, given that A is false, it is a matter of indifference what the counterfacts are under the supposition that A. More formally:

Restricted Actualism $\forall Y_A \in W_A, V(\bar{A}, Y_A) = V(\bar{A})$

Like Prospect Actualism, Restricted Actualism does not seem plausible for counterfactual supposition because what would have been the case were A true can be relevant to one's evaluation of $\neg A$. But its credentials as a condition on evidential supposition seem much stronger. Compare the counterfactual claim 'He would have won the race if he had turned left' to the factual claim 'If he turns left, he has won the race'. The former, if true, makes the fact that he turned right all the more regrettable. This is not true for the latter claim, for in this case the supposition that he turns left is rendered inert by the fact that he turns right and so the indicative conditional as a whole conveys no information.

This claim is important because it turns out that, in the presence of Supposition Independence, the principle of Restricted Actualism is sufficient to characterise desirability under evidential supposition in the sense that it implies both Thesis 6.4 and Thesis 7.4. For, in virtue of Restricted Actualism, it follows from the axiom of desirability and Centring that

$$V(W, X_A) = V(A, X_A) \cdot P(A|X_A) + V(\bar{A}, X_A) \cdot P(\bar{A}|X_A)$$
$$= V(A \cap X) \cdot P(A|X_A) + V(\bar{A}) \cdot P(\bar{A}|X_A)$$

But then, by Supposition Independence and the fact that the desirability axioms imply that $V(W) = V(A) \cdot P(A) + V(\bar{A}) \cdot P(\bar{A}) = 0$, it follows that

$$V(W, X_A) = V(A \cap X) \cdot P(A) + V(\bar{A}) \cdot P(\bar{A})$$
$$= V(A \cap X) \cdot P(A) - V(A) \cdot P(A)$$
$$= V(X|A) \cdot P(A)$$

So:

1. *(Bradley's Thesis)* Recall that the content of the indicative conditional $A \mapsto X$ is just the ordered pair (W, X_A). By hypothesis, indicative conditionals should be interpreted evidentially. So it follows from the above that
$$V([A \mapsto X]) = V(X|A) \cdot P(A)$$

2. *(Conditional Evidential Desirability)* As we are concerned with evidential supposition, let us denote the 'marginal' of V on Ω_A by V_A^+. By the Ramsey Test for desire, $V(W, X_A) = k_A V_A^+(X)$, for all $X_A \in \Omega_A$. So $k_A V_A^+(X) = V(X|A) \cdot P(A)$. To satisfy Thesis 6.4, we simply choose $V(\cdot|A)$ as our measure of the desirability under the evidential supposition that A and $P(A)$ as the 'discount' factor commensurating unconditional and conditional degrees of desire.

8.6 PARTITIONING CONDITIONALS

Up to this point we have been able to restrict attention to a single supposition. But to examine the conditions under which the final pair of theses advanced in the previous chapters – Belief Independence and Value Independence – will hold in the multidimensional possible-worlds framework we need to extend

TABLE 8.5. *Three-Dimensional Possibility Space*

Counteractual not A-worlds	Counteractual A-worlds			
	w_1	w_2	w_3	
w_4	$\langle w_1,w_1,w_4\rangle$ $\langle w_4,w_1,w_4\rangle$	$\langle w_2,w_2,w_4\rangle$ $\langle w_4,w_2,w_4\rangle$	$\langle w_3,w_3,w_4\rangle$ $\langle w_4,w_3,w_4\rangle$	$B_{\bar A}$
w_5	$\langle w_1,w_1,w_5\rangle$ $\langle w_5,w_1,w_5\rangle$	$\langle w_2,w_2,w_5\rangle$ $\langle w_5,w_2,w_5\rangle$	$\langle w_3,w_3,w_5\rangle$ $\langle w_5,w_3,w_5\rangle$	$C_{\bar A}$
	B_A		$\bar B_A$	

our toy model by allowing for more than one supposition. To illustrate the main results it suffices to consider just two: the supposition that A and the supposition that $\neg A$. In this case the elementary possibilities will be world-triples specifying the facts and the counterfacts under the supposition that A and $\neg A$.

A three-dimensional space of worlds is somewhat harder to represent graphically, but we will assume that Centring holds and therefore that, in every world-triple $\langle w,w_A,w_{\bar A}\rangle$, either $w = w_A$ or $w = w_{\bar A}$ depending on whether w is an A-world or not. This reduces the number of elementary possibilities from the $5 \times 3 \times 2 = 30$ in the cross-product of W, W_A and $W_{\bar A}$ to just the $2 \times 3 \times 2 = 12$. Table 8.5 gives a two-dimensional representation of this three-dimensional space of elementary possibilities by displaying all triples consistent with any combination of suppositions.

Prospects in this extended model are just sets of world-triples, with $(A, B_A, C_{\bar A})$ being the prospect that A is the case, B is/would be the case if A is/were and C is/would be the case if A is/were not. The contents of conditionals with antecedent A are given by the rows of Table 8.5 and those of antecedent $\neg A$ by the columns. The intersection of the two gives the contents of partitioning conditionals. For instance, the content of the partitioning conditional $(A \rightarrow B)(\neg A \rightarrow C)$ is the prospect $(W, B_A, C_{\bar A})$ – the bottom-left and bottom-middle cells in the table – because this sentence is true at the set of world triples $\langle w,w_A,w_{\bar A}\rangle$ such that $w_A \in B$ and $w_{\bar A} \in C$.

As before, we can define a joint density p and utility u on the space of world-triples and corresponding probability P and desirability V on the space of prospects, with marginals P_A^* and $P_{\neg A}^*$ and associated desirability functions, V_A^* and $V_{\neg A}^*$, on the counterfacts under the supposition that A and that $\neg A$. So, as before, the conditions of Probabilism, Desirabilism, Suppositional Belief and Suppositional Desire and the Ramsey Tests for belief and desire all hold with respect to these functions.

Under what conditions will an agent's degrees of belief and desire in partitioning conditionals satisfy Theses 7.5 and 7.6? The short answer is: when they regard counteractual worlds under disjoint suppositions as probabilistically and desirabilistically independent of each other. In the simple

TABLE 8.6. *Probability and Desirability under Counterfactual Independence*

	w_1	w_2	w_3	
w_4	$p_A(w_1) \times p_{\bar{A}}(w_4)$ $u_A(w_1) + u_{\bar{A}}(w_4)$	$p_A(w_2) \times p_{\bar{A}}(w_4)$ $u_A(w_2) + u_{\bar{A}}(w_4)$	$p_A(w_3) \times p_{\bar{A}}(w_4)$ $u(w_3) + u(w_4)$	$B_{\bar{A}}$
w_5	$p_A(w_1) \times p_{\bar{A}}(w_5)$ $u_A(w_1) + u_{\bar{A}}(w_5)$	$p_A(w_2) \times p_{\bar{A}}(w_5)$ $u_A(w_2) + u_{\bar{A}}(w_5)$	$p_A(w_3) \times p_{\bar{A}}(w_5)$ $u(w_3) + u(w_5)$	$C_{\bar{A}}$
	B_A		$\bar{B}_A$	

three-dimensional case this condition is formally expressed by the following constraint on joint probabilities and desirabilities of worlds:

$$p(\langle w_A, w_{\bar{A}} \rangle) = p_A(w_A) \times p_{\bar{A}}(w_{\bar{A}})$$

$$u(\langle w_A, w_{\bar{A}} \rangle) = u_A(w_A) + u_{\bar{A}}(w_{\bar{A}})$$

When this condition is satisfied the agent's attitudes to the counterfacts can be represented compactly as in Table 8.6.

This condition of probabilistic and desirabilistic independence of worlds under disjoint suppositions is equivalent to a corresponding partition-independent condition that says that the desirabilities of conditional prospects under mutually exclusive suppositions are independent of one another (proof in the Appendix as Theorem A.23). Formally:

Counterfact Independence $V(B_A, C_{\bar{A}}) = V(B_A) + V(C_{\bar{A}})$

Now, in the three-dimensional possible-worlds model, the content of the indicative conditional $(A \mapsto B)(\neg A \mapsto C)$ has as its content the prospect $(B_A, C_{\bar{A}})$. So it follows that:

1. (*Value Independence*) $V([(A \mapsto B)(\neg A \mapsto C)]) = V(B_A, C_{\bar{A}}) = V(B_A) + V(C_{\bar{A}}) = V([A \mapsto B]) + V([\neg A \mapsto C])$. Hence, by the definition of conditional desirability,

$$V([A \mapsto B] \mid [\neg A \mapsto C]) = V([A \mapsto B])$$

   i.e. an indicative conditional is desirabilistically independent of the counterfacts under the supposition that its antecedent is false.
2. (*Belief Independence*) By Theorem A.4, Counterfact Independence implies that $P(B_A, C_{\bar{A}}) = P(B_A) \cdot P(C_{\bar{A}})$. Hence $P([(A \mapsto B)(\neg A \mapsto C)]) = P([A \mapsto B]) \cdot P([\neg A \mapsto C])$ and so, by the definition of conditional probability, $P([A \mapsto B] \mid [\neg A \mapsto C]) = P([A \mapsto B])$ – i.e. an indicative conditional is probabilistically independent of the counterfacts under the supposition that its antecedent is false.

TABLE 8.7. *Model Conditions and Rationality Properties*

Model conditions	Rationality thesis
Basic definitions	(1) Probabilism (2) Desirabilism (3) Suppositional Belief (4) Suppositional Desire (7) Ramsey Test for belief (9) Ramsey Test for desire
+ Centring	Regularity
+ Supposition Independence	(5) Conditional Evidential Belief (8) Adams' Thesis
+ Restricted Actualism	(6) Conditional Evidential Desire (10) Bradley's Thesis
+ Counterfact Independence	(11) Belief Independence (12) Value Independence

8.7   CONCLUDING REMARKS

In this chapter I have shown that the various rationality theses discussed in the previous part of the book are satisfiable in a natural way in the kind of multidimensional possible-worlds model employed here. Some of these theses are immediate consequences of the way in which measures of agent's degrees of belief in, and desire for, both factual and conditional prospects are constructed in this framework: most notably, perhaps, the Ramsey Test hypotheses for belief and desire. Others depend on the characterisation of evidential supposition through various independence and ethical actualist hypotheses. The relations between these hypotheses and the rationality theses examined in the previous part of the book are summarised in Table 8.7.

It might also be helpful at this point to refer back to Table 7.1 at the end of the previous part of the book to link these model assumptions to the characteristic axioms for the rationality claims. Note in particular the relationships between Supposition Independence and the axiom of Suppositional Rigidity, between Restricted Actualism and the axiom of $\mapsto$-Betweenness, and between Counterfact Independence and the axiom of $\mapsto$-Separability.

Finally, it is worth mentioning some of the logical relationships that hold between the various conditions examined in this chapter. (Further details and proofs for all the claims can be found in Bradley & Stefánsson, 2015.) For simplicity I assume Centring throughout. First, the independence conditions on belief. As we noted before, Fact–Counterfact Independence is equivalent to World Independence. Both are also equivalent, given Centring, to the conjunction of Supposition Independence and the probabilistic independence of counterfacts under disjoint suppositions (the implication of Counterfact Independence mentioned above). In summary:

$$\text{World Independence} = \text{Fact–Counterfact Independence}$$

$$= \left[\begin{array}{c} \text{Supposition Independence} \\ + \text{Counterfact Independence} \end{array} \right]$$

Something similar is true for the combinations of independence and value actualist assumptions on belief and desire. World Actualism and World Independence are jointly equivalent to Prospect Actualism and Fact–Counterfact Independence. Restricted Actualism is weaker than Prospect Actualism but it neither implies nor is implied by World Actualism. On the other hand, the conjunction of Supposition Independence and Restricted Actualism is implied by the conjunction of World Actualism and World Independence (this follows from Theorem A.22 in the Appendix). Finally, Prospect Actualism and Fact–Counterfact Independence are jointly implied by the combination of Restricted Actualism, Counterfact Independence and Supposition Independence. In summary:

$$\left[\begin{array}{c} \text{World Independence} \\ + \text{World Actualism} \end{array} \right]$$

$$= \left[\begin{array}{c} \text{Fact–Counterfact Independence} \\ + \text{Prospect Actualism} \end{array} \right]$$

$$= \left[\begin{array}{c} \text{Supposition Independence} \\ + \text{Restricted Actualism} \end{array} \right] + [\text{Counterfact Independence}]$$

9

Taking Action

9.1 DECISION PRINCIPLES

Decision theory addresses two questions. What states of mind or sets of judgements are rationally permissible? And what choices should we make, given our state of mind or judgement? The previous chapters have proposed an answer to the first question and enough is now in place for us to turn to the second.

There is broad agreement amongst decision theorists that, under the conditions characterised by subjective uncertainty about the state of the world and in which the agent has determinate degrees of belief in, and preference for, all relevant prospects, actions should be chosen in accordance with the expected subjective benefit of so doing. In contrast, there is a good deal of variety in the way that this idea is cashed out. Much of this is driven by differences in the way that decision problems, and especially the objects of choice, are conceived and formally represented. But some of it emanates from more substantial differences both in the interpretation of the principle of maximising expected benefit and in the formal assumptions made for the purposes of deriving it. One aim of this chapter will be to explore the relationship between these theories. Another will be to defend a particular one.

As the theory of rationality presented in the previous chapter is essentially an extension of Richard Jeffrey's decision theory, it is natural to start by reminding ourselves about what he says about choice of action. In Jeffrey's theory, an action is simply a prospect that can be made true at will (i.e. which is a genuine option for an agent to exercise) and its choice-worthiness is identified with its desirability as defined earlier – i.e. as the conditional expectation of utility given the performance of the action. We can therefore express his theory of choice thus:

Jeffrey Of all the prospects that constitute options for you, pick one with greatest desirability.

Although Jeffrey's theory has been very influential in philosophy, it has few contemporary supporters,[1] with the majority of philosophical decision theorists accepting the objections to it lodged by causal decision theory. In economics and the (extra-philosophical) decision sciences, on the other hand, Jeffrey's theory is largely unknown, and it is the version of decision theory due to Leonard Savage (1974/1954) that finds near-universal acceptance for the treatment of decision problems under subjective uncertainty, supplemented by the work of von Neumann & Morgenstern (2007/1944) and Anscombe & Aumann (1963) to treat, respectively, decision problems under risk and combinations of risk and uncertainty. (More recently, several non-Bayesian theories of decision making under ambiguity have been gaining influence, but we defer discussion of this to the last part of the book.)

Two types of questions now arise. Firstly, there is the relationship between these different theories of decision making under uncertainty. Are they just notational variants of the same theory? Are some special cases of one another? Or are they rival theories that make different predictions? There are two dimensions along which these questions can be posed: the relationship between the different, but broadly Bayesian, theories of choice under uncertainty (such as those of Ramsey, Savage, Jeffrey and causal decision theorists) and the relationship between these Bayesian theories and theories of decision making under risk. For the most part, these relationships remain ill-understood. Although rarely explicitly articulated, it is generally held, for instance, that von Neumann and Morgenstern's theory is the special case of Savage's subjective expected utility theory that is obtained when probabilities are objective or given. As we shall see, however, this view is not quite correct. Similarly, there is a common view that Savage, Ramsey and Jeffrey merely present variants of a single Bayesian decision theory, employing slightly different vocabulary and making slightly different assumptions. On the other hand, it is also often held that causal decision theory is a rival to Jeffrey's theory, but not to Savage's. It is difficult to see how both views could be true.

Secondly, there is the question of which of these is the most adequate as a normative theory of choice. Across the large body of decision theoretic literature, it is possible to extract two main variants of the claim that Jeffrey's theory is not. One variant argues that Jeffrey is wrong in thinking that actions are just like other prospects. Actions are not ordinary prospects or states of the world, goes this line of argument, but interventions in the world; a feature better captured by Savage's representation of them as functions from states to consequences. A second variant argues that Jeffrey is wrong in thinking that the choice-worthiness of an action is measured by its desirability. The desirability of an action is a measure of how good it would be to learn that the action will be performed – its auspiciousness or news value – rather than a measure of how effective it is in bringing about good consequences – its causal or instrumental efficacy. But it is the latter rather than the former that

---

[1] Notable exceptions include Eells (1982) and Ahmed (2014).

should guide our choices. So, what makes actions distinct as prospects is not so much their form but the fact that they are susceptible to a different kind of evaluation. The first line of criticism has not, to my knowledge, been explicitly articulated anywhere, but I suspect that one of the reasons why Jeffrey's theory has made little inroad into economics and the social sciences lies in the fact of widespread tacit acceptance of it. In contrast, the second line of criticism, which is at the heart of the debate over causal versus evidential decision theory, has been extensively discussed.

To investigate both of these questions, I will present a broad framework within which these different (and potentially rival) theories can be formulated and compared, with a view to identifying the conditions under which they yield the same prescriptions and those in which they do not. In sections 9.3 and 9.4, I will show that the choice rules of von Neumann and Morgenstern and of Savage alike are simply special cases of the rule of maximisation of desirability applied to the particular objects of choice that these theories postulate and which are normatively compelling only when certain additional assumptions are made. In later sections I will address the relationship between causal and evidential (Jeffrey) decision theory and, in particular, the prospects for a measure of choice-worthiness that is distinct from desirability and reflective of causal efficacy. The first task, however, is to clarify, at a more general level, the relationship between rational preference and choice.

9.2 PREFERENCE-BASED CHOICE

In its most abstract form, a decision problem consists of two elements: the alternatives or options the agent must choose amongst and the resources she can bring to bear in making her decision. Resources will be construed here simply as judgements or attitudes: the beliefs and preferences that the agent can use to evaluate the options before her. A decision rule is then a way of determining, for any given decision problem (a pair of options and judgemental resources), a non-empty subset of the available options, understood to be those options that are permissible choices given the evaluation determined by the agent's judgements.

More formally, let $\mathcal{D} = \langle O, S \rangle$ be a decision situation with O a non-empty set of options drawn from some background set Υ of alternatives and $S = \langle \Gamma, \unrhd, \succsim \rangle$ the agent's judgemental state, where $\unrhd$ and $\succsim$ are, as before, her credibility and preference relations on an algebra of prospects $\Gamma = \langle Y, \models \rangle$. A choice rule C is a mapping from any decision situation to a non-empty subset of O, called her 'choice set'. Intuitively, the choice set $C(\mathcal{D})$ is the subset of options available to the agent that she is rationally permitted to choose given her judgemental state.

What kind of choice rule is appropriate in a decision situation will depend on what the set of options looks like and how rich the agent's judgemental set is. In models of decision making under 'mild' uncertainty of the kind that we will study in this chapter, it is assumed that the credibility relation $\unrhd$ is complete over the set of states and the preference relation $\succsim$ is complete over the set of consequences. In contrast, in the models of decision making under ambiguity that we will look at later on, $\unrhd$ is not assumed to be complete over the set of states, but the assumption that $\succsim$ is complete over consequences is typically retained. By weakening the last assumption, a model would allow for severe evaluative uncertainty as well.

In this chapter we will assume that the agent's credibility and preference relations are defined on a Boolean algebra of prospects and satisfy rationality conditions sufficient to determine a Jeffrey representation of them – i.e. a pair of probability and desirability measures $\langle P, V \rangle$ expressing her factual and evaluative uncertainty. In some decision problems, the agent's counterfactual uncertainty about the effects of the exercise of one of her options will also be relevant, requiring that her judgement set determine not just measures of her degrees of belief and desire but also measures of her suppositional beliefs and desires. In these cases we will take her credibility and preference relations to be defined on a conditional algebra of prospects so that her counterfactual uncertainty can be equated with her uncertainty regarding the truth of counterfactual conditionals.

Recall that the Choice Principle (introduced in Chapter 2) dictates that the agent should choose the option she most prefers from amongst those that are available. More exactly, to allow for indifferences, the agent's choice set should consist of those options that she most prefers. This criterion of rational choice is usually rendered as the requirement that agents choose only *optimal* elements, these being the options which are at least as preferred as all others that are available (see, for instance, Arrow, 1959, and Sen, 1971). Formally, for all $\alpha \in O$:

Optimal Choice $\alpha \in D(O) \Leftrightarrow \forall \beta \in O, \alpha \succsim \beta$

The norm expressed by Optimal Choice is adequate when an agent's preferences over options are complete. But if an agent is unable or unwilling to compare some of the options, perhaps because she lacks important information about relevant features of the world or because of conflicting value considerations, then there may be no optimal elements available. So all that we should require of her is that her choice set consist of the *maximal* elements of her option set. This is to say that something is a permissible choice from a set of options if no other available option is strictly preferable to it. Formally, for all $\alpha \in O$:

Maximal Choice $\alpha \in D(O) \Leftrightarrow \neg \exists \beta \in O$ such that $\beta \succ \alpha$

When the preference relation is complete, the maximal alternatives are all optimal. But, when it is not, there can be no alternative strictly preferred to some option α without it being the case α is as least as preferred as all other options. For example, suppose $O = \{\alpha, \beta, \gamma\}$ and the agent's preferences are such that $\alpha \succ \beta$ and $\gamma \succ \beta$ but α and γ are incomparable. Then α is a maximal choice but not an optimal one. Since rationality does not require completeness of preference, Maximal Choice is the choice criterion that should be endorsed.

If an agent's choices are formally based on her preferences in the sense of conforming with the principle of Maximal Choice, we will typically want to say that her preferences provide the reasons for her choice, both in being the cause of her action and in providing justification for it. (Only typically, because an agent's choice might be formally preference-based, but not substantially so, if the conformity is just an accident.) There is a lively debate in the social sciences as to the extent to which human action is preference-based, with factors such as emotions, social and moral norms, and habits or customs often being proposed as alternative reasons for, and/or causes of, actions. I will not enter into these questions, but it is worth making some remarks about how preference-based choice relates to the more general notion of reason-based choice.

It is obvious that preferences are often based on a wide variety of reasons. I prefer to take the bus than to take a taxi because it's cheaper; I prefer the woollen jacket to the cotton one because it's warmer; I prefer not to deal with him because I believe him to be dishonest; and so on. Drawing attention to these reasons serves not to displace preference-based explanation of choice but to deepen it.[2] The more radical question is whether one can have an all-things-considered reason to make a choice that is contrary to preference. Consider the following variant of an example that Sen (1977) draws on to support this claim. Your host at a party offers you cake. There are three slices on the plate of different sizes. Being greedy, you prefer the largest. But, if you were to choose it, you would violate a norm of politeness. So you pick the middle-sized piece instead, apparently contrary to your preferences. It would seem thus that two factors determine your (non-maximal) choice: your preferences and a social norm of politeness.

There are, however, two alternative ways of describing this situation that are consistent with regarding the choice of the middle-sized piece as preference-based. The first is to take your option set to consist of taking either the middle-sized or the small-sized cake slice, the option of taking the large-sized one having been excluded by the politeness norm. On this view, social and moral norms are side constraints on action, determining which options are permitted in much the same way as physical constraints determine which are feasible. On a second description of this situation, your preferences over the available actions depend both on your preferences over cake slice sizes and your preferences for norm conformity, together with relevant background beliefs. On this reading your choices *are* preference-based, but your preferences over actions have determinants of different kinds. If either of these are viable interpretations of it, then Sen's example cannot be used to argue that counter-preferential choice can be rational.

---

[2] See Dietrich & List (2013a,b) for a sophisticated discussion of the relation between preferences, choices and reasons.

Let me finish with a rather basic point, which nonetheless has some important implications. Even if rationality requires maximising choice on the basis of one's all-things-considered preferences, it does not follow that preferences always suffice to explain choice. An agent can be indifferent between some alternatives, for instance, or her preferences may be incomplete. In such cases her preferences will not completely determine what she should choose and she must look to other considerations to help her decide. So preference-based explanations and rationalisations of choice must necessarily be limited in scope; able to explain why certain options are not chosen, but not always why a particular one was. To take up the slack, explanations must draw on factors other than preference: psychological ones such as the framing of the choice problem or the saliency of particular options, or sociological ones such as the existence of norms or conventions governing choices of the relevant kind. Rationalisations too must look to other kinds of reasons for choice.

It follows from this that observations of actual choices will only partially constrain preference attribution. For instance, that someone chooses a banana when an apple is available does not allow one to conclude that the choice of an apple was ruled out by her preferences, only that her preferences ruled the banana in. In this simple observation lies a serious obstacle to the ambition of Revealed Preference theory to give conditions on observed choices sufficient for the existence of a preference relation that rationalises them. For the usual practice of inferring the completeness of the agent's preferences from the fact that she always makes a choice when required to is clearly illegitimate if more than one choice is permitted by her preferences.

9.3 DECISIONS UNDER RISK

Let us now confine our attentions to cases in which the agent has complete preferences over the set of prospects, leaving questions of rational choice under incompleteness to the last part of the book. In this case, we have seen that the agent's preferences are numerically representable by a utility function and, in this sense, rational preference-based choice will be utility-maximising. Indeed, the prescription to maximise utility is nothing other than the prescription to make optimal choices relative to one's preferences. I now want to explore the implications of the prescription to maximise utility when the utility measure takes the specific form of a desirability function – i.e. is a numerical function on a Boolean algebra of prospects satisfying the axioms of Normality and Averaging (see page 78). In this section, I will consider rational choice under conditions of risk or objective uncertainty; in the next, rational choice under conditions of subjective uncertainty.

What is required by desirability maximisation will depend, of course, on the nature of the options under consideration. In the von Neumann & Morgenstern (2007/1944) (hereafter vN–M) framework, which is the standard one for the treatment of risk, the options amongst which the agent must choose are taken to be lotteries, these being objective probability distributions over a given set of outcomes (called the 'prizes'). The vN–M framework is rather different in appearance from the one developed here, so our first task will be to develop a way of representing lotteries within a propositional framework. We do so by drawing on the treatment of chance propositions introduced in Part II of the book.

As before, let $\mathcal{Z} = \{Z, \models\}$ be a Boolean subalgebra of prospects for which chances are defined, $\Pi = \{ch\}$ be the set of all probability functions on $\mathcal{Z}$ and $\Delta = \wp(\Pi)$ be the set of all chance prospects. Let $\mathcal{X} = \{X_1, ..., X_n\}$ be an n-fold partition of Z, with the X_i being the factual prospects that constitute 'prizes' or, more generally, the outcomes of a random process. Let $Ch(X) = x$ denote the proposition that the chance of X is x (and the chance of $\neg X$ is $1 - x$), with $x \in [0,1]$. Let $\bigwedge_{i=1}^{n}(Ch(X_i) = x_i)$ denote the conjunction of the n propositions $Ch(X_1) = x_1$, $Ch(X_2) = x_2$, ..., and $Ch(X_n) = x_n$, where the x_i are such that $\sum_{i=1}^{n} x_i = 1$. A proposition $\bigwedge(Ch(X_i) = x_i)$ thus expresses the chances of realising each of the 'prizes' represented by the propositions X_i, thereby serving as the propositional equivalent, in this framework, of a vN–M lottery.

Let $\mathcal{L}$ be the set of all such lottery propositions. Let L_x be $\bigwedge(Ch(X_i) = x_i)$ and L_y be $\bigwedge(Ch(X_i) = y_i)$ and let $z_i^\lambda := \lambda x_i + (1 - \lambda)y_i$ for some real number $\lambda \in [0,1]$. Define the λ-linear combination of lottery propositions L_x and L_y by

$$\lambda L_x + (1 - \lambda)L_y := \bigwedge(Ch(X_i) = z_i^\lambda)$$

Note that $\bigwedge(Ch(X_i) = z_i^\lambda)$ is itself a lottery proposition. It follows that the set $\mathcal{L}$ of lottery propositions is closed under linear combination as is required by the von Neumann and Morgenstern framework.

In a decision problem under risk, the option set $O_\mathcal{L}$ is just a subset of $\mathcal{L}$. For the corresponding decision problem $(O_\mathcal{L}, \langle P, V \rangle)$, the vN–M theory recommends that:

von Neumann and Morgenstern Of all the lotteries available to you, you should choose the one that maximises the objective expectation of desirability given the lottery. Formally:

$$L_x \succsim L_y \Leftrightarrow \sum_i V(X_i) \cdot x_i \geq \sum_i V(X_i) \cdot y_i$$

What is the relation between the rule of desirability maximisation and the prescription to maximise the objective expectation of desirability? It turns out that the two rules are consistent and that the desirabilities of lotteries go by the expected desirabilities of their 'prizes', in accordance with the vN–M rule, iff the Principal Principle holds and chances are neutral in a practical or desirabilistic sense, a condition that I will call Chance Neutrality. More formally, the latter condition says:

Chance Neutrality For any $X \in \mathcal{X}$ and $\bigwedge(Ch(X_i) = x_i) \in \mathcal{L}$:

$$V(\bigwedge(Ch(X_i) = x_i) \mid X) = 0$$

The intuition expressed by this condition is that chances do not matter to us intrinsically; they only matter instrumentally, as means to getting the outcomes we care about intrinsically. We care about the chances of getting cancer, for instance, because we care about cancer, about the chances of rain because we care about staying dry, and so on. Once it is settled whether or not we will get cancer or stay dry, then the chances no longer matter.[3]

We are now in a position to state the main claim of this section:

Theorem 9.1 *(Lottery Theorem)* (Stefánsson & Bradley, 2015) *Let $S = \langle P, V \rangle$ be an agent's judgemental state, with P and V both defined on a Boolean algebra of prospects containing $\mathcal{L}$. Suppose that P respects the Principal Principle and V respects Chance Neutrality. Then*

$$V(L_x) \geq V(L_y) \Leftrightarrow \sum_i V(X_i) \cdot x_i \geq \sum_i V(X_i) \cdot y_i$$

The Lottery Theorem tells us that an agent who maximises desirability thereby values lotteries by their expected desirability in accordance with the theory of von Neumann and Morgenstern iff her degrees of belief and desire, respectively, respect the Principal Principle and Chance Neutrality. If these principles are sound, then this result may be read as a vindication of vN–M expected utility theory as a guide to rational decision making under risk. But are they?

The usual objections to the Principal Principle don't seem applicable here. But I do have doubts about Chance Neutrality. Consider the following two examples (drawn from Stefánsson & Bradley (2015)).

Example 1 (Taking Risks). *Ann is an experienced mountain climber. She does not directly seek death or severe injury, but nonetheless tackles climbs that put her at risk of both. She does this because part of the value of climbing, according to her, stems from the confrontation of risk. Ann is not completely foolhardy and will not undertake climbs if the chances of death or injury are too high. But for her the activity is of little worth if there is no associated chance of death or injury, even though death and injury are outcomes that she would strongly prefer to avoid. Indeed, for her there is an optimal region of risk, where the chances of death or injury are high enough to require courage of the climber, but not so high as to make the activity foolish.*

Example 2 (Equal Treatment). *Bob's father has been given a watch as a present, but since he already has one he decides it should go to one of his children. Bob doesn't particularly want or need the watch, but nonetheless is outraged when his father announces his intention to give the watch to Bob's*

---

[3] Chance Neutrality should not be confused with the notion of risk neutrality. A risk-averse agent (someone who prefers the certainty equivalent of a lottery to the lottery itself) can still respect Chance Neutrality, for instance.

sibling. Seeing Bob's reaction, his father decides to toss a coin to settle who gets the watch, which leaves Bob satisfied.

These two examples are very different, but share the common feature of a protagonist who finds the chance (or risk) of some outcome desirable even though s/he does not desire the outcome itself. Ann does not value the risk of death and injury associated with climbing as a means to these highly undesirable outcomes. Rather, she values them as intrinsic features of an activity that she considers worthy. We do not have to share Ann's penchant for danger to recognise this type of value. It is often the case that the difficulties associated with some activity or project (generically the risk of failure) are part of what makes them worthy as activities or projects. This can be as true for extreme crocheting as it is for mountaineering. And, in these cases, there is an important difference between achieving some outcome when the chances of failing were high and achieving it when success was pretty much guaranteed. Bob, on the other hand, does not value the chance of a watch because of its instrumental relation to the outcome of receiving the watch, since he does not even want the watch. Nor does he view the chances neutrally: for him it is far worse not to get the watch when he never had a chance of getting it than to not get it when he had a fair shot of doing so. So both Ann and Bob violate Chance Neutrality with respect to certain goods.

These examples do not show that there is no description of Ann and Bob's decision problem to which the condition of Chance Neutrality could be applied. Perhaps there are 'ultimate' goods to which it does. The important point is simply that if we want to use the vN–M theory then care must be taken to ensure that the decision problem is framed in such a way as to ensure that 'prizes' that are taken to be the goods at stake in the decision must be such that the agent takes chance-neutral attitudes to them. It is an open question whether this is always both possible and desirable. But, even if, as I suspect, the most natural representation of many decision problems under risk will not permit application of the vN–M theory, we are always able to follow the prescription to maximise desirability. For desirability maximisation does not presuppose Chance Neutrality.

9.4 DECISION MAKING UNDER UNCERTAINTY

The central tenet of Bayesian decision theory is that, under conditions of uncertainty, rationality requires choice of an action that maximises the expectation of subjective benefit. But there are significant differences in the way that this idea is cashed out by different versions of Bayesian decision theory both with respect to how actions are conceived and with respect to how exactly their choice-worthiness is measured. In these sections, I will focus on a three-way comparison between Jeffrey, Savage and causal decision theory.

Recall that Savage construes actions as functions from states to consequences and requires choice of action in accordance with:

Savage Of all the actions that constitute options for you, pick one with
greatest subjective expected utility

We noted earlier that Savage's theory correctly evaluates actions only
when three conditions are met: each action – state pair determines a unique
consequence, the states of the world are probabilistically independent of the
available actions, and the consequences of these actions are desirabilistically
independent of the state of the world in which they are realised. The last of
these conditions, though important for Savage's representation theorem, is
easily dispensed with when his theory is formulated in our framework. But the
other two are not.

Jeffrey took the fact that his theory was applicable even when states are
probabilistically dependent on the acts to be a singular advantage of it.
Changing the state of the world is precisely the point of acting, he argued
(if you like, the consequences of actions *are* changed states). Choice-worthy
actions are those which make the greatest positive *difference* to the probability
of the states of the world that one desires. In essence, this is the rationale
for desirability maximisation, for, in making true the most desirable of the
prospects that are options for you, you maximise the expected improvement
in the state of the world.

Probabilistic dependencies are also the source of a major difficulty for
Jeffrey's theory. For the improvement brought about by making a prospect
true is measured in Jeffrey's theory by calculating the (weighted average
of the) differences between the unconditional probabilities of the possible
states of the world and their conditional probabilities given the truth of the
prospect. But the causal efficacy of the exercise of an option in bringing
about some consequence is not always accurately measured by the conditional
probability of the consequence, given the option. Recall the example of the
relationship between smoking, lung cancer and breathing difficulties illustrated
in Figure 6.1. In that example, lung cancer and smoking have a common
genetic cause, making them probabilistically correlated. Since the probability
of lung cancer given smoking is higher than the unconditional probability of
lung cancer, and lung cancer is undesirable, it follows that smoking makes
a negative difference to the expected state of the world. But, by assumption,
smoking is not a cause of lung cancer, so it would be irrational to refrain from
smoking simply because it was evidence for the gene that is the real cause of
the lung cancer.

Causal decision theorists argue that maximisation of desirability leads to
poor choices in situations in which probabilistic correlations between actions
and outcomes do not track the causal relationships between the two. They
propose instead that the choice-worthiness of an action be measured by its
causal efficacy in bringing about desirable outcomes, a feature that may be
termed its causal expected utility. This gives us a third broadly Bayesian
decision principle to consider:

Causal Decision Theory Of all the prospects that constitute options for you,
pick one with greatest causal expected utility

9.4.1 Savage Acts

Subjective expected utility and causal expected utility are both, on the face
of it, different measures of the choice-worthiness of a prospect from its
desirability. But the precise relationships between these three concepts are
difficult to discern because of differences in the vocabulary and background
assumptions employed by the three Bayesian theories in which they occur. To
clarify them, let us first formulate them within the framework developed in
the previous part of the book (hereafter referred to as Desirabilism), in which
an agent's judgemental state is given by a pair of probability and desirability
measures $\langle P, V \rangle$, unique up to linear transformation of V, and defined on
a regular conditional algebra $\Gamma = \langle Y, \models, \rightarrow \rangle$ based on a Boolean algebra of
factual prospects $\Omega = \langle X, \models_X \rangle$.

Jeffrey's theory is just the 'inner core' of this framework: the restriction
of $\langle P, V \rangle$ to Ω together with the prescription to maximise desirability relative
to the subset of factual prospects that can be made true at will. To give a
corresponding formulation of Savage's decision theory we need to do two
things. First, we must identify prospects within the conditional algebra with
a structure equivalent to Savage's actions; and, second, we must show how to
ascribe to them Savage's measure of choice-worthiness.

As before, let consequences be factual prospects that are maximally specific
with regard to all that matters to the agent (in that decision situation) and
states be prospects that are maximally specific with regard to all possible
features of the environment relevant to the determination of a consequence.
Savage's actions, in this framework, are then identified by partitioning
indicative conditionals of the form $(S_1 \mapsto C_1) \wedge (S_2 \mapsto C_2) \wedge \ldots \wedge (S_n \mapsto C_n)$,
where the S_i are mutually exclusive states and the C_i are the corresponding
consequences. For this, partitioning conditional says that if S_1 is the state of the
world then C_1 is the consequence, if S_2 is the state of the world then C_2 is the
consequence, and so on – i.e. it describes the function assigning consequences
to states that is constitutive of a Savage act. Note that, since in our theory
(unlike Savage's) the distinction between states and consequences is pragmatic
rather than metaphysical, the subset of partitioning conditionals that should
be treated as actions will not be fully identifiable from their logical form alone.
But this will not pose any difficulty so long as we know how to evaluate every
partitioning conditional, whether or not the agent regards it as an option that
she can exercise. And it has the advantage of separating the problem of option
uncertainty from that of identifying which options are available.

The next step is to identify the conditions, if any, under which maximisation
of desirability and maximisation of subjective expected utility coincide in their
evaluation of acts. When they do, the following version of the Subjective
Expected Utility (SEU) hypothesis must be true.

SEU Hypothesis Let $\{S_i\}_{i=1}^n$ be an n-fold partition of the set of prospects X into states, $\mathcal{C} = \{C_i\} \subseteq X$ be a set of consequences and $\bigwedge_i (S_i \mapsto C_i)$ be a conjunction of the (non-contradictory) indicative conditionals $(S_1 \mapsto C_1)$, $(S_2 \mapsto C_2)$, ... and $(S_n \mapsto C_n)$. Then:

$$V(\bigwedge_i (S_i \mapsto C_i)) = \sum_{i=1}^{n} V(S_i \wedge C_i) \cdot P(S_i)$$

The SEU Hypothesis as stated here does not place any constraints on the set of consequences or on the set of states, other than that they should partition the algebra of prospects. Furthermore, no requirement of state independence is built into the hypothesis: the desirability of the possible consequences of the action can depend on the state of the world in which they are located simply because the hypothesis uses the desirabilities of the $S_i \wedge C_i$, and not the C_i alone, to evaluate the desirability of the partitioning conditional. So, in this general form, the SEU Hypothesis expresses a stronger claim than Savage's theory. But, of course, by appropriate specification of constraints on the interpretation of states and consequences the hypothesis can be weakened. And if the general version can be derived within the framework of Desirabilism then so too can any weaker one. In particular, Savage's theory is derivable if consequences are required to be maximally specific with regard to all that matters to the agent (in which case $V(S_i \wedge C_i) = V(C_i)$).

Although this version of the SEU hypothesis dispenses with the assumption of state independence, it retains a second condition essential to Savage's theory: that states of the world be probabilistically independent of acts. For it turns out that if choice of action in accordance with the principle of desirability maximisation is to imply maximisation of subjective expected utility then the partitioning conditionals representing acts must be probabilistically independent of the antecedents of the conditionals that occur in them. This explains why Savage's acts must be represented by partitioning *indicative* conditionals and not by counterfactuals. For the probabilistic independence of a conjunction of conditionals from the antecedents of each is an implication of two conditions – Adams' Thesis and Belief Independence – that I previously argued must be satisfied by indicative conditionals, but not by counterfactuals.

We are now in a position to give a precise statement of the conditions under which desirability maximisation implies maximisation of subjective expected utility:

Theorem 9.2 *(Desirability of Actions Theorem) (Bradley & Stefánsson, 2015)[4] Assume Desirabilism; then the SEU hypothesis is equivalent to the conjunction of Supposition Independence, Restricted Actualism and Value Independence*

---

4   In Bradley & Stefánsson (2015) it is called the EU equivalence theorem.

Recall that, to obtain a desirability representation satisfying both Supposition Independence and Restricted Actualism, preferences for conditionals must have the Indicative property and satisfy $\mapsto$-Betweenness and that to obtain one satisfying Value Independence they must satisfy both $\mapsto$-Separability and $\mapsto$-Betweenness. This follows from the representation results for indicative conditionals (proved as Corollary A.17) and for partitioning conditionals (proved as Corollary A.18) given in Chapter 7. The $\mapsto$-Separability condition is doing the same work here as Savage's Sure-Thing Principle, namely to ensure that counterfacts under disjoint suppositions are separable under the preference relation. More interesting is the role of the Indicative property, which, as we saw in Chapter 7, implies (within a conditional algebra) the probabilistic independence of conditionals from their antecedents (equivalent, within a regular logic of conditionals, to Adams' Thesis). Within Savage's framework, this condition follows from the much stronger postulate P3, which requires the state independence of utility. The $\mapsto$-Betweenness condition, which, together with the Indicative property, ensures that the desirability function satisfies the condition of Restricted Actualism, is implicitly assumed by Savage when he identifies the qualitative probability relation, derived with the help of P5, with the agent's relational beliefs.

The significance of the Desirability of Actions theorem is twofold. Firstly, it shows that the differences between Jeffrey's and Savage's theory of choice do not stem from the fact that they work with rival concepts of value, for both can be construed as advocating desirability maximisation as the principle of choice. What distinguishes their theories is the set of prospects that play the role of objects of choice – in Jeffrey's case, the factual prospects that can be made true at will; in Savage's, partitioning indicative conditionals – and the associated conditions imposed on them.

Secondly, the theorem makes transparent how much more demanding the requirement to maximise subjective expected utility is than the requirement to maximise desirability. For an agent may choose prospects in accordance with their desirability without thereby maximising subjective expected utility if her options don't have the specific form of Savage acts or if her degrees of desire don't conform to Supposition Independence, Restricted Actualism and Value Independence. Jointly, these three conditions are very demanding. In particular, as we noted at the end of Chapter 8, they jointly imply that the facts and counterfacts are probabilistically independent (Fact–Counterfact Independence) and that the strict counterfacts don't matter (Prospect Actualism). So the normative validity of the SEU Hypothesis, within Desirabilism, depends not only on the probabilistic independence of states from acts but also on a strong version of value actualism.

9.5 BAYESIAN DECISION THEORY

We have seen that within the broad framework of Desirabilism the theories of Jeffrey, von Neumann and Morgenstern and Savage can all be viewed as advocating desirability maximisation under constraints of one kind or another. From this perspective, these theories are not distinguished by the standard of value they endorse but on the question of how the prescription to maximise desirability should be applied: in particular, which prospects should play the role of objects of choice and what conditions preferences for them should satisfy. Jeffrey advocated desirability maximisation with respect to any prospect that could be made true at will, a liberalism not shared by other decision theorists. Followers of von Neumann and Morgenstern, for instance, will insist that the objects of choice should be construed as lotteries and that Desirabilism should be supplemented with Chance Neutrality (and the Principal Principle), so that desirability maximisation equates with maximisation of the objective expectation of utility. Likewise, followers of Savage will insist that in situations of uncertainty the objects of choice should be construed as Savage-style acts (i.e. as partitioning indicative conditionals) and will want to supplement Desirabilism with Supposition Independence, Restricted Actualism and Value Independence, so that desirability maximisation equates with maximisation of subjective expected utility.

Causal decision theory too can be fitted into this mould. On the formulation of causal decision theory due to Brian Skyrms (1981), for instance, the choice-worthiness of an act is the unconditional expectation of its desirability calculated relative to the 'right' partition: a partition of the space of prospects whose elements are maximally specific with regard to features of the world outside the agent's causal influence but upon which the causal efficacy of the action depends. Suppose that $\{\kappa_i\}$ is such a partition. Then, for Skyrms, the choice-worthiness of an action α is its (causal) expected utility U, or its expected desirability calculated with respect to the partition $\{\kappa_i\}$. Formally:

$$U(\alpha) = \sum_i V(\alpha \wedge \kappa_i) \cdot P(\kappa_i)$$

David Lewis (1981) offers a very similar formulation of causal decision theory, except that for him the right partition consists in dependency hypotheses: maximally specific propositions about how the possible consequences of acting depend causally on the choice of act.

On either formulation, the causal expected utility of an act α will equal the desirability of the conjunction of indicative conditionals $\bigwedge_i(\kappa_i \mapsto \alpha)$ with antecedents drawn from a partition $\{\kappa_i\}$ satisfying the conditions of Supposition Independence, Restricted Actualism and Value Independence. For then, as we have seen

$$V(\bigwedge_i(\kappa_i \mapsto \alpha)) = \sum_i V(\alpha \wedge \kappa_i) \cdot P(\kappa_i) = U(\alpha)$$

So these causal decision theorists, like Savage, should not be construed as disagreeing with Jeffrey about the standard of value that applies in choice (desirability) but about the appropriate objects of such choice. They too will restrict the objects of choice to 'acts': partitioning indicative conditionals with antecedents drawn from the right kind of partition, namely 'states' that are maximally specific conjunctions of causal factors or dependency hypotheses. But they need not insist, as Savage does, that the action–state pairs (the $\alpha \wedge \kappa_i$) pick out maximally specific consequences, so causal decision theory applies even in situations of option uncertainty when exact consequences are not known (recall the discussion in Chapter 3).

Let's take stock. It would seem that a broad Bayesian framework for tackling decision problems under uncertainty can be discerned, whose core consists of three principles.

1. *Choice Principle*: Rational choice is a matter of picking an option with maximum desirability (an option being a prospect that can be made true at will).
2. *Framing Principle*: Options should be represented as 'acts': partitioning indicative conditionals relating states to consequences.
3. *Evaluation Principle*: The desirabilities of such acts satisfy the SEU Hypothesis.

Let us call the conjunction of these three principles the *Bayesian synthesis*. Jeffrey accepted only the first principle, of course. And Savage didn't work with as rich a value framework as that provided by Desirabilism, so he cannot strictly be construed as accepting the Choice Principle. But we have seen that versions of both Savage's theory and of causal decision theory can be constructed which satisfy all three principles (and a similar exercise could be performed with Ramsey's). Indeed, because my formulation of the Bayesian synthesis leaves open the interpretation of states and consequences, it provides a framework broad enough to encompass a very wide variety of theories.

The central question that we now need to address is: does the Bayesian synthesis present an adequate theory of rational choice? We can break this into two subsidiary questions. First, can and should situations of decision making under uncertainty be framed as choices between options satisfying the Framing Principle? And, second, is the theory normatively valid given the Framing Principle – i.e. in choices between options of this kind, is it normatively binding on agents to choose the option that maximises subjective expected utility, as implied by the conjunction of the Choice and Evaluation Principles?

The answer to the first question depends to large extent on what further conditions are placed on the interpretations of states and consequences by versions of the theory. Savage, for instance, packs a lot of the content of the theory into the conditions that characterise the prospects that serve in these roles. I have already argued that these conditions are more restrictive than is ideal given that it may not always be desirable or even possible for particular agents to frame actions as functions from maximally specific states to maximally specific consequences. But I want to set this issue aside and focus on the question of whether the desirability that the SEU hypothesis attaches to Savage-style acts truly measures their choice-worthiness.

TABLE 9.1. *Allais' Paradox*

Actions	Ticket numbers		
	1	2–10	11–100
f	$1,000,000	$1,000,000	$1,000,000
g	$0	$5,000,000	$1,000,000
f'	$1,000,000	$1,000,000	$0
g'	$0	$5,000,000	$0

The main challenge in this regard is to deal with the well-known objections to Savage's theory deriving from the Allais and Ellsberg Paradoxes (and which will carry over to any version of the Bayesian synthesis). Some defendants of Savage argue that the problematic preferences exhibited in these set-ups are simply irrational. But this conclusion sits uncomfortably with the stubbornness with which people stick to them even after their putative irrationality has been explained. So let's see if we can say something in defence of the theory on the assumption that these preferences are not irrational. I will comment only briefly on Allais' Paradox since the contours of the response to it are well known. The Ellsberg Paradox is more difficult to deal with and so I will develop a response in some detail.

9.5.1 Allais' Paradox

Allais' set-up is reproduced again here in Table 9.1. Recall that the problem arises because subjects frequently report preferences for action f over action g and for g' over f', in violation of $\mapsto$-Separability. (In the transcription of Savage offered here, the action g is represented by the partitioning indicative conditional 'If ticket 1 is drawn then you win nothing, if tickets 2–11 are drawn then you win $5,000,000, ...', and so on.)

There are two reasonable responses one might make to this challenge if one regards this pattern of preferences as rationally permissible. The first is to conclude that, although rationality requires maximisation of desirability, it does not always require maximisation of subjective expected utility. Allais (1979) himself thought that Value Independence was not a requirement of rationality because there could be 'complementarities' between different states of the world – i.e. ways in which the evaluation of one state is influenced by the properties of the others. In particular, the overall distribution of outcomes across different states could affect how risky the action is perceived to be. While options f and f', for example, have the same consequence in the event ticket numbers 1 through 11 are drawn, the fact that f has this very same consequence for all other ticket draws makes it the case that it yields a desirable consequence *with certainty*, a feature not shared by f'. Recently arguments of this kind have served to motivate several (closely related) rival theories to

Savage's, including the non-expected utility theory of Machina (1982), the rank-dependent utility theory developed by Quiggin (2012) and its philosophical counterpart, the risk-weighted expected utility model of Buchak (2013).

Others have suggested instead that Value Independence fails because of the regret or disappointment that subjects experience when the outcome of a choice is worse than it would have been had they chosen differently. On this latter diagnosis, it is not the 'horizontal' complementarities between the outcomes of the same act in different states that are the problem but the 'vertical' complementarities between the consequences of different acts in the same state. This argument has motivated the regret theory developed by Loomes & Sugden (1982), which departs more radically from Savage's theory in allowing for intransitivities in preferences over actions resulting from the fact that agents' attitudes to them will depend on the set of alternatives.

In this large literature on Allais' Paradox and alternatives to subjective expected utility theory, there is very little recognition of the distinct roles played by the separability conditions and value actualist ones or, more informally, between the idea that the counterfacts under orthogonal suppositions are probabilistically and desirabilistically independent and the idea that the strict counterfacts don't matter. It is the latter idea – Value Actualism – that is at stake in much of the literature on Savage's theory, with critics offering examples of the kind mentioned above of how counterfacts matter and defendants retorting that if the counterfacts matter then they must do so because they have some effect in the actual world. For instance, if the fact that an action is risky is a reason not to prefer it, then there must be some effect of its riskiness that matters to the agent. If it is the disappointment that she will feel when the outcome of the choice of the risky act is a bad one, then this disappointment must be considered to be part of the consequence of that act in the 'unlucky' state. What distinguishes acts g and g′ in Allais' experiment, on this hypothesis about why riskiness matters, is that, if ticket 1 is drawn, the agent who has chosen g will experience a good deal of disappointment because *ex ante* she was almost certain to win more, while the agent who has chosen g′ will experience only a small amount of disappointment because *ex ante* she was unlikely to win anything anyway.

Now, proper application of Savage's theory requires that consequences be maximally specific with regard to all that matters to the agent. So whatever the reasons for the counterfacts mattering they must be written into the description of the consequences. If we do so in Allais' experiment, the violation of Value Independence will disappear. Properly described, for instance, acts g and g′ will no longer have the same consequence when ticket 1 is drawn. As John Broome (1991, p. 107) puts it, 'Plainly, therefore, the case against the Sure-thing Principle is absurd. It depends on making a distinction on the one hand, and denying it on the other.' For either consequences should be distinguished by the counterfacts, in which case the acts available in Allais' set-up are not properly represented by Table 9.1, or they should not, in which case the Allais preferences are indeed irrational.[5]

This defence of Savage and, more generally, the Bayesian synthesis amounts to two claims: first, that a partitioning indicative conditional cannot serve to identify an act unless the descriptions of consequences it gives are maximally specific with regard to all that matters to the agent; and, second, when descriptions of consequences are maximally specific, then the counterfacts will not matter. So the viability of this line of defence depends on whether it is always possible to specify consequences in such a way that Value Actualism is satisfied. If the agent cares about the counterfacts because of how she will actually feel, then clearly it will be possible to do so. But, even if no non-modal property of the consequence serves to capture the relevance of the counterfacts, some modal property will.

Suppose, for instance, that the desirability of X in state S_1 depends on whether or not it would have been the case that Y if state S_2 had occurred. Then the desirability of the indicative conditional $(S_1 \mapsto X)(S_2 \mapsto Y)$ should be equated with that of the conditional $(S_1 \mapsto (X(S_2 \to Y))(S_2 \mapsto Y(S_1 \to X))$, which has the counterfacts written in and, hence, which satisfies Value Independence. (Note the occurrence of two different kinds of conditionals: indicatives at the 'base' level and counterfactuals embedded in the consequences.) Is there a threat of infinite regress because we are now required to refine the consequence $X(S_2 \to Y)$ to reflect the counterfact that $Y(S_1 \to X)$? No, because there is no suggestion that it matters in S_1 that had it been S_2 then it would have been the case that had it been S_1 then it would have been X. These 'second-order' counterfacts don't matter.

This move doesn't entirely rescue Savage's theory, for, as we noted in Chapter 4, it leads to conflict with the Rectangular Field assumption. No such difficulties plague the Bayesian synthesis as formulated here. Nonetheless, there are limits to this strategy of reindividuation. Firstly, it makes the Framing Principle extremely demanding because consequences may need to be very specific indeed. Secondly, if the counterfactual dependencies are of the 'vertical' kind – where what matters is what would have occurred if a different action had been chosen, rather than if a different state had been the actual one – then the reindividuation strategy would imply that we cannot say what the consequence of any action is until we know what actions it is being compared with. This is clearly unsatisfactory: we should be able to individuate options independently of the set of alternatives available at any one time. So if the vertical counterfacts matter then the Evaluation Principle will have to be weakened and desirability maximisation will no longer amount to maximisation of subjective expected utility.

5 In making this argument, one need not follow Broome in seeing a role for rationality in determining whether the counterfacts matter. It suffices that, as a matter of fact, they do matter to the agent making the Allais choices.

TABLE 9.2. *The Ellsberg Paradox*

	Red	Black	Yellow
L_1	$100	$0	$0
L_2	$0	$100	$0
L_3	$100	$100	$0
L_4	$0	$100	$100

9.5.2 Ellsberg's Paradox

The Ellsberg Paradox is generally taken to present Bayesian decision theory with a much more difficult challenge than Allais' Paradox. Not only are the troublesome preferences that are often observed in experiments like it incompatible with Savage's theory, but they also violate causal decision theory and many of the aforementioned rival theories to Savage's designed to deal with the Allais Paradox. Consider a slightly simplified version of Daniel Ellsberg's set-up, with an urn that contains three balls: one red and the remaining two black or yellow in an unknown proportion. When asked to choose between the options displayed in Table 9.2, many subjects express a preference for L_1 over L_2 and for L_4 over L_3, in violation of the Sure-Thing Principle. Ellsberg's explanation for these observations was that agents are averse to what he called *ambiguity*, this being the lack of information as to the precise probability distribution over the state space. In the first choice situation that the subjects find themselves in they are given information which makes it reasonable for them to put the probability of drawing a red ball at one-third, but with regard to the probability of a black ball, they know only that it is no more than two-thirds. In view of this, many subjects, he conjectured, would 'play it safe' and opt for the lotteries with a known probability of paying out over one in which there is a good deal of uncertainty about this probability.

Ambiguity aversion of the kind postulated by Ellsberg has been established in numerous experiments involving set-ups similar to his, a fact which has fuelled a search for rival decision models to the Bayesian one. Although these models differ in various ways, they all take it as given that ambiguity aversion is inconsistent not only with Savage's Sure-thing Principle but also with the view that the individuals are probabilistically sophisticated in that they base their decisions on precise probabilities for the contingencies upon which the consequences of their choice depend. But, if Ellsberg's conjecture about how subjects perceive the decision problem they face is correct, then Table 9.2 does not provide the correct representation of their decision problem. For Savage, a properly specified decision problem is one in which the descriptions of states are maximally specific with regard to the presence or absence of all factors relevant to the determination of consequences that are causally independent of the actions available to the agent and in which descriptions of consequences are

TABLE 9.3. *The Reframed Ellsberg Paradox*

	RBB	RBY	RYY
L_1	$\frac{1}{3}$	$\frac{1}{3}$	$\frac{1}{3}$
L_2	$\frac{2}{3}$	$\frac{1}{3}$	0
L_3	$\frac{1}{3}$	$\frac{2}{3}$	1
L_4	$\frac{2}{3}$	$\frac{2}{3}$	$\frac{2}{3}$

maximally specific with regard to features that matter to the agent's evaluation of their desirability. Now, according to Ellsberg – correctly, I believe – agents will regard the distribution of balls in the urn as relevant to the determination of the consequences: the monetary gains attendant on a draw. This means that, strictly speaking, the states of the world are not draws of red, black or yellow but combinations of distributions of balls in the urn plus the draw from it, such as 'The urn contains 30 red balls, 25 black balls and 35 yellow balls; a yellow ball is drawn' and 'The urn contains 30 red balls, 60 black balls and no yellow balls; a red ball is drawn'.

Does the distribution from which a ball is drawn matter, given the specification of the ball actually drawn? To answer this question consider an alternative (coarse-grained) representation of the Ellsberg paradox in which states are individuated by distributions of balls in the urn and consequences are chances of monetary gains. This is illustrated in Table 9.3 in which the cell entries are the chances of winning \$100. Thus an entry of x indicates that making that choice when the world is in that state gives the agent a chance of x of winning \$100 and a chance of $1 - x$ of winning nothing.

In this representation the betting acts L_1 to L_4 are functions from possible ball distributions to chances. Now, let us suppose that our agent is a subjective expected utility maximiser *à la* Savage and, moreover, that she regards the two possible states of the world as equally likely (perhaps in virtue of symmetry considerations). In this case, a preference for L_1 over L_2 reveals that for the agent $V(\frac{1}{3}) > V(\frac{2}{3}) + V(\frac{1}{3}) + V(0)$, while a preference for L_4 over L_3 reveals that for the agent $V(\frac{2}{3}) > V(\frac{1}{3}) + V(\frac{2}{3}) + V(1)$. Together, these imply that

$$V\left(\frac{1}{3}\right) - V(0) > V\left(\frac{2}{3}\right) - V\left(\frac{1}{3}\right) > V(1) - V\left(\frac{2}{3}\right) \qquad (9.1)$$

So we can conclude that an agent who maximises subjective expected utility *can* have Ellsberg preferences over betting acts, provided she values gains in chances of monetary payoffs less as the minimum/maximum chance rises – i.e. if the chances of money have diminishing marginal utility for her.

This characteristic of an agent's degrees of desire is ruled out by the theory of von Neumann and Morgenstern. Recall that according to the vN–M theory a lottery has a value equal to a linear average of the utilities of its prizes. So, in particular, lotteries of the kind appearing in the Ellsberg set-up which yield $100 with a certain chance x are valued at $x \times V(\$100)$. From which it follows of course that

$$V\left(\frac{1}{3}\right) - V(0) = \frac{1}{3}V(\$100) = V(\$100) - \frac{2}{3}V(\$100) = V(1) - V\left(\frac{2}{3}\right) \quad (9.2)$$

in contradiction to Equation 9.1. So we can conclude that the Ellsberg preferences are not inconsistent with Savage's theory but that they are inconsistent with the combination of this theory and that of von Neumann and Morgenstern.

This conclusion will be very surprising for those who take it that Savage's theory is simply a generalisation of vN–M that allows for subjective uncertainty. But this is not the case. Savage's theory, strictly speaking, says nothing at all about how chances should be valued. In contrast, as we saw earlier, the vN–M theory requires that agent's desires satisfy the Chance Neutrality condition. I argued before that Chance Neutrality was doubtful as a condition on rational degrees of desire, and this discussion serves only to strengthen the case against it. For, if a Bayesian wants to make room for the type of caution exhibited by the Ellsberg preferences, she must reject Chance Neutrality.[6]

9.6 CAUSAL DECISION THEORY

On the view presented in the previous sections, a theory of rational evaluation needs to be supplemented by principles which regulate the application of the theory to choice problems. The central one is the requirement that we formulate our options with reference to a partition of states and consequences that captures everything of relevance to the decision problem. The advantage of doing things this way is that we can get along with a single value measure – desirability – in order to make our choices. The disadvantage is that we need to be able to formulate our options in the right way in order to apply the theory, and this can be very demanding on the decision maker. So there is interest in looking at the possibility of a partition-independent formulation of causal decision theory based on a notion of value that applies to all prospects but coincides with desirability on the appropriate class of objects.

As causal decision theorists emphasise, the distinctiveness of causal efficacy from that of desirability derives from the kind of supposition involved: causal expected utility is the expectation of utility on the interventional supposition that the action was performed, rather than the conditional expectation of utility given the (evidence of the) action. So, while the desirability of any action α is given by

$$V(\alpha) = \sum_{\omega \in \alpha} u(\omega) \cdot P(\omega|\alpha) \quad (9.3)$$

---

[6] Or, strictly speaking, the Principal Principle. But this condition does not seem to be central to questions of the rationality of cautious preferences, and I very much doubt that denying it will solve the dilemma that the Ellsberg preferences present.

its causal expected utility will be given by

$$U(\alpha) = \sum_{\omega \in \alpha} u(\omega) \cdot P^*_\alpha(\omega) \tag{9.4}$$

These two quantities will be equal only when $P^*_\alpha = P(\cdot|\alpha)$ – i.e. when the mode of supposition of the performance of the action is evidential. And it precisely because evidential supposition leads to poor choices in cases when probabilistic and causal dependence diverge that causal decision theory looks to alternative modes of supposition to characterise the kind of suppositional reasoning necessary to determine how choice-worthy an action is.

Unfortunately, causal expected utility, as defined by (9.4), cannot serve as a rival concept of value to desirability, at least if we require that such a concept be partition-invariant. To see this, note that partition invariance would require that, for all prospects β,[7]

$$U(\alpha) = U(\alpha\beta) \cdot P^*_\alpha(\beta) + U(\alpha\neg\beta) \cdot P^*_\alpha(\neg\beta)$$

From which it would follow that

$$U(\alpha) = P^*_\alpha(\beta)[\sum_{\omega \in \alpha\beta} u(\omega) \cdot P^*_{\alpha\beta}(\omega)] + P^*_\alpha(\neg\beta)[\sum_{\omega \in \alpha\neg\beta} u(\omega) \cdot P^*_{\alpha\beta}(\omega)]$$

But from (9.4) we also have

$$U(\alpha) = \sum_{\omega \in \alpha} u(\omega) \cdot P^*_\alpha(\omega)$$

$$= \sum_{\omega \in \alpha\beta} u(\omega) \cdot P^*_\alpha(\omega) + \sum_{\omega \in \alpha\neg\beta} u(\omega) \cdot P^*_\alpha(\omega)$$

And these can be the same (non-accidentally) only if

$$P^*_{\alpha\beta}(\omega) \cdot P^*_\alpha(\beta) = P^*_\alpha(\omega)$$

And hence, in particular, that

$$P^*_\beta(\omega) = \frac{P(\omega)}{P(\beta)} = P(\omega|\beta)$$

In recognition of this, James Joyce (1999) argues that the causal utility of a prospect should not be construed as a form of value different from desirability, but as desirability from a particular *perspective*, namely that achieved by supposing that an action were performed. To get a handle on how his proposal solves the problem, note firstly that the multidimensional possible-worlds framework introduced in the previous chapter supports the

---

7 In view of the background assumption of World Actualism, we can, without loss of generality, confine attention to factual prospects.

construction of a whole family of desirability-like measures of value. The desirability of a prospect, as defined in Equation 9.3, is the conditional expectation of the utility, as measured by u, given the truth of the prospect – i.e. for any β, $V(\beta) = \mathbb{E}_p(u|\beta)$. In this case, as indicated by the subscript, the expectation is calculated relative to the probability density p measuring the agent's unconditional degrees of belief in worlds (or world n-tuples). Similarly, V_α^*, the measure of desirability under the supposition that α, is the conditional expectation of utility, as measured by u_α, calculated relative to the probability density p_α, where u_α and p_α, respectively, represent the agent's degrees of desire for, and belief in, worlds on the supposition that α – i.e. for any β, $V_\alpha^*(\beta) = \mathbb{E}_{p_\alpha}(u_\alpha|\beta)$. But we can of course calculate the conditional expectation of utility given the truth of a prospect relative to any given utility and probability pair. In particular, we can use $\mathbb{E}_{p_\alpha}(u|\cdot)$, the conditional expectation of u relative to the probability p_α, to define a function $V_{p_\alpha}(\cdot)$, such that

$$V_{p_\alpha}(\beta) = \mathbb{E}_{p_\alpha}(u|\beta) = \sum_{\omega \in W} u(\omega) \cdot P_\alpha^*(\omega|\beta)$$

Note that $V_{p_\alpha}(\cdot)$, although not strictly a desirability function since it is not normalised relative to the tautology (i.e. it does not satisfy axiom V1), is a desirability-like measure of value in that it conforms to axiom V2. It is also a partition-independent value measure, since, for any partition $\{\beta_i\}$ of β,

$$V_{p_\alpha}(\beta) = \sum_i V_{p_\alpha}(\beta_i) \cdot P_\alpha^*(\beta_i|\beta)$$

Apart from these formal properties, why is the function $V_{p_\alpha}(\cdot)$ of any interest? Because

$$V_{p_\alpha}(\alpha) = \sum_{\omega \in W} u(\omega) \cdot P_\alpha^*(\omega|\alpha) = \sum_{\omega \in W} u(\omega) \cdot P_\alpha^*(\omega) = U(\alpha)$$

So $V_{p_\alpha}(\alpha)$ would seem to be just the partition-independent measure of the causal expected utility of α that we were looking for.

Joyce argues that causal expected utility, as defined by $U(\alpha) = V_{p_\alpha}(\alpha) = \mathbb{E}_{p_\alpha}(u|\alpha)$, is the correct measure of the choice-worthiness of an action and that, so defined, is the desirability of an action on the supposition of its performance. The first claim I agree with, but not the second. The desirability of an action under the supposition of its performance is *not* a good measure of its choice-worthiness because, if we suppose that an action will be performed, we in effect bracket out our attitude to it. For instance, if I consider how probable it would be that I could afford a holiday in Barbados if I were to win the lottery, I bracket out the improbability of my winning the lottery. Similarly, when considering the desirability of taking the holiday if I were to win, I bracket out the desirability of the lottery win itself. But then, under the supposition that I am going to perform an action, its performance is neither desirable nor undesirable. It's just a given. That is why, on the account developed in Chapter 6, desirability under the supposition that α is measured by $\mathbb{E}_{p_\alpha}(u_\alpha|\alpha)$ (i.e. V_α^*), not $\mathbb{E}_{p_\alpha}(u|\alpha)$ (i.e. not V_{p_α}). But it is the latter that Joyce correctly identifies as the best measure of choice-worthiness.

Let me put the point slightly differently. Note that, because $P_\alpha^*(\omega|\alpha) = P_\alpha^*(\omega|\top)$, the conditional expectation of utility given α, calculated from the perspective of the supposition that α is true, must be the same as the conditional expectation of utility, given the tautology, from that perspective. But the truth of the tautology is neither desirable nor undesirable (under any supposition), so it is wrong to use the desirability of α under the supposition on its truth as a measure of its choice-worthiness. What seems to lead Joyce to think that the two can be equated is an argument of the following kind. Because the desirability of the current status quo is just that of the tautology – it being by definition what is expected and therefore lacking news value – and because an action's desirability can be thought of as the desirability of the status quo that the action produces, it follows that the desirability of the action can be equated with the desirability of the tautology from the standpoint induced by the supposition of the action's performance.

But this reasoning is mistaken. The status quo is a variable prospect (indexed by expectations) that, at each supposition-induced standpoint, has a desirability equal to the tautology, *from that standpoint*. But what the status quo is from my current perspective is not the status quo from the perspective induced by a supposition. So the desirability of what is currently the status quo is not the same, from any standpoint, as that of the induced status quo. More formally, let SQ be a function that associates each state of expectation with a corresponding prospect, intuitively the prospect describing what the agent expects to be the case. Let s be the agent's current state of expectation and $s^{*\alpha}$ be the state of expectation induced by the supposition that α. Then the current desirability of the current status quo is $V(SQ(s))$, and that of the status quo induced by the supposition that α is $V(SQ(s^{*\alpha}))$. Now

$$V(SQ(s)) = V(\top) = V_\alpha^*(\top) = V_\alpha^*(SQ(s^{*\alpha}))$$

But

$$V(SQ(s^{*\alpha})) \neq V_\alpha^*(SQ(s^{*\alpha}))$$

So the current desirability of the status quo induced by the supposition that α cannot be equated with the desirability of that state, under the supposition that α.

It is the former quantity, $V(SQ(s^{*\alpha}))$, that is of interest to the causal decision theorist, for it measures the desirability of the expected status quo consequent on the intervention α. So we should expect that

$$V_{p_\alpha}(\alpha) = V(SQ(s^{*\alpha})) \tag{9.5}$$

In the previous section, we examined the manner in which the right-hand side of this identity is unpacked within the Bayesian synthesis, namely by identifying prospects that satisfy this equation. The trick was to identify actions with the state of expectation that they induce. In particular, the prospect whose desirability is just that of action α's causal utility is the prospect that w_1 is the case if w_1 would be the case if α were, w_2 is the case if w_2 would be the case if α were ... , for all ω consistent with α. So it is natural to define $SQ(s^{*\alpha})$ by

$$SQ(s^{*\alpha}) := \bigwedge_\omega [(\alpha \to \omega) \mapsto \omega]$$

Then, given Supposition Independence, Restricted Actualism and Value Independence and the fact that the $\alpha \to \omega$ are mutually exclusive, it follows that

$$V(SQ(s^{*\alpha})) = \sum V(\{\omega\}) \cdot P(\alpha \to \omega) = \sum u(\omega) \cdot P_\alpha^*(\omega) = V_{p_\alpha}(\alpha)$$

in accordance with Equation 9.5.

The left-hand side of Equation 9.5 gives us Joyce's formulation of causal expected utility in terms of the conditional expectation of utility calculated relative to a suppositional probability. If this is not the desirability of the action under the supposition of its performance, then what is it? The suggestion I want to make is that an action is choice-worthy insofar as the desirability of what we can expect supposing its implementation exceeds that of what we currently expect – i.e. insofar as the desirability of the status quo it induces exceeds that of the current status quo (as Equation 9.5 suggests).

To state this more formally, let's define the change in desirability in prospect β induced by the supposition that α as the difference between the conditional expectation of utility, given β, relative to the supposition that α and the conditional expectation relative to the supposition of $\top$. More precisely:

$$\Delta_\alpha^*(V(\beta)) := \mathbb{E}_{P_\alpha^*}(u|\beta) - \mathbb{E}_P(u|\beta) = \sum_{\omega \in \beta} u(\omega) \cdot [p_\alpha^*(\omega|\beta) - p(\omega|\beta)]$$

It follows in particular that

$$\Delta_\alpha^*(V(\top)) = \sum_{\omega \in \top} u(\omega) \cdot [p_\alpha^*(\omega) - p(\omega)]$$

$$= \sum_{\omega \in \beta} u(\omega) \cdot p_\alpha^*(\omega)$$

in virtue of the fact that by V1, the axiom of normality, $\sum u(\omega) \cdot p(\omega) = 0$. This makes $\Delta_\alpha^*(V(\top))$ the change in desirability of the status quo induced by the supposition that α; and the required measure of the causal expected utility of α. We have now the partition-independent causal decision principle that we have been looking for. It is:

Causal Decision Value Of the actions available to you, you should perform the one with the greatest desirability gain induced by its performance – i.e. the action α which maximises the quantity $\Delta_\alpha^*(V(\top))$

10

The Learning Agent

In the first part of the book we developed a theory of rational prospective agency based around a set of claims about what rationality requires of the judgements of agents: consistency between her beliefs, desires and preferences; between her conditional and unconditional attitudes; and between her attitudes and her choices. This theory can reasonably be criticised both for being too demanding and for being insufficiently so. Too demanding because real agents face resource constraints which make it impossible to form opinions about all possible contingencies and unwise that they attempt to do so. Insufficiently demanding, on the other hand, because it requires of agents only that their beliefs and desires be consistent, and not that they be responsive to what is true or valuable in any way. In the next part of the book we will take up the first challenge. Our task in this chapter will be to respond to the second.

There are two directions in which one might seek to constrain the sorts of attitudes that a rational agent can permissibly hold. First, one might put forward conditions that go beyond consistency but which fall short of simply demanding that the agent believe only truths and desire only what is good (which would be to demand what cannot generally be achieved). Second, one might require that the agent improves her opinions over time by responding appropriately to information acquired through interaction with the environment or through deliberation. Many conditions of both kinds are to be found in the epistemological literature, for different conceptions of belief, but I will focus on those deriving from the widely shared thought that an agent's beliefs should be constrained by the *evidence* she holds about the state of the world and, more generally, that her attitudes should reflect what she has learnt from experience.

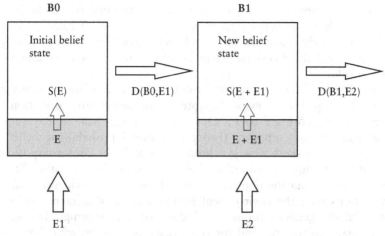

FIGURE 10.1. Evidence-Based Constraints on Belief

Let us restrict attention to belief for the moment and return to the other attitudes later. Evidence-based constraints on belief can take both a synchronic form – that an agent's beliefs at a particular time should appropriately reflect the evidence that she holds at that time – and a diachronic one – that she should revise her beliefs appropriately in response to the acquisition of new evidence. Both kinds of constraint can be captured more formally by a rule or function:

- a synchronic belief formation rule $\mathcal{S}$ that specifies for any given body of evidence E the set of permissible belief states, $\mathcal{S}(E)$;
- a diachronic belief revision rule $\mathcal{D}$ which specifies for any given belief state B and (new) evidence E_N, the set of permissible new belief states $B' = \mathcal{D}(B, E_N)$.[1]

Schematically, we can picture this evidence-constrained process of belief formation and revision as in Figure 10.1. The agent is in an initial belief state $B0$ that is constrained by the evidence E that she holds at that time in accordance with the rule $\mathcal{S}$. She then receives new evidence $E1$, to which she responds by revising her belief by application of rule $\mathcal{D}$, which takes as arguments the evidence $E1$ and her initial belief state $B0$, thereby arriving at new belief state $B1$.

This simple picture of learning from experience can be filled out in different ways depending on how its basic elements – belief states, evidence and

---

[1] It is to be expected that the formation and revision rules would be closely related. For instance, if the rules $\mathcal{S}$ and $\mathcal{D}$ are consistent with one another, then the belief state mandated by the revision rule $\mathcal{D}$ when taking initial belief state B and new evidence E_N as arguments should be permissible according to the formation rule $\mathcal{S}$ given the agent's total evidence E_T, the combination of her initial evidence and new evidence – i.e. it should be the case that $B' \subseteq \mathcal{S}(E_T)$.

inference rules – are conceived. For instance, if one conceives of both belief states and evidence as sets of propositions, then classical logic provides a natural set of synchronic constraints on belief states (logical consistency with the evidence) and the AGM theory of belief revision a corresponding set of diachronic rules.

In this chapter, however, we will work with the quantitative conception of belief developed in previous chapters and investigate formation and revision rules appropriate to it. Our starting point will be Bayesianism: the predominant contemporary theory of rational probabilistic belief and belief revision. Bayesians view synchronic rationality as conformity of one's degrees of belief with the laws of probability and diachronic rationality as a matter of drawing out the implications of information or evidence generated by interactions with the environment by application of inference rules that are probabilistic generalisations of Modus Ponens. Although Bayesianism comes in many forms, the foil for our discussion will be what I will term *Classical Bayesianism*, an epistemological doctrine that consists of four specific propositions.[2]

1. *Certain Evidence*: The agent's evidence consists of the set of propositions that she regards as certain. Learning canonically takes the form of coming to believe a true proposition to degree one.
2. *Precise Probabilism*: A rational agent's belief state is a single probability measure on the set of prospects
3. *Consistency with Evidence*: An agent's belief state at any time must be consistent with her evidence at that time (the synchronic condition)
4. *Conditioning*: The agent is required to change her beliefs by conditionalisation on new evidence (the diachronic condition).

Although none of these propositions is undisputed amongst Bayesians, historically it is the synchronic condition that has been the subject of most debate, specifically on the question of whether the condition is strong enough or, instead, whether rationality imposes further a (non-evidential) constraints on our belief state. I will have little to say about it however because, in my opinion, this issue is overshadowed by the question of whether agents must adopt a full probability function as their belief state in the first place (a question addressed in the next chapter). The focus of this chapter will instead be on rules for attitude revision. In the first section I will consider what experience is, arguing that it should be construed far more broadly than is allowed by the principle of Certain Evidence. This will open space for examination of a greater variety of revision rules than just Classical Bayesian conditioning, appropriate

---

[2] Bayesian decision theory, the topic of the previous chapter, adds a fifth proposition: that rational agents maximise subjective expected utility in their choices.

to different of types of experience. In the final sections I will turn to the issues of desire and preference change.

10.2 EXPERIENCE AND LEARNING

The learning agent changes her beliefs and desires in the light of what she gleans from experience. In doing so she takes experience to place constraints on her attitudes that go beyond those of the consistency conditions considered in the previous chapters, constraints that encode what she has learnt from her interaction with the environment. So much is common ground; the question is: what to make of it? For many Bayesians, the process of learning from experience is precisely where rationality begins to have some bite, providing a royal route out of the confines of the static view of rationality as consistency. No matter how idiosyncratic agents' initial states of belief, they conjecture, exposure to sufficient common evidence will drive them towards a consensus based on the truth. This view does not, it seems to me, survive proper analysis of the concept of evidence and that of experience upon which it is based.

There are two competing senses of the notion of experience that need to be carefully distinguished. On the one hand, there is what happens to the agent in the world: she makes contact with objects, light rays strike her retina, sound waves bombard her eardrums, and so on. Let us call this experience$_1$. On the other hand, there is the agent's understanding or interpretation of what has happened to her. Let us call this experience$_2$. The two need not coincide: what the world delivers to the agent, what the objective informational content of her experience$_1$ is, can be different from what she takes from her interaction with the world and the significance she attaches to it – i.e. the content of her experience$_2$. Consequently, two agents may share an experience of the first kind, while having quite different experiences of the second (the opposite is possible too).

Consider, for example, an English-language speaker with no knowledge of French hearing a cry of 'Feu!' that has been produced in response to a fire of which she is unaware. The sound waves striking her eardrums carry the information that there is a fire: there is a causal chain leading from the fire to the passage of the sound waves that makes their reception a reliable indicator of the presence of the fire. But this is not what she takes from this experience because she is unable to interpret the words she hears; her experience is not of a warning of *fire*. What is true in this particular example is true in general. Even when signals do not arrive in foreign languages, our senses do not simply transmit informational inputs from the world to our brain. The contents of our perceptions, at least by the time they are available as judgements, are interpretations of the world, based on largely automatic inferences that the sensory organs and the brain make when stimulated by the outside world.

Classical Bayesianism inherits the empiricist contention that what we *take* from our interaction with the world through experience$_2$– our evidence –

can and should be decomposed into that which is *given* by the world in experience$_1$ – that which the agent *knows* to be true – and that which is inferred from it. Upon this rests the assurance that learning washes out the subjectivity of the agent's prior beliefs. But this contention must be rejected for two reasons. Firstly, because this decomposition is not possible: we cannot separate out from the content of our experience$_2$ that part of it which faithfully records experience$_1$ because the constraints we adopt don't come with in-built certification as 'given as true', 'inferred', and so on. And, secondly, because our experience$_2$ need not take the form of a set of things about which we are certain. For instance, we may come to regard a proposition as more probable than before, or to believe that one proposition is as likely to be true as another, or that the conditional probability of one proposition given another is one.

> Probabilistic judgement may be appropriate as a direct response to experience, underived from sure judgment that the experience is of such and such a character. (Jeffrey, 1992, p. 45)

In rejecting the possibility of extracting the objective 'given' from experience$_2$, I don't mean to deny that we sometimes describe our experience in terms of a set of things that we take to be true. We can, for instance, often list the phenomenal properties of the world in terms of the way things *seem* to us. But such 'seemings' don't provide the kind of certainties about the world required for an objective conception of learning, for they have no special connection to the way things actually are. If an agent is properly tuned to environment then things will tend to seem to her just the way they in fact are. But nothing about the character of something seeming one way or another carries with it the certification of such faithfulness to the true state of the world.

The distinction between concepts of experience has obvious implications for the justificatory role of experience in belief formation and change. When we consider whether our English speaker should form the belief that there is a fire, we confront two reasonable answers. Yes, because she should believe what is true. But also no, because she does not have any reason (or indeed any access to a reason) for believing this. She is required$_1$ to believe there is a fire, but not required$_2$ to do so. To put it differently, our agent does have a fault, namely that of not being properly tuned to the information source in virtue of not speaking French. But she is not at fault from the point of view of rationality, as her beliefs are not out of line with what she takes herself to have learnt from her experience.

When I ask what someone is required to believe on the basis of experience or of the evidence garnered from it, it is to experience$_2$ that I refer, because my interest is in principles that the person can actually apply, given her state of knowledge. Principles that require that she be correctly tuned to the external world so that she draws from experience just what it contains by way of information cannot be applied by her. So her failure to satisfy them is not one she can address directly. (She can address it indirectly, of course, by taking steps to tune herself better to the environment – e.g. by learning the local language.) On the other hand, the requirement that an agent make the correct inferences from what she takes herself to have learnt is one that she *can* hope to fulfil.[3]

The upshot of this is that learning will be conceived here in terms of what van Fraassen (1989) calls the 'Voluntaristic' model, in which both the constraints deriving from experience and the principles governing how they are dealt with are imposed by agents upon themselves. In this model the information or evidence that an agent acquires through experience$_2$ *is* simply the set of new constraints on her attitudes that she adopts as a direct result of her interaction with the environment. Correspondingly, her total evidence at any particular moment of time is simply the sum total of constraints that she adopts on her attitudes at that time. The agent's evidence thus consists of those features of her attitudinal state that derive directly from experience$_2$ distinct, on the one hand, from the signals that reach her from the environment and, on the other, from the attitudes that she acquires by inferences that take these constraints as arguments.

This treatment of evidence has a number of implications. Firstly, it implies that the distinction between the agent's evidence and the attitudes that it supports or permits is one that is *internal* to the agent's state of mind. So the fact that an agent responds appropriately to the evidence that she holds about the world cannot provide justification$_1$ of her belief state. Indeed, whether rational belief change leads to true or accurate beliefs depends on whether the constraints that the agent adopts as a result of experience truly or accurately reflect the state of the world. No such guarantee of veracity is afforded by rationality.

Secondly, contrary to the tenets of Classical Bayesianism, the agent's evidence need not be restricted to just the propositions she regards as certain, nor need learning be restricted to cases in which some proposition comes to be regarded as such. For she may adopt constraints of many different forms as a result of experience. This opens the door to consideration of a much wider range of revision problems, characterised by different kinds of adopted constraints, and to a correspondingly wider range of inference rules that provide solutions to these problems.

In this chapter, four particular types of learning experience will be studied and corresponding rules of attitude revision defended. In line with the quantitative models developed in previous chapters, let the agent's initial or prior state of belief be represented by a probability function P and her (posterior) state of belief after an experience E by a probability function Q. Belief revision rules will be treated as mappings from P to Q as a function of

---

3 Normally anyhow. Some inferences may be too complicated for her to make and so she cannot be required to make them.

the experience E, the latter being represented by the constraints that it imposes on Q.

The question to be addressed in each of the four cases is *how* an agent should revise her attitudes, *given* the constraints that she has adopted. In this regard there are two points that can be made immediately. Firstly, how she should respond will of course depend on the form that the constraints take – i.e. on the type of experience she has. And, secondly, for any given type of constraint there will typically be many ways in which they can be accommodated consistently. This is because experience will never be so rich as to completely determine what attitudes the agent must adopt. So, if we are to say anything very specific about how the agent should revise her beliefs, we will need to introduce considerations that go beyond that of respecting consistency. In common with many other models of attitude revision, I will adopt the principle that the agent's revisions in response to evidence should be *minimal*, in the sense of involving the least departure from her current attitudes consistent with the constraints she accepts. The basic motivation for the principle is that an agent should not make any revisions to her attitudes that are not warranted by experience, both because she risks error in adopting new attitudes and because the attitudes they replace are a resource that should not be squandered needlessly. These claims will be made more precise in due course.

10.3   CLASSICAL BAYESIAN CONDITIONING

Let us start with the simplest and most-studied case, namely when the agent learns that some proposition α is true, and so adopts as a constraint on her posterior probabilities the requirement that $Q(\alpha) = 1$. Classical Bayesianism says that, in these circumstances, your new degrees of belief in any proposition γ should simply be your old degrees of belief in γ, on the (evidential) supposition that α. This is equivalent to obtaining your new degrees of belief, Q, from your old by Bayesian conditioning on the truth of α, where this is defined as follows.

Bayesian conditioning New degrees of belief Q are obtained from (old) degrees of belief P by

$$Q(\cdot) = P(\cdot | \alpha)$$

A very large number of different kinds of arguments have been given in favour of this claim, including (i) *pragmatic arguments*, of which perhaps the best known is the dynamic Dutch Book argument, which shows that Bayesian conditioning is the only belief revision policy that does not leave one vulnerable to the possibility of a sure loss at the hands of an unscrupulous bookie (see Teller, 1973), (ii) *epistemic utility arguments* (such as those of Leitgeb & Pettigrew, 2010a,b) which aim to establish Bayesian conditioning as the rule that maximises the expectation of epistemic utility, defined by goals such as truth or accuracy; (iii) *minimal change arguments*, which show that Bayesian conditioning minimises the distance between prior and posterior probability function, where distance can be measure in a variety of different ways without affecting the argument (see Diaconis & Zabell, 1982); and (iv) *symmetry arguments*, such as those of van Fraassen (1989) and Grove & Halpern (1998), which show that Bayesian conditioning is the only revision rule that preserves salient structural features of the revision problem.

Collectively, these arguments amount to a powerful case in favour of the claim that we should revise our beliefs by Bayesian conditioning in the case where our learning experiences are exhaustively described by our coming to believe with certainty that some proposition is true. Rather than evaluate them in detail, however, let's look at a more direct argument for Bayesian conditioning, since it will reveal many of the difficulties that all these arguments face. Bayesian conditioning may profitably be regarded as a probabilistic generalisation of a familiar form of argument. Here is how it goes. Suppose that you come to believe that some proposition α is true. Suppose also that you believe that if α then β. By Modus Ponens, from α and $\alpha \rightarrow \beta$ you can infer that β, and from α and $\alpha \rightarrow \neg\beta$ you can infer that $\neg\beta$. So it would seem that, if you learn that α, you should believe that β iff you believe that $\alpha \rightarrow \beta$.

An important qualification: this argument assumes that your belief that if α then β (or its contrary, if α then $\neg\beta$) is not overturned by learning that α. Consequently, even though Modus Ponens is a valid principle of inference, it is not a requirement of *rationality* that you believe its conclusion if you believe its premises. Rationality requires only that you do not believe both the premises and the negation of the conclusion. But you are free not to believe the conclusion and instead give up your belief in one of the premises. So the requirement that you believe that β in circumstances in which you believe that $\alpha \rightarrow \beta$ and come to learn that α presupposes that you regard both that α and that $\alpha \rightarrow \beta$ as constraints on your beliefs.

All of this carries over to the probabilistic case. Suppose that your prior degree of belief in β, on the evidential supposition that α, equals x. Then, if you learn that α is indeed true, you should come to believe β to degree x. For the degree of belief that you afford β, under the supposition that α is in fact true, encodes your judgement of the probabilistic implications of the truth of α, in the light of your other beliefs. So they should imply, together with the truth of α, just this degree of belief for β. Note once again an important qualification: this argument presupposes that learning that α is true does not change your conditional degrees of belief for β, given that α. More exactly, it presupposes the *rigidity of your evidential suppositional degrees of belief given* α in the face of this experience – i.e. that:

Rigidity $Q_\alpha^+ = P_\alpha^+$

Recall that our treatment of evidential supposition entails that, for any factual propositions α and β, $P_\alpha^+(\beta) = P(\beta|\alpha)$. So, together with the constraint imposed by experience in the case under consideration, namely that $Q(\alpha) = 1$, Rigidity implies classical Bayesian conditioning. *Proof:* By the law of total probability, $Q(\beta) = Q(\beta|\alpha) \cdot Q(\alpha) + Q(\beta|\neg\alpha) \cdot Q(\neg\alpha) = Q(\beta|\alpha)$. But, by Rigidity, $Q(\beta|\alpha) = P(\beta|\alpha)$. Hence $Q(\beta) = P(\beta|\alpha)$.[4]

Some of the arguments for Bayesian conditioning seem to imply that Rigidity is generally satisfied. But this is not true. The problem in a nutshell is that when we learn the truth of some proposition we may learn more than just that proposition's truth. Bob has been ill recently. Conditional on him being at work I believe him to have fully recovered. And, typically, if I in fact learn that he is at work I will infer his recovery. But if learn that he has come to work today by discovering him dead in his office, I will (quite reasonably) not adopt my prior conditional belief that he is fully recovered, given that he is at work.

We can put the point slightly differently. What my degrees of belief on the evidential supposition that α represent is my expectation of truth given that α. The expectation reflects in part my uncertainty about how α could be true, an uncertainty that could be resolved (or partially resolved) in the learning of α's truth. For example, suppose that I believe to degree 0.8 that I will be soaked in the event of rain, a judgement that reflects my uncertainty as to how hard it will rain (if it does). If I learn about the rain by it falling on me, then its intensity will be revealed along with its presence, and so I should not slavishly adopt my prior conditional degrees of belief, which were based on an uncertainty that has now been resolved.[5]

To cover these cases we might say that Rigidity applies only when α is the strongest proposition learnt. But this too is not sufficient. For, as Richard Jeffrey has pointed out, interaction with the environment may lead to a redistribution of one's belief over some set of prospects without one becoming certain of any of them. For instance, suppose that someone recounts some event (α) to you, and that being told about α vaguely reminds you of something that you saw or heard in the past (β). Then, the total effect of your experience will be represented both by the constraint that $Q(\alpha) = 1$ and that $Q(\beta) = b \neq 1$. Then unless $Q(\beta|\alpha)$ just happens to equal b, you cannot, on pain of inconsistency, revise your beliefs by Bayesian conditioning on α.

This observation suffices to undermine the famous dynamic Dutch Book argument for Bayesian conditioning. This argument shows that commitment to a belief revision policy other than Bayesian conditioning leaves one vulnerable

---

4 It also allows for a derivation of rules for revising beliefs in conditional prospects, and indeed, as we shall see, revision rules for cases in which we come to learn the truth a conditional prospect – both extensions to the classical Bayesian framework. Indeed, it reveals that classical Bayesian conditioning is *not* appropriate when revising beliefs in conditional prospects, for, as we discovered earlier, $P_A^+(B \to C) = P(C|AB) \neq P(B \to C|A)$.

5 This fact explains the famous Monte Hall paradox. See (Jeffrey, 1992, pp. 122–4) for a discussion.

to a sure loss at the hands of an unscrupulous bookie who knows what your policy is. But it does not follow from this that one should actually adhere to such a policy. For, as we have just argued, in cases when the truth of α does not represent all that one has learnt, conditioning on α will lead one into inconsistency. Given that one cannot rule out such eventualities, one should not commit oneself to any general belief revision policy for all cases of learning that α, Bayesian or otherwise. All the Dutch Book argument establishes is that it would be inconsistent to *expect* oneself to revise one's beliefs in any way other than by adopting one's current suppositional degrees of belief. But, as we pointed out before, this expectation reflects uncertainty about how one will learn the truth of α. And this uncertainty may be resolved in the actual learning of it in a way which requires one *not* to condition on the truth of α.

The upshot is this. Rigidity holds only when the adopted constraint exhaustively describes the effect of interaction with the environment; in particular, when α is all and everything that is learnt. With this restriction, there is a compelling argument for Rigidity. Suppose that α is indeed all and everything that is learnt, that all changes to the agent's partial belief are rational effects of her learning that α but that her new degrees of belief are not her old degrees of belief conditional on α. Then her conditional degrees of belief given α must have changed. But the truth of α is not itself a reason to change one's conditional beliefs given α, so something more than α must have been learnt. But that is contrary to the supposition that α is all that is learnt.

The crucial step in this argument is the claim that learning that α is not in itself a reason to change one's conditional degrees of belief given that α. Now, learning something gives one reason to change one's attitudes when what one has learnt constrains one's attitudes in some way. For instance, learning that it will rain tomorrow clearly constrains what attitudes I can take to tomorrow's weather, but it does not seem relevant to the question of whether my house will be flooded if it rains. So, in learning that it will rain (and no more than this), I am not thereby constrained in the conditional probability that I attach to my house being flooded given that it will rain.

To make this more precise, note that a probability function P can always be decomposed in the following way, provided that $P(\alpha\beta)$, $P(\alpha\neg\beta)$ and $P(\neg\alpha)$ are non-zero:

$$P(\cdot) = [P(\cdot|\alpha\beta) \cdot P(\beta|\alpha) + P(\cdot|\alpha\neg\beta) \cdot (1 - P(\beta|\alpha))] \cdot P(\alpha) + P(\cdot|\neg\alpha) \cdot (1 - P(\alpha))$$
$$(10.1)$$

Let us call this the decomposition of P in terms of α and β, and the various terms appearing in it the $\alpha\beta$-factors of P. Now, an experience can be said to be a reason to change our attitude to any of the $\alpha\beta$-factors just in case we can fix the value of all the other factors and vary the one under consideration without constraint – i.e. consistently with the experience. This means that, if we fix the values of all the other factors, then for each possible value for the factor under consideration, there corresponds a probability function consistent with the experience assigning just this value.

Suppose that I learn that α. This experience is relevant to the factors $P(\alpha)$ and $P(\neg\alpha)$ as I cannot vary them consistently with the truth of α, whatever value I give to the other factors. But is it relevant to the factor $P(\beta|\alpha)$? It is not. For let it take any value $k \in [0,1]$, set $P(\alpha) = 1$ and fix the values for the other factors. Then a probability function P' is determined by Equation 10.1 as follows:

$$P'(\cdot) = P(\cdot|\alpha\beta) \cdot k + P(\cdot|\alpha\neg\beta) \cdot (1 - k)$$

And P' is such that

$$P'(\alpha) = k \cdot P(\alpha|\alpha\beta) + (1 - k) \cdot P(\alpha|\alpha\neg\beta) = 1$$

So P' is consistent with the experience, irrespective of the value of k. We can conclude that learning that α places no constraint on our conditional probability for β, given that α. And hence that it gives us no reason to change it. Since Bayesian conditionalisation is the only rule for experiences of this type (when some proposition α is learnt to be true) that preserves conditional probabilities given the learnt proposition, it follows that it is also the unique revision rule for such experiences that conforms to the requirement that no revisions should be made where none are required.[6]

10.4 JEFFREY CONDITIONING

Bayesian conditioning on the proposition that α is appropriate if this proposition represents all the evidence gleaned from experience. But people are often not aware of all that they have learnt or are unable to adequately represent it. For example, suppose I overhear a conversation in a foreign language and from the sounds of words and the mannerisms of the speakers I conclude that they are most likely, say, Spanish, but perhaps Catalan or even French. They seem to be assenting to each other's remarks by utterances of 'Si', but perhaps I am mishearing. Could the 'Si' be the French denial of a negated assertion? There is no hope here of producing a sentence that summarises all and only the facts learnt (and believed with probability one) in the encounter. Relevant evidence is not entirely indubitable, many of the cues never make it into consciousness (and perhaps cannot) and I don't have well-defined conditional degrees of belief for the speaker's language, given the bits of evidence that do.

Such cases of uncertain evidence, as we might call them, can be accommodated by the classical Bayesian conditioning model if one's new probabilities for the propositions about what one has perceived or remembered are themselves products of conditioning on some 'deeper' evidence proposition

---

[6] A more precise and detailed version of this argument can be found in Dietrich *et al.*, 2016.

of which one is certain. The logical possibility of such a proposition is somewhat beside the point, however, since in practice we rarely have any idea what this proposition might be and don't have well-defined conditional probabilities given its truth that we can use to apply the classical conditioning rule. So it's better to consider how to respond directly to probabilistic evidence.

For cases where one acquires uncertain evidence – when memory or perception may be mistaken – Richard Jeffrey (1992) offers a rule of conditioning that generalises the Bayesian one. Specifically, suppose that $\mathcal{A} = \{\alpha_i\}$ is a partition of propositions and that, as a result of interaction with the environment, the agent's probabilities for each of the α_i changes from $P(\alpha_i)$ to $Q(\alpha_i)$ (hereafter called a Jeffrey experience). Then:

Jeffrey Conditioning New degrees of belief, Q, are obtained from old degrees of belief, P, by *Jeffrey conditioning* on the partition $\mathcal{A}$ just in case, for all $\beta \in \Omega$:

$$Q(B) = \sum_i P(B|\alpha_i) \cdot Q(\alpha_i)$$

Note that Jeffrey conditioning is a generalisation of Bayesian conditioning because the latter is yielded as a special case when the probability of one element of the partition $\mathcal{A}$ goes all the way to one.

When you revise your beliefs by Jeffrey conditioning on some partition, you keep your conditional probabilities given the partition elements fixed and then calculate your new probabilities using the law of total probability – i.e. you average your old conditional probabilities with your new unconditional probabilities for the partition elements. This is illustrated in Figure 10.2, where the computation of a new probability for prospect β is broken down into two steps. In step 1, new probabilities for each cell element for the table displaying the possible conjunctions of the elements of the partition $\{\beta, \neg\beta\}$ with those of the partition $\{\alpha_1, \alpha_2, \alpha_3\}$ are calculated by multiplying the new probabilities for the α_i with the old conditional probabilities for β and $\neg\beta$ given the α_i. In step 2, a new probability for β is calculated by summing over these cell probabilities.

Adequacy As the computation reveals, Jeffrey conditioning on $\mathcal{A}$ is appropriate whenever redistribution of belief across the α_i leaves the agent's degrees of conditional belief given the partition elements unchanged – i.e. when the Rigidity condition applies to each of the α_i. This too is just a matter of probability theory. *Proof*: By the law of total probability, $Q(B) = \sum_i Q(\beta|\alpha_i) \cdot Q(\alpha_i)$. But, by Rigidity, $Q(\beta|\alpha_i) = P(\beta|\alpha_i)$. Hence $Q(\beta) = \sum_i P(\beta|\alpha_i) \cdot Q(\alpha_i)$.

When would we expect the Rigidity condition to apply to the elements of $\mathcal{A}$? Just as with classical Bayesian conditioning, the answer is: whenever all and everything gleaned from experience is represented by the redistribution of belief across it. Suppose that an agent's degrees of beliefs are represented at time t_0 by probability P and that at t_1 she undergoes some interaction with

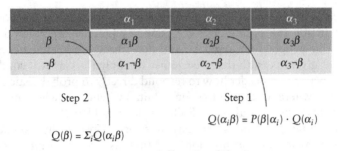

FIGURE 10.2. Jeffrey Conditioning

the environment that yields information about the partition $\mathcal{A}$ and only about this partition. Soon thereafter, at time t_2, she learns the true member of the partition. Suppose that in the interval between t_0 and t_2 she learns nothing about propositions in any other partition. Let Q represent her degrees of belief at t_1 and R her degrees of belief at t_2. But, if Bayesian conditioning is appropriate when α_i is learnt with certainty, then it is required both that $R(\cdot|\alpha_i) = P(\cdot|\alpha_i)$ and that $R(\cdot|\alpha_i) = Q(\cdot|\alpha_i)$. So it is required that $Q(\cdot|\alpha_i) = P(\cdot|\alpha_i)$ – i.e. that Q be obtained from P by Jeffrey conditioning.[7]

Objections Let us turn to consideration of some objections to Jeffrey conditioning. Its most controversial feature is the fact that it is non-commutative: the effect of Jeffrey conditioning on one constraint followed by another is not necessarily the same as the effect of Jeffrey conditioning on the second and then the first. Many commentators, including Döring (1999) and Lange (2004), have argued that the order in which evidence is received should not make any difference to what one believes on the basis of it and hence that the non-commutativity of Jeffrey conditioning is a flaw. But these commentators are mixing up experience₁ and experience₂; while the order of experience₂ can matter, that of experience₁ clearly shouldn't.

Suppose, for instance, that I expect an essay from a student. I arrive at work to find an unnamed essay in my pigeonhole with familiar writing. I am 90 per cent sure that it is from the student in question. But then I find that he left me a message the day before saying that he thinks that he may well not be able to bring me the essay in the next couple of days. In the light of all that I have learnt, I now lower to 30 per cent my probability that the essay was from him. Suppose now I got the message before the essay. The final outcome should be the same, but I will get there a different way: perhaps by my probabilities for the essay coming from him initially going to 10 per cent and then rising to 30 per cent on finding the essay. The important thing is that this reversal of the order of experience does not

---

7   This argument was first made by Skyrms (1987).

produce a reversal of the order of the probabilities: I do not think it 30 per cent likely that I will get the essay after hearing the message and then revise it to 90 per cent after checking my pigeonhole. *The same experiences have different effects on my probabilities depending on the order in which they occur.* (This is, of course, just a particular application of the rule that my posteriors depend both on the priors and the inputs.) On the other hand, when we compare a situation in which the probability of γ is revised to x_1 and then to x_2 with that in which it is revised to x_2 and then to x_1, we are generally dealing with cases where the experiences causing the probabilities to take these values must be different. So they will not be instances of order reversal of experience$_1$.[8]

A different challenge is posed by Weisberg (2009), who argues that the Rigidity condition is in tension with the thought that all learning is fallible. In particular, a sequence of experiences may initially lead one to a conclusion that subsequent experience reveals to be unsound, because the later experience undermines the basis for the inferences supporting the conclusion. Weisberg imagines a poorly lit room in which one sights a reddish-looking jelly bean, an observation that leads one to attach high probability to the proposition that (J) the jelly bean is red. Subsequently one learns that (R) the light in the room was itself red, undercutting the basis for the inference from the reddish appearance of the jelly bean to it being red. Weisberg argues that, since the first observation should not have had any effect on R, Rigidity makes it impossible to revise one's initial inference to J. But here Weisberg is quite wrong. The agent who admits the possibility of the light being red must increase her probability for it being so when she makes her first observation. In other words, the partition she should take as the basis for Jeffrey conditioning is not $\{J, \neg J\}$ but $\{J \wedge R, J \wedge \neg R, \neg J \wedge R, \neg J \wedge \neg R\}$.

This is not the end of the issue, however. As Weisberg points out, there may be any number of potential 'underminers' associated with any observation, and so the conditioning partition will have to be very fine-grained indeed if one wants to avoid problems down the line. In the context of the highly idealised models typical of work on belief revision, in which it is assumed that agents have inexhaustible computational resources, this is perhaps not a problem. But resource-bounded agents have to make some judgement about what coarse-grained partition to use as the conditioning base, running the risk that potential underminers are not represented in it. If they are subsequently surprised by evidence of the truth of an unanticipated underminer, they will not be able to condition themselves out of trouble. But this problem – the problem of unforeseen contingencies – is not unique to Jeffrey conditioning; it affects all belief revision models that take the space of possible evidences as given to the agent. We return to this problem later.

---

[8] See Wagner (2002) for an excellent discussion of this issue.

10.5 ADAMS CONDITIONING

Conditional beliefs, like the unconditional ones we have been investigating, change as a result of a number of different kinds of interaction with our environment, including observation, experimentation and testimony.[9] Some examples.

1. *Observation*: Testing for the sex ratios of different species of fish in some particular lake by means of random sampling can lead to revisions to one's probability for a fish being male, given that it is a particular kind of fish.
2. *Experimentation*: I am having trouble opening a lock on the front door, but know I have the right key. After much fiddling, I conjecture that, if I pull the door towards me while turning the key, the lock will open. After repeated trials, I am sure that I am basically right, even though it does not work on every attempt. My conditional probabilities for someone opening the door given that they follow this procedure rise near to one.
3. *Testimony*: I consult an oracle to find out whether I will succeed at my driving test tomorrow. The oracle tells me that if the examiner has a moustache then I will pass. Since I trust the oracle completely, I assign a conditional probability of one to passing given that the examiner has a moustache.

Classical Bayesians must represent such changes in conditional probabilities as consequences of learning the truth of some set of evidence propositions. This is plausible in some cases. The belief change induced by the sampling of fish to test for sex ratios of pike, for instance, might be the result of conditioning on a data proposition believed to degree one concerning the proportion of pike in the sample that was found to be male. The other cases are less amenable to treatment of this kind, however. Although I believe the conclusion I reach about how to open the door is right in virtue of certain facts about the position of the door, the state of the lock, and so on, I do not know what these facts are. There are also some facts of which I am aware, in particular those relating to my actions and their outcome, that no doubt played a role in my coming to believe that the lock will open if I pull the door while turning the key. But, though my judgement is made in light of these facts, it is not necessitated by them in any apparent way.

These examples belong to a particularly interesting set of cases in which the interaction with the environment gives us cause to change one or more of our conditional beliefs, given some possibility, without it giving us cause to change our probabilities for the possibility itself. In these cases the salient form of

---

9 Here I am summarising the discussion in Bradley (2007a), which contains a lengthier explanation of these claims.

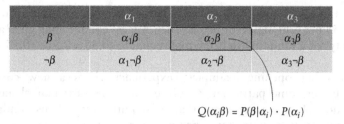

$$Q(\alpha_i\beta) = P(\beta|\alpha_i) \cdot P(\alpha_i)$$

FIGURE 10.3. Adams Conditioning

revision is Adams conditioning.[10] Let $\mathcal{A} = \{\alpha_i\}$ and $\mathcal{B} = \{\beta_j\}$ be partitions of propositions and suppose that the agent has an experience (hereafter called an Adams experience) which changes her conditional degrees of belief for the β_j given the α_i from $P(\beta_j|\alpha_i)$ to $Q(\beta_j|\alpha_i)$. Formally:

Adams Conditioning New degrees of beliefs Q are obtained from old degrees of belief P by *Adams conditioning* on the partitions $\mathcal{A}$ and $\mathcal{B}$ just in case:

$$Q(\gamma) = \sum_i \left[\sum_j P(\gamma|\alpha_i\beta_j) \cdot Q(\beta_j|\alpha_i) \right] \cdot P(\alpha_i) \qquad (10.2)$$

When Adams conditionalsing on a redistribution of conditional probabilities, the agent calculates new conditional probabilities for γ given the α_i by using the new conditional probabilities for the β_j given the α_i yielded by experience and her old conditional probabilities for γ given the $\alpha_i\beta_j$, and then averages these with her old probabilities for the α_i to get a revised probability for γ. Equivalently, as is illustrated in Figure 10.3, she can calculate new probabilities for the cells of the intersection of $\mathcal{A}$ and $\mathcal{B}$ using her prior probabilities for the α_i and the revised conditional probabilities for β and $\neg\beta$ given the α_i that were yielded by experience, and then Jeffrey conditionalising on these new probabilities for the $\alpha_i\beta_j$ as per Figure 10.2.

In canonical cases experience is relevant only to the conditional probabilities for some particular prospect β given the α_i. Then Equation 10.2 reduces to the much simpler

$$Q(\gamma) = \sum_i P(\gamma|\alpha_i\beta) \cdot Q(\beta|\alpha_i) \cdot P(\alpha_i)$$

In particular, $Q(\beta) = \sum_i Q(\beta|\alpha_i) \cdot P(\alpha_i)$. In our first example, for instance, sampling of lake fish produced new conditional probabilities for being male given that the fish is of a certain type. Suppose that the lake contains pike

[10] I introduced this term in Bradley (2005b), motivated by the affinity between this rule of belief revision and Ernst Adams' theory of conditionals. But Adams himself never held the view that what I dubbed Adams conditioning is the correct way to revise conditional belief. Douven & Romeijn (2011) use the term slightly differently to describe a principle for revising beliefs in response to learning the truth of conditional sentences.

and perch in equal quantities and that sampling yields 53 per cent males for pike and 55 per cent males for perch. Then Adams conditioning gives a revised probability for a random fish being male equal to $(0.53 \times 0.5) + (0.55 \times 0.5) = 0.54$.

In the other opening examples, experience yields a new conditional probability for some particular β only given some particular element α of the partition $\mathcal{A}$, a weaker input than is required for Adams conditioning. But it is natural to assume that, when no information is yielded regarding the conditional probability of β given the other elements of $\mathcal{A}$, that these should not change.

Example 10.1 (Judy Benjamin). *Bas van Fraassen (1981) imagines private Judy Benjamin, the protagonist of the eponymous movie, being parachuted into combat with her platoon into a territory that is divided into (R) Red and (¬R) Blue areas and by the company (H or ¬H) occupying it. They receive the radio message 'If you are in the Red area then the probability you are in H-company territory is 0.75'. Van Fraassen asks what Judy's new probabilities for being in the Red area should be. Suppose that in response to the message Judy adopts as new conditional probabilities $Q(H|R) = 0.75$ and $Q(¬H|R) = 0.25$, but does not change her conditional probabilities given ¬R (about which the message was silent). Then by Adams conditioning she obtains*

$$Q(\gamma|R) = P(\gamma|HR) \cdot Q(H|R) + P(\gamma|\neg HR) \cdot Q(H|R)$$
$$= 0.75 \cdot P(\gamma|HR) + 0.25 \cdot P(\gamma|\neg HR)$$
$$Q(\gamma|\neg R) = P(\gamma|\neg R)$$

In particular:

$$Q(R) = [0.25P(R|HR) + 0.75P(R|\neg HR)] \cdot P(R) + P(R|\neg R) \cdot P(\neg R) = P(R)$$

So Judy's new probabilities for being in the Red area are the same as her prior probabilities.[11]

What makes Adams conditioning particularly salient is the fact that, in a certain sense, it is the exact complement of Jeffrey conditioning. For in Adams conditioning it is the conditional probabilities with respect to elements of a partition that change while the probabilities of the elements themselves remain unchanged, rather than the other way round. Consequently, study of this kind of revision offers the possibility of modelling cases where interaction with the environment affects both the agent's unconditional beliefs and her conditional

---

[11]   This is what van Fraassen suggests is the intuitively correct answer. But he shows that no distance minimisation rule yields it.

ones, by representing them in terms of combinations of Jeffrey and Adams conditioning.

When is Adams conditioning the right way to change your beliefs in response to an Adams experience? When the total effect of the interaction with the environment on your conditional beliefs for the β_j given α_i is representable by a redistribution of probability over the partition $\{\alpha_i\beta_j\}$ satisfying:

1. *Independence*: $Q(\alpha_i) = P(\alpha_i)$
2. *Rigidity*: $Q(\cdot|\alpha_i\beta_j) = P(\cdot|\alpha_i\beta_j)$

So much is just a matter of probability theory.[12] But under what conditions should we expect these conditions to apply? Here one might invert the argument for classical and Jeffrey conditioning and argue that, just as changes in one's degrees of belief concerning some partition $\{\alpha_i\}$ do not in themselves give one reason to change one's conditional beliefs for prospects given the α_i, so too if what one learns from an interaction with the environment is appropriately represented by a shift in one's conditional degrees of belief for the β_i given α, and by nothing more and nothing less than these shifts, then the interaction has furnished no reason for a change in one's unconditional degrees of belief for α. Or else the effect of the interaction was not properly represented in the first place.

Suppose your prior degrees of belief are given by P, that you have a Jeffrey experience on the partition $\{AB, A\neg B, \neg A\}$ and that as a result you adopt posterior degrees of belief R. In the previous section we argued that R should be obtained from P by Jeffrey conditioning on the redistribution of probability across that partition induced by the experience. Now the redistribution can be decomposed into a redistribution of unconditional belief across the partition $\{A, \neg A\}$ and a redistribution of conditional belief given A between B and $\neg B$ – i.e. we can think of it as composed of a Jeffrey experience plus an Adams experience. Suppose these experiences are separated in time so that you have the Jeffrey experience on the partition $\{A, \neg A\}$ first, leading you to revised degrees of belief Q, and then the Adams experience taking you all the way to degrees of belief R. By the argument of the previous section, Q must be obtained from P by Jeffrey conditionalisation. It follows that $R(\neg A)$ must equal $Q(\neg A)$ and $R(\cdot|AB)$, $R(\cdot|A\neg B)$ and $R(\cdot|\neg A)$ must equal $Q(\cdot|AB)$, $Q(\cdot|A\neg B)$ and $Q(\cdot|\neg A)$. But we have just seen that these equalities characterise Adams conditioning. So if Jeffrey conditioning is the correct revision rule for Jeffrey experiences then Adams conditioning is the correct rule for Adams experiences.

---

[12] For a proof of both the necessity and sufficiency of Independence and Rigidity for Adams conditioning, see Bradley (2007a).

10.6 PREFERENCE CHANGE

Preferences, like other attitudes, can and often do change as a result of interaction with the environment – in response to observation, experimentation and verbal testimony, for instance – as well as a result of internal processes, both cognitive and biological – in response to deliberation or physical maturation, for instance. An understanding of how preferences change, or should change, as result of these processes is critical to a range of questions, both empirical and normative. Despite this, work on preference revision is rather heterogeneous. Until the recent growth of behavioural economics, it was largely a neglected question in neoclassical economics, whose models typically treated preferences as both exogenously given and stable. In philosophy the situation is not dissimilar, with little attention being given to rational preference or value revision compared to, say, the wealth of work on belief revision.[13]

Broadly speaking, there are two main types of preference change: belief-driven and desire-driven changes. The former include such variants as the following.

1. Re-evaluations of preferences over options in the light of new information about the state of the world – e.g. when you change your attitude to going to the beach when you learn that it is likely to rain.
2. Changes in attitude to an event in the light of information about someone's intentions or actions – e.g. when learning that your neighbour is planning a barbecue makes you prefer that it rains that day.
3. Adaptation of preferences for outcomes as their realisation becomes more or less probable – e.g. when you lose interest in a sporting competition when it becomes clear that you have no hope of winning it.

The first case is easily handled within classical Bayesian theory because the change in preference can be explained by a change in the expected utility of going to the beach induced by conditionalising on the new information. The second case is much more difficult to model in this framework, because this time the preferences that change are those that are directed at the state of the world rather than at actions. But, as we shall see, they can be dealt with elegantly in our model. Case 3 differs from the other two in that the belief change causes the preference change without being a reason for it: the fact that I am unlikely to win a competition is no reason to regard winning it as less valuable than winning at one in which my prospects are good. Nonetheless, such cases can be modelled in the same way as the other two, since, formally

---

[13] Important exceptions are Hansson (1995, 2001), Van Benthem & Liu (2007) and Dietrich & List (2013a,b).

speaking, a change in the probability of a prospect can serve as a driver for preference change in the same way as other kinds of reasons.[14]

Some types of desire-driven preference change are as follows.

1. Maturation of tastes with age – e.g. in types of music.
2. Conditioning or cultivation of taste by habituation or exposure – e.g. weaning infants onto cows' milk or acquiring a taste for olives.
3. Discovery of the value of things through experience – e.g. when you learn that relationships require discretion as well as honesty, or that red wine is best drunk with cheese.
4. Changing preferences for an activity as the amount of anticipated pleasure deriving from it changes with increasing skill – e.g. learning how to play the piano, or mastering the crossword puzzle.

None of these cases are reducible in any simple way to belief changes, let alone to informational changes. In habituation cases repeated experience of something leads to a re-evaluation of it (typically unconsciously) despite the fact that any informational gains are made only in the early repetitions. One can grow tired of a foodstuff, for instance, not because of anything one learns about it, but simply because of the jading of one's palate. In cases of value discovery, some kind of learning is involved, but it seems to be of a different nature to that involved in the improvement of belief. When one learns that a particular wine is a good companion to a particular cheese, one does of course learn something about the two products. But what one learns about them is how they stand in relation to one's tastes – a discovery that must give rise to an improved evaluation of the products in combination, before it gives rise to a new (and improved) belief about them. Cases of skill change seem to have aspects in common with both habituation and value learning. What is special about them, however, is that the value is created rather than learnt or acquired; it is the mastery of the skill that makes the activity pleasurable.

The rest of this chapter will be devoted to developing a model of these different kinds of changes; a model that has a very broad range of applications. What falls outside their scope are cases in which preferences are formed (or withdrawn) rather than revised. Sometimes we initially lack any preference at all between various alternatives but come to form them through inquiry or deliberation – e.g. when we develop a preference for one hotel over another by comparing the reviews on each or visiting each in turn. And sometimes entirely novel alternatives become available and we need to make a first assessment of them – e.g. new goods come onto the market, you meet someone for the first time or you are told about a restaurant that you had not heard of before. Treatment of such cases will be left to a later chapter.

---

[14] Indeed there are arguably cases in which adaptation of this kind *is* reason-based – e.g. if I realise that wishing for something that is not true makes me unhappy.

10.6.1 Generalised Conditioning

Let us start with the most general effect of experience that we will consider, namely when the initial perturbation to the agent's state of mind resulting from experience is exhaustively described by a redistribution of probability and desirability across a particular partition $\mathcal{A} = \{\alpha_i\}$ of the space of prospects. For instance, such a set of constraints might represent the outcome of a tasting of a range of wines, leading to a new set of desirabilities for their consumption, or of a debate with someone on the effects of different taxation policies, leading to new desirabilities for their implementation. Our treatment here of these cases will be quite abstract and formal, but in subsequent sections we will look at applications to more concrete situations.

To extend the framework adopted earlier in the chapter, let the agent's initial or prior state of mind be represented by a pair of probability and desirability functions $\langle P, V \rangle$, defined on a Boolean algebra of prospects $\Omega = \langle X, \models \rangle$, and her state of mind after an experience E (her posterior state of mind) by a pair of probability and desirability functions $\langle Q, W \rangle$ on the same algebra. Revision rules are just mappings from $\langle P, V \rangle$ to $\langle Q, W \rangle$ as a function of the experience E. Now suppose that as a result of experience the agent changes her degrees of belief for the $\alpha_i \in \mathcal{A}$ from $P(\alpha_i)$ to $Q(\alpha_i)$ and her degrees of desire for them from $V(\alpha_i)$ to $W(\alpha_i)$. Then we say:

Generalised Conditioning New degrees of belief and desire $\langle Q, W \rangle$ are obtained from old degrees of belief and desire $\langle P, V \rangle$ by *generalised conditioning* on $\mathcal{A}$ iff for all prospects β and for all $\alpha_i \in \mathcal{A}$ such that $P(\beta|\alpha_i) > 0$

$$Q(\beta) = \sum_i P(\beta|\alpha_i) \cdot Q(\alpha_i) \tag{10.3}$$

$$W(\beta) = \sum_i [V(\beta|\alpha_i) + W(\alpha_i)] \cdot Q(\alpha_i|\beta) \tag{10.4}$$

We can think of generalised conditioning as involving the two steps illustrated in Figure 10.4, which exhibits the intersection of example partitions $\{\beta, \neg\beta\}$ and $\{\alpha_1, \alpha_2, \alpha_3\}$. In the first step, revised probabilities and desirabilities for the cells are obtained from the prior conditional probabilities and desirabilities for β and $\neg\beta$, given the α_i, and the new probabilities and desirabilities for the α_i yielded by experience, by setting $Q(\alpha_i\beta) = P(\beta|\alpha_i) \cdot Q(\alpha_i)$ and $W(\alpha_i\beta) = V(\beta|\alpha_i) + W(\alpha_i)$. In the second step, the revised probabilities and desirabilities for β and $\neg\beta$ are calculated from the cells by applying the laws of probability and desirability.

It is relatively straightforward to establish that Q and W, as defined by Equations 10.3 and 10.4, are, respectively, a probability and desirability

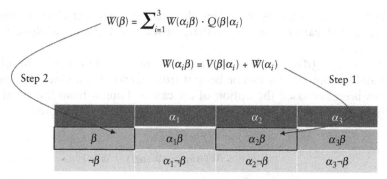

FIGURE 10.4. Generalised Conditioning

function, so that $\langle Q, W \rangle$ is indeed a rational state of mind.[15] The interesting question is whether, or under what conditions, $\langle Q, W \rangle$ is the *uniquely* rational state of mind to adopt after revision when the effect of experience is correctly described by the redistribution of probability and desirability over the partition in question. The answer is: *whenever interaction with the environment leaves the agent's conditional attitudes, given the α_i, undisturbed.* More precisely, new degrees of belief and desire $\langle Q, W \rangle$ that accord with Equations 10.3 and 10.4 are completely determined by the redistribution of probability and desirability across $\mathcal{A}$ and the requirement that for all prospects β such that $P(\beta|\alpha_i) > 0$ and all $\alpha_i \in \mathcal{A}$:

Rigidity of Conditional Desire $W(\beta|\alpha_i) = V(\beta|\alpha_i)$

So much is just a matter of the rules of probability and desirability.[16] Again the important question is: under what conditions will Rigidity of Conditional Desire hold? A Bayesian should say: whenever the redistribution of probability and desirability over the α_i describes all and everything that is learnt by the agent as a result of interaction with the environment and all changes to the agent's partial attitudes are rational effects of what she learns. This claim can be buttressed by showing that an agent who violates Rigidity is vulnerable to a money pump.

A Money Pump Argument for Rigidity Consider firstly the limiting case in which experience teaches the agent that α. For the purposes of the exercise let us suppose that the option of making prospects true can be bought and sold in some market so that, in an appropriate currency, $V(\beta)$ and $W(\beta)$ give the fair prices for the agent, before and after learning that α, of the option to make true the prospect β. (A fair price is the one at which an agent is prepared to

[15] Proof in Bradley (1999).
[16] Proof in (Bradley, 2007a, Lemma 22).

both buy and sell.) Suppose firstly that the agent commits herself to a revision policy in case of learning that α such that, for some β, $V(\beta|\alpha) \neq W(\beta|\alpha)$. There are two cases.

(i) $V(\beta|\alpha) > W(\beta|\alpha)$. In this case the agent can be sold the option of $\alpha\beta$ for $V(\alpha\beta)$ and the option of α can be bought from her for $V(\alpha)$. Once the option of α has been exercised the option of $\alpha\beta$ can be bought from the agent for $W(\alpha\beta)$. So the agent has spent $V(\alpha\beta) - V(\alpha) - W(\alpha\beta)$. By assumption

$$V(\alpha\beta) - V(\alpha) = V(\beta|\alpha)$$
$$> W(\beta|\alpha) = W(\alpha\beta) - W(\alpha) \quad \bullet$$

But $W(\alpha) = 0$ since the fair price for α after learning that α is 0. So $V(\alpha\beta) - V(\alpha) > W(\beta|\alpha)$. Hence $V(\alpha\beta) - V(\alpha) - W(\alpha\beta) > 0$ – i.e. the agent makes a loss.

(ii) $W(\beta|\alpha) > V(\beta|\alpha)$. In this case the option of $\alpha\beta$ can be bought from the agent for $V(\alpha\beta)$ and the option of α sold for $V(\alpha)$. Once the option of α has been exercised the option of $\alpha\beta$ can sold back to the agent for $W(\alpha\beta)$. So in this case the agent has spent $V(\alpha) - V(\alpha\beta) + W(\alpha\beta)$. By assumption $W(\alpha\beta) = W(\beta|\alpha) > V(\beta|\alpha) = V(\alpha\beta) - V(\alpha)$. Hence $V(\alpha) - V(\alpha\beta) + W(\alpha\beta) > 0$. So in this case too she makes a loss. It follows that an agent who commits to a revision policy for cases of learning that α that fails to satisfy Rigidity of Conditional Desire will be vulnerable to a sure loss.

To extend this argument to the more general case of a revision policy for any redistribution of probability over a partition $\{\alpha_i\}$ consider a two-stage revision process. At the first stage, interaction with the environment induces the agent to adopt new probabilities for the elements of the partition $\{\alpha_i\}$, without the probability of any one of them going to one. In the second stage the agent learns which of the α_i is the truth. Suppose that this process leads to a transformation of her state of mind from $\langle P, V \rangle$ to $\langle Q, W \rangle$ and then to $\langle R, U \rangle$. By our previous argument for Rigidity in the context of learning the truth of a particular prospect, $U(\cdot|\alpha_i) = V(\cdot|\alpha_i)$ and $W(\cdot|\alpha_i) = V(\cdot|\alpha_i)$, since both the revisions from $\langle P, V \rangle$ to $\langle R, U \rangle$ and that from $\langle P, V \rangle$ to $\langle Q, W \rangle$ fall under its scope. It follows that $W(\cdot|\alpha_i) = V(\cdot|\alpha_i)$ for any of the α_i and hence that Rigidity of Conditional Desire holds for pure probabilistic shifts as well.

The Scope of Rigidity Money pump arguments, like their close relatives the Dutch Book arguments for probabilistic constraints, show that failure to satisfy some condition or other renders the agent vulnerable to exploitation. It does not follow without further argument that rigidity of conditional attitudes is a general requirement of rationality. After all, one can render oneself invulnerable to money pumps by simply not declaring a revision policy. Indeed, this would seem to be a sensible precaution, since there are cases in which one's attitudes may change as a result of interaction with the environment but not (entirely) because of the information that one acquires during it, simply because the manner in which something is learnt has some non-rational effect on one's attitudes. If, for instance, one learns of the consequences of excessive alcohol consumption by doing the drinking oneself, or of the presence of a poisonous snake in the house by standing on it, there is every possibility that other attitudes will be altered in the process and in a manner not representable as conditioning on what has been learnt. Unless the manner in which information is acquired can be controlled somehow (as perhaps it is in scientific experiments), it would be unwise to commit oneself to a revision policy in the manner required by the money pump argument. The money pump argument is not therefore conclusive.

On the other hand, should an agent's conditional desires change, either directly as a result of experience or by inference from it, the net effect on the agent's attitudes can be expressed by a set of constraints on some more refined partition. For instance, if she reasons as above, so that not only her degrees of desire for α change but also her conditional degrees of desire for β given α, then she could adopt as her posterior constraint a redistribution of probability and desirability over the partition $\{\alpha\beta, \alpha\neg\beta, \neg\alpha\}$ rather than the initial partition $\{\alpha, \neg\alpha\}$. Then we can ask whether Rigidity of Conditional Desire is satisfied relative to this more refined partition. Crucially, there will always be some level of refinement at which Rigidity of Conditional Desire will be satisfied. More exactly:

Theorem 10.2. *(Bradley, 2009) Assume that Ω is countable. Let $\langle P, V \rangle$ and $\langle Q, W \rangle$ be, respectively, an agent's prior and posterior states of mind. Then there exists some partition of Ω such that $\langle Q, W \rangle$ is obtained from $\langle P, V \rangle$ by generalised conditioning on this partition.*

Theorem 10.2 tells us that any revision of a state of mind is representable as an instance of generalised conditioning as long as the revision produces a new state of mind that is internally consistent. It does not follow, of course, that agents, or even just the rational ones, always revise by this method, or indeed that they should. But we can conclude they always revise 'as if' by generalised conditioning on a particular partition. Furthermore, from the agents' point of view, if they can adequately capture the immediate import (for them) of their experience in the form of a revision of their attitudes to the elements of some particular partition of prospects, then generalised conditioning is the uniquely rational method for propagating the implications of this experience to the rest of their attitudes.

Foundations Let's finish by considering what the characteristic features are of preference change through generalised conditioning. To do so, let us say that preference change as a result of some experience is *rigid*, relative to a partition $\mathcal{A}$ of the space of prospects, just in case the agent's conditional preferences, given the elements of $\mathcal{A}$, do not change as a result of the experience. Let $\succsim$ be the agent's initial or prior preferences over a Boolean algebra of prospects $\Omega = \langle X, \models \rangle$ and let $\succsim^*$ be her posterior or revised preferences after the experience. Now the defining condition for rigid preference change is the following:

Invariance of Conditional Preference For all $\alpha_i \in \mathcal{A}$ and for all $\beta, \gamma \in X$ such
that $\gamma\alpha_i \neq \perp \neq \beta\alpha_i$:

$$\beta \succsim_{\alpha_i} \gamma \Leftrightarrow \beta \succsim^*_{\alpha_i} \gamma$$

When an agent's preferences are rigid relative to $\mathcal{A}$, in the sense of satisfying
the Invariance of Conditional Preference condition with respect to it, she can
be represented as revising her attitudes in the face of experience by generalised
conditioning on $\mathcal{A}$. More precisely:

Representation of Generalised Conditioning (Bradley, 2007a, theorem 13)
Let $\succsim$ and $\succsim^*$ be a pair of continuous preference relations that satisfy
Invariance of Conditional Preference with respect to the partition $\mathcal{A} =$
$\{\alpha_i\}$. If the pairs $\langle P, V \rangle$ and $\langle Q, W \rangle$, respectively, represent $\succsim$ and $\succsim^*$ then
$\langle Q, W \rangle$ is obtained from $\langle P, V \rangle$ by generalised conditioning on $\mathcal{A}$.

This theorem has both normative and methodological significance. Normatively the lesson is that an agent who regards her experience as irrelevant to
her conditional preferences given the α_i must revise her attitudes in response
to this experience by generalised conditioning on $\mathcal{A}$. Methodologically, that
satisfaction of the Invariance condition with respect to a particular partition
serves as a criterion for when the effects of experience on an agent can
be localised to that partition – i.e. for when the significance of experience,
according to the agent, is exhausted by its implications for its elements.

10.6.2 Belief-Driven Preference Change

Generalised conditioning is demonstrably rational whenever the Rigidity
condition holds. But in order to revise one's preference in this way it is
necessary to start with a rather rich input, namely an exhaustive specification
of the effects of experience on one's distribution of probability and desirability
across a particular partition. In this section we shall attempt to go a bit
further than this and consider how a rational agent should revise her attitudes
when the conditioning base is less rich. The two most salient cases are when
experience gives the agent immediate cause to revise her degrees of desires,
but not her beliefs, and when it gives her cause to revise her beliefs, but
not her desires. In these cases the basis for conditioning takes the form of a
redistribution of either probability or desirability (but not both) across some
partition of the space of prospects. We will look at the latter in this section
and the former in the next.

Intuitively, there are two kinds of effects of belief change that are especially
relevant to preference. The first kind is the effect on the instrumental value of
some prospect of a change in the conditional probability, given its realisation,
of prospects that matter to the agent. Thus, if I learn that drinking red wine,
but not white, reduces the chances of a heart attack, I may as a result come to
prefer drinking red wine to white.

The second kind of belief change relevant to preference is when a change in the probability of some possibility α makes the prospect of β more attractive, not because of any probabilistic dependence between the two but because of the desirabilistic dependence of β on α. For example, if I have planned to take my children to the park if I can get away from work early enough, then learning that no rain is forecast for later in the day will make the prospect of getting off work early more attractive. This is not because the forecast affects the likelihood of getting off work, but because I prefer not to go to the park in the rain.

To analyse these effects of belief change on preference we once again take the rigidity conditions as our compass for directing minimal revision. But a further principle is required in order to determine the effect of belief changes on desires. In general, a change in belief concerning some prospect should and will affect the desirability of other prospects. But a change in the degree to which one believes that β does not in itself give one reason to change one's attitude to the desirability of β itself. And, more generally, changes in the relative probability of the elements of some partition of prospects do not rationalise changes in their relative desirability. For instance, suppose that I strongly prefer to teach an advanced course in decision theory than an introductory course in logic. Then getting wind of the Head of Department's intentions regarding teaching allocations should not make any difference to the degree to which my preference for the former alternative exceeds the latter. To be sure, changes in belief regarding some prospects can *cause* a change in its desirability without being a reason for it. This is what happens in cases of adaptive preference change (such as that illustrated by Case 3 in our list of preference change types). Such adaptations are ruled out by the following condition.

Local Independence of Preference from Belief   A change in an agent's degrees of belief over a partition $\{\alpha_i\}$ should have no effect on her preferences for the α_i – i.e.

$$\alpha_i \succsim^* \alpha_j \Leftrightarrow \alpha_i \succsim \alpha_j .$$

We now examine the implications of this principle in the context of the various types of belief change examined in the previous chapter.

New Information   Suppose that as a result of some such interaction with the environment an agent learns that α, and nothing more than that α, so that the initial effect of this interaction is exhaustively described by the constraint on her posterior state of mind that $Q(\alpha) = 1$. In these circumstances, classical Bayesian conditioning requires that $Q(\cdot) = P(\cdot|\alpha)$. But how should the agent revise her desires? In this case, Rigidity of Conditional Desire alone implies that the agent's new degrees of desire should equal her prior conditional degrees of desire given that α – i.e. that

$$W(\cdot) = V(\cdot|\alpha) \tag{10.5}$$

Proof: $W(\beta) = W(\alpha\beta) \cdot Q(\alpha|\beta) + W(\neg\alpha\beta) \cdot Q(\neg\alpha|\beta) = W(\alpha\beta)$, since $Q(\alpha|\beta) = 1$. In particular, $W(\top) = W(\alpha\top) = 0$. Hence $W(\alpha\beta) = W(\beta|\alpha) + W(\alpha) = V(\beta|\alpha)$ in virtue of Rigidity of Conditional Desire.

It follows from Equation 10.5, that whenever an opinionated agent revises her beliefs and desires in this manner – by conditioning on the information that α – her new preferences between any prospects β and γ will go by her old preferences between the prospects $\alpha\beta$ and $\alpha\gamma$ – i.e. that

$$\beta \succsim^* \gamma \Leftrightarrow \alpha\beta \succsim \alpha\gamma$$

This is just what we would expect whenever learning that α is the case has a purely informational effect on the agent's preferences. For example, in Case 2 of our initial list of causes of preference change, my prior preference for my neighbour holding a barbecue in rainy weather to his holding it in sunny weather is what explains my acquisition of a preference for rainy over sunny weather upon receipt of the information that he has planned a barbecue.

New Probabilities Suppose now that as a result of experience the agent redistributes her degrees of belief across some partition $\mathcal{A} = \{\alpha_i\}$ of the algebra of prospects, a circumstance in which Jeffrey conditioning is the appropriate form of rigid belief change. But now the Rigidity condition alone does not determine new desirabilities for all prospects since it does not determine new desirabilities for the elements of the partition $\mathcal{A}$. The Local Independence condition in effect restricts any changes to desirabilities of the α_i, however, following the change in their probabilities, to those that are mere renormalisations of the prior values.

What minimal renormalisation is required by the axioms of desirability? It is the following. As before, let $k = \sum_i V(\alpha_i) \cdot Q(\alpha_i)$ measure the overall desirability gain to the agent as result of the change in probabilities over $\mathcal{A}$ (informally, we can say that it expresses the amount by which the world has been revealed, by the experience inducing the belief change, to be a better or worse place than initially believed). Then set

$$Q(\alpha_i) = V(\alpha_i) - k \qquad (10.6)$$

It is obvious that the Local Independence condition is satisfied by this renormalisation. And a small amount of algebra will confirm that the axioms of normality and desirability are too.[17] Then Rigidity of Conditional Desire allows application of the equations of generalised conditioning to yield

$$W(\beta) = \sum_i V(\alpha_i\beta) \cdot Q(\alpha_i|\beta) - k \qquad (10.7)$$

In other words, up to renormalisation by k, new desirabilities are obtained by averaging the old cell desirabilities for the $\alpha_i\beta$ by the new probabilities

---

[17] See Bradley (2004) for details.

induced by Jeffrey conditionalisation on A. (The normalisation factor drops out in preference comparisons.) For example, when I acquire new probabilities for rain, the desirability of going to the park is obtained by weighting the old desirabilities for going to the park in the rain and going in the sunshine with the new probabilities for rain and sunshine.

New Conditional Probabilities Suppose that as a result of experience the agent adopts new conditional probabilities for the elements of a partition $B = \{\beta_j\}$ given the elements of partition A, a circumstance in which Adams conditioning is the appropriate form of minimal belief change. As we observed before, Adams conditioning can be thought of as Jeffrey conditioning on the joint partition $A \times B = \{\alpha_i \beta_j\}$ subject to the additional constraint that the probabilities of the α_i do not change. Given this, and the assumption that the Rigidity condition applies to $A \times B$, it will follow by application of Equation 10.7 that the agent's new degrees of desire for any prospect γ should be obtained from her old by averaging her old degrees of desire for the $\gamma \alpha_i \beta_j$ by her new conditional degrees of belief for the $\alpha_i \beta_j$ given γ (and renormalising). More formally:

$$W(\gamma) = \sum_i \left[\sum_j V(\gamma \alpha_i \beta_j) \cdot Q(\alpha_i \beta_j | \gamma) \right] - k \qquad (10.8)$$

As in the earlier wine example, we are often interested in the relevance of the change in the conditional probabilities of the β_j given some particular α to the desirability of α itself. In this case, Equation 10.8 reduces to

$$W(\alpha) = \sum_j V(\alpha \beta_j) \cdot Q(\beta_j | \alpha) - k$$

In words: new desirabilities are obtained by reweighting old ones by the new conditional probabilities yielded by experience. For instance, to assess the new desirability of drinking red wine, we should weight our old desirabilities for red wine with and without a heart attack with the new conditional probabilities for heart attack or not, given red wine consumption. And this will differ from the prior desirability of red wine drinking just to the extent that these new conditional probabilities differ from the old.

10.6.3 Desire-Driven Preference Change

Intuitively, there are two kinds of effects of desire changes that are relevant to preference revision. The first kind is the effect on the desirability of some prospect of a change in the desirability of its possible consequences. For instance, if an agent's taste in music 'matures' over time, so that her high regard for rock music and low regard for classical music is replaced by a low regard for the former and a greater appreciation of the latter, then her earlier preference for a night in a club over an evening at a concert hall is likely to be reversed at some point in time. The second is the effect of a change in the

conditional desirabilities of a set of alternative prospects, given the presence
of some condition, on the desirability of the condition itself. For instance,
discovering that strawberries taste even better if eaten with cream may lead
one to value cream more highly and to purchase it more often.

To derive these intuitive conclusions concerning the effect on an agent's
preferences of desire or 'taste' changes from the model of generalised
conditioning, we need to draw on another principle concerning the effect of
desire change on belief: that agents should not engage in wishful thinking.
A person who engages in wishful thinking allows her judgements about how
probable certain prospects are influenced by how desirable or undesirable she
finds them. When a jilted lover convinces himself that his estranged partner
will realise the error of her ways and return to him, or a driver thinks she is
far less likely to have an accident on the road then the statistics suggest, they
are engaging in wishful thinking. So too the famous fox of Aesop's fable, who
lost his taste for grapes he could not reach. Wishful thinking, though common,
is regarded by many as epistemically irrational (see, for instance, Elster, 1985
and Binmore, 2008). The following principle guards against it.

Local Independence of Belief from Desire   A change in an agent's degrees of
   desire over a partition $\{\alpha_i\}$ should have no effect on her degrees of belief
   for the α_i

Local Independence of Belief from Desire prohibits changes in an agent's
degrees of desires for some prospect from affecting her degrees of beliefs
regarding it. As it stands, the condition is too strong to apply with complete
generality. If the desirability of some prospect increases, I might well infer that
I will try and secure its realisation. This will make the probability of my taking
certain actions greater, for if I am rational then my actions will be guided by
the expected desirability of their consequences. So I should violate the Local
Independence condition.

There are two ways one might deal with this problem. One is to qualify the
condition so as to exclude cases such as these; this will be straightforward if
exceptions are restricted to specific categories. A second is to break revision
into two steps: a first stage in which Local Independence is applied to revision
in response to taste change and a second stage in which beliefs are revised in
response to the changes in the expected desirabilities of actions induced by the
first-stage revisions. This second option is the one that I would advocate.

Suppose that as a result of experience the agent redistributes her degrees of
desire across some partition $\mathcal{A} = \{\alpha_i\}$ of the algebra of prospects. The effect
of this redistribution on the instrumental value of any prospect β will depend
on how probable each of the α_i are given β. To measure this effect, let $l_\beta :=$
$[W(\alpha_i) - V(\alpha_i)] \cdot P(\alpha_i|\beta)$ be the sum of the changes in the desirabilities in the
α_i weighted by their probabilistic dependence on β. Then Local Independence
of Belief from Desire implies that rigid attitude change will respect, for any

prospect β

$$W(\beta) = \sum_i [V(\beta|\alpha_i) + W(\alpha_i)] \cdot P(\alpha_i|\beta)$$

$$= \sum_i [V(\beta\alpha_i) - V(\alpha_i) + W(\alpha_i)] \cdot P(\alpha_i|\beta)$$

$$= V(\beta) + l_\beta$$

The agent's new preferences will differ from her old, in virtue of such a change of taste, as a function of the magnitude of these gains and losses. In particular, her new preferences over prospects can be derived from her old plus the magnitudes of desirability transmitted, in virtue of the fact that

$$\beta \succeq {}^*\gamma$$
$$\Leftrightarrow W(\beta) \geq W(\gamma)$$
$$\Leftrightarrow V(\beta) + l_\beta \geq V(\gamma) + l_\gamma$$
$$\Leftrightarrow V(\beta) - V(\gamma) \geq l_\gamma - l_\beta$$

It follows that a preference reversal between β and γ will occur just in case the difference in the magnitudes of the desirability gains to each as a result of the taste change is greater than the prior difference in their desirability. For instance, to pick up our previous example, if the α_i refer to listening to different types of music and β and γ to going to classical concert and a rock club, respectively, then an early preference for the latter over the former will be reversed just in case the sum of the desirability gains associated with listening to classical music and the desirability loss associated with listening to rock exceeds the initial desirability difference between the evening at the concert hall and the night at the club.

PART IV

RATIONALITY WITHIN BOUNDS

11

Imprecise Bayesianism

11.1 INTRODUCTION

The second and third parts of the book developed a model of rational decision making based on a number of idealisations. Firstly, that agents' attitudes – their beliefs, desires and preferences – are *consistent* both in themselves and with respect to one another. Secondly, that agents are *logically omniscient*, in that they believe all logical truths and endorse all the logical consequences of their attitudes. Thirdly, that they are *fully aware* of all relevant prospects, including what options they have, what the possible states of the world are and what the potential consequences of exercising their options are. And, lastly, that agents are *maximally opinionated*, in that they have determinate belief, desire and preference attitudes to all prospects under consideration.

These assumptions had a purpose: they simplified the discussion enough to allow the presentation of a comprehensive theory of rationality. There is no doubt, however, that they are unrealistic. All of us violate some of them most of the time and most of them some of the time. Anyone interested in a descriptively accurate account of human reasoning and decision making will therefore want to dispense with them. And indeed there has been a great deal of work in this area over the last few decades, broadly under the banners of bounded rationality and behavioural economics, which has emphasised the extent to which our thinking and choosing are affected by factors such as context, framing, emotions and biases.[1]

Normative theories of rationality also need to dispense with these assumptions; not in order to become more descriptively accurate (that is not their goal) but in order to provide prescriptions relevant to real agents. The aim is not to abandon full rationality for a bounded variant that more accurately

[1] See Samson (2014) for a useful introduction to this literature.

corresponds to the way human decision makers are, but to characterise (full) rationality for agents who are bounded in the sense that they are not maximally opinionated, logically omniscient, fully aware or even completely consistent. So each of these assumptions will need to be relaxed to some degree, both in order to examine what modifications are required to the theory of rationality developed so far and in order to say something about the rational way to behave when you know yourself to be bounded in these ways.

While all four assumptions are descriptively unrealistic, their normative statuses are rather different. Consistency has the strongest case for being a core rationality requirement and something that agents should aim for. At the other end of the spectrum, the case for maximal opinionation is very weak; indeed, it can be argued that in many circumstances we should *not* be opinionated on everything. The same can be said for full awareness: while it may be desirable that agents are aware of all contingencies, it cannot reasonably be said that it is irrational of them not to be so.[2] Logical omniscience has some claim to be a requirement of rationality, but it is weaker than that of consistency because it is more demanding.

In the next couple of chapters I will extend our treatment of rationality, first to agents who are not fully opinionated and then to those who are not fully aware or logically omniscient. This is not just a matter of retreating from some of the very strong rationality claims made in previous chapters (though there will be some of this); it also involves enriching the account developed in them, in particular regarding attitudinal change. In essence, the focus of the account of rationality will shift from the description of an ideal judgemental state to a description of the process by which agents can manage their limitations by seeking to improve their judgements.

This chapter will examine a model of non-opinionated rationality called Imprecise Bayesianism, offering foundations for it in a theory of incomplete preference and raising several challenges to it. In subsequent chapters I turn to the treatment of unawareness and to the question of how less than fully opinionated agents should change their minds and how they might make decisions. First, however, I will give an informal sketch of the model of bounded rationality I will be defending in these next chapters.

11.2 BOUNDED MINDS

11.2.1 Unawareness and Incompleteness

Let us start out by making use of the simple qualitative models of categorical belief introduced earlier, in which an agent's state of belief is given by a set of sentences or propositions, understood to be the propositions that they

---

[2] One may of course learn that one was uaware of something by becoming aware it. But there are no checks for unawareness.

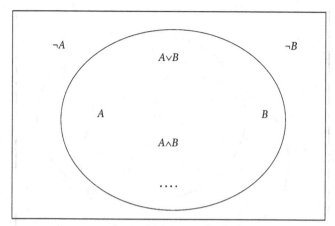

FIGURE 11.1. Opinionated Belief State

believe true. In such models, the states of minds of fully opinionated agents are typically represented by *maximal* sets of propositions – i.e. by sets containing, for every proposition X in the domain of belief, either X or $\neg X$. Such a model is partially illustrated in Figure 11.1 , where the propositions inside the oval are those believed true by the agent and the ones outside it those that are believed false. Only partially, because the figure doesn't display all the propositions that the agent believes (or doesn't believe), just a relevant selection of them. Note that, in contrast to standard practice in belief revision theory, there is no assumption that the maximal sets are either logically closed or consistent (for reasons that will become clear below).

Our task is to extend such modelling to other kinds of cases. It is usual to represent non-opinionated states of mind by sets of propositions that fall short of maximality. But this will not do for present purposes, because it doesn't allow for a distinction between not having an opinion on some proposition and being unaware of it. So here a non-opinionated belief state will be represented by a *set* of fully opinionated belief states – i.e. by a set of maximal sets of propositions – where, intuitively, each set represents a state of full opinionation that the agent regards as permissible given her current beliefs. In contrast, the state of mind of an agent who is unaware of certain propositions (but has opinions regarding those she is aware of) will be represented by a single, but non-maximal, set of propositions.

In the case of less than full opinionation illustrated by Figure 11.2, the agent believes that A, B and $A \rightarrow B$ but has no opinion on C. Consequently, her belief state consists of two intersecting sets of propositions corresponding to the two states of full opinionation that she regards as permissible, one containing C and the other $\neg C$, and both containing A, B and $A \rightarrow B$. In contrast, in the case illustrated by Figure 11.3, the agent is simply unaware of C, rather than

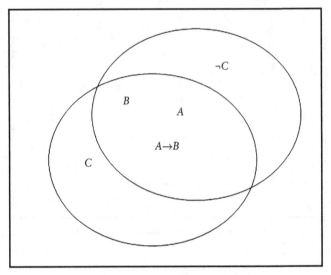

FIGURE 11.2. Non-Opinionation

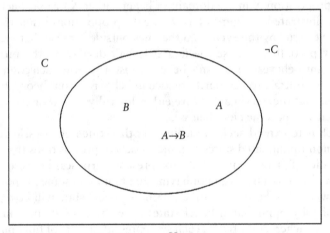

FIGURE 11.3. Unawareness

having no opinion on it, but is fully opinionated on *A*, *B* and that *A* → *B*. Consequently, her belief state is represented by a set of propositions that contains neither *C* nor its negation.

The thought underlying this treatment of the distinction between non-opinionation and unawareness is that an agent who has not made up her mind about some prospect of which she is aware is in a state in which she regards as permissible a variety of opinions about it, in the sense that she is not ready to exclude them as opinions she could hold. If I have not reached a judgement on whether it will rain today, both the opinion that it will do so and the opinion that it will not are open to me. My state of mind is thus akin to that of a group made of individuals holding a variety of opinions and in which it has not been settled as to which should be adopted by the group as a whole. (I will call these 'members' of the agent her *avatars*.) In contrast, an agent who is unaware of a proposition regards no opinion on it as either permissible nor impermissible.

Different kinds of transitions between these three belief states are possible. An agent may initially be unaware of the possibility of a hail storm, have her attention drawn to this possibility by a friend and then go and read a weather report in order to form an opinion as to whether one will occur or not. In doing so, the agent moves from the kind of belief state represented by Figure 11.3, through the kind represented by Figure 11.2 and ends up in the kind represented by Figure 11.1. The trajectory is reversed in the case of someone who initially believes in ghosts, suspends opinion on the matter after acquiring evidence that undermines his beliefs and then forgets all about the matter. Such transitions are the topic of the next chapter.

11.2.2 Types of Inconsistency

Non-opinionation and unawareness are conditions of agents that reflect their cognitive and informational limitations. To be bounded in either of these ways is not to be irrational. But there are normatively problematic cases in which an agent fails to hold an opinion on a proposition. In cases of failure of *logical omniscience*, for instance, the agent lacks the opinion on a proposition that is logically implied by other opinions that she holds and so fails to hold an opinion that, in some sense, she should. In cases of *non-extendability*, on the other hand, the agent cannot form an opinion without running into inconsistency.

To make these notions more precise, let us compare them to some familiar failures of rationality. An agent is *strictly inconsistent* when her belief set contains both a proposition and its negation. Strict inconsistency can and should be avoided. An agent who lacks logical omniscience is not (thereby) guilty of strict inconsistency, but there is nonetheless something normatively problematic about her state of mind. What this is can be brought out by observing that she fails to regard as impermissible certain strictly inconsistent belief states – i.e. she has at least one inconsistent avatar. So she tolerates inconsistency even if she does not embrace it.

More problematic than failures of logical omniscience are cases of what will be called *implicit inconsistency*.[3] An implicitly inconsistent agent is one who cannot satisfy logical omniscience without violating strict consistency. So an

---

3   I take this term from Gold & List (2004), where this property is applied to majority decisions.

agent is implicitly inconsistent when she believes a proposition whose negation is a logical consequence of other propositions she believes.

Stating these three conditions more formally will help clarify the connection between them. Let Ω be a background set of propositions that is closed under negation and potentially containing conditional propositions of the form $A \rightarrow B$. The avatars of the agent are just maximal subsets of Ω and her belief set $\mathcal{B}$ is the intersection of them: the subset of Ω containing all and only the propositions she believes. Let $Cn(\mathcal{B})$ be the closure of her belief set under a consequence operation Cn. A consequence operation is just a function that assigns a set of propositions to every set of propositions, intuitively the former being the propositions that are valid consequences of the latter. For our purposes, we don't need to say much about the properties of the consequence operation, but let's take it to be reflexive, idempotent and monotonic and to satisfy Modus Ponens (i.e. $A \in Cn(\mathcal{B})$ and $A \rightarrow B \in Cn(\mathcal{B})$ implies that $B \in Cn(\mathcal{B})$).[4] Now we say that the belief set $\mathcal{B}$ of the agent is

- *strictly inconsistent* iff there exists some proposition $A \in \Omega$ such that $A \in \mathcal{B}$ and $\neg A \in \mathcal{B}$
- *logically non-omniscient* iff there exists some proposition $A \in Cn(\mathcal{B})$ such that $A \notin \mathcal{B}$
- *implicitly inconsistent* iff there exists some proposition A such that $A \in Cn(\mathcal{B})$ and $\neg A \in Cn(\mathcal{B})$

For example, consider someone who believes both that A and that if A then B. If she does not believe that B then her failure is one of a lack of logical omniscience. On the other hand, if she believes that $\neg B$ then her failure is one of implicit inconsistency. Finally, if she believes both B and $\neg B$ then she is strictly inconsistent. These three cases are illustrated in Figure 11.4.

Since $\mathcal{B} \subseteq Cn(\mathcal{B})$ an agent who is inconsistent must also be implicitly inconsistent and an agent who is implicitly inconsistent, but not strictly inconsistent, must fail to be logically omniscient. Implicit inconsistency is compatible with full opinionation, however. Indeed, someone who is fully opinionated, but not logically omniscient, is forcibly implicitly inconsistent, because the latter implies there must exist some proposition $A \in Cn(\mathcal{B})$ such that $A \notin \mathcal{B}$ and the former that if $A \notin \mathcal{B}$ then $\neg A \in \mathcal{B}$.

11.2.3 Coherent Extendability

A second type of failure that is associated with incomplete states of mind is that of non-extendability. An agent's opinions are *coherently extendable* when she can form an opinion on every proposition (with regard to which she previously held no opinion) without finding herself with an inconsistent set

---

4   A relation R on a set $\{\alpha, \beta, ...\}$ is reflexive iff $\alpha \in R(\alpha)$, idempotent iff $\beta \in R(\alpha) \Leftrightarrow \beta \in R(R(\alpha))$ and monotonic iff $\alpha \subseteq \beta \Rightarrow R(\alpha) \subseteq R(\beta)$.

Inconsistency   Implicit inconsistency

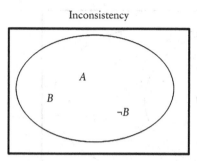
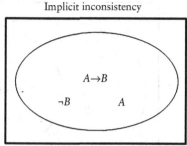

Non-omniscience

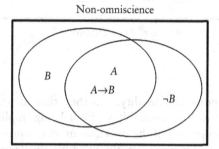

FIGURE 11.4. Three Failures of Belief

of judgements. Correspondingly, an agent's judgements are non-extendable when she cannot develop an opinion on some proposition without inducing implicit inconsistency; hence cannot form an opinion and draw the logical consequences of so doing without finding herself with strictly inconsistent opinions.

More precisely, let C be a completion of B iff C is a maximal subset of Ω such that if $A \in B$ then $A \in C$. Then an agent's beliefs or opinions are coherently extendable iff there exists a strictly consistent completion of her belief set, and non-extendable otherwise – i.e. the belief set B of the agent is

- *non-extendable* iff if C every completion of B is strictly inconsistent

For instance, in the case illustrated in Figure 11.5, the agent does not have strictly inconsistent beliefs. But she lacks an opinion as to whether A is true or not, and were she to acquire such an opinion she would be required by Modus Ponens to infer that B, in contradiction with her belief that $\neg B$. So her judgements are non-extendable.

While implicit inconsistency means that the agent cannot satisfy logical omniscience without violating strict inconsistency, non-extendability means that the agent cannot satisfy *completeness* without implicit inconsistency. So these notions are conceptually distinct. They are nonetheless closely related. When the consequence operation is monotonic, as we assumed, implicit

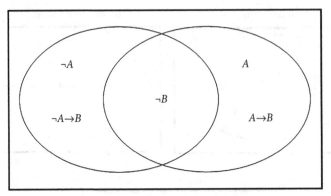

FIGURE 11.5. Non-Extendability

inconsistency implies non-extendability. On the other hand, whether or not a set of beliefs can fail to be extendable without being implicitly inconsistent depends on both the properties of the background consequence operation and how rich the set Ω is. For instance, in the example displayed in Figure 11.5, each avatar of the agent is implicitly inconsistent, but the intersection of them is not. So the agent's beliefs are not coherently extendable even though they are not implicitly inconsistent. On the other hand, when Ω is a full Boolean algebra and the consequence operation includes classical logic then the two notions coincide, in the sense that a belief set will be coherently extendable iff it is implicitly consistent.

This leaves us with a hierarchy of failures ranging from strict inconsistency through implicit inconsistency to non-extendability and failure of logical omniscience. Where to draw the line between what rationality requires and what it does not is bound to be controversial. My own view is that coherent extendability is a requirement of rationality, but that logical omniscience is not. Logical omniscience guarantees extendability; useful, no doubt, but more than is strictly required to stay out of trouble. But I grant that this view is likely to be contentious. On the other hand, neither completeness nor full awareness are plausibly general requirements of rationality, nor is inconsistency plausibly not.

Each type of 'failure' requires a different response from the agent. The inconsistent agent must develop new avatars by withdrawing her opinion on some propositions; the non-omniscient one must eliminate strictly inconsistent avatars by forming an opinion through deliberation; and so on. To model these responses, as well as the transitions between states of unawareness/awareness and non-opinionation/opinionation, we need a more general theory of attitude change than the one developed in Chapter 10. This will be the task of the next chapter. For now we focus on non-opinionation in quantitative models of belief and desire, and in particular on its relation to the requirement of coherent extendability.

11.3 IMPRECISE BAYESIANISM

From a descriptive and a normative point of view, perhaps the most dubious of the characteristics attributed to rational agents in the first parts of the book is that of maximal opinionation. Real people often do not, cannot or will not reach judgements about some prospects. Furthermore, there is little plausibility to the claim that rationality requires them to do so. When we lack the information or the skills required to reach a satisfactory judgement or find ourselves torn between conflicting considerations, it is not irrational for us to suspend or defer judgement.

In the rest of this chapter and the ones that follow, I will ask what the requirements of rationality are on a less than fully opinionated agent. This will involve addressing the same three questions considered before: what are the properties of a rational state of mind? How should a rational agent change her mind in response to experience? How should a rational agent decide what to do? For each I will give an answer that involves a modification – sometimes minor, sometimes not – of the one tailored to maximally opinionated agents, but which is nonetheless continuous with it.

The main aim of this chapter is to spell out an answer to the first question, which will be termed Imprecise Bayesianism. It is little more than an application to quantitative models of agents' states of mind of the position sketched out in the previous section, namely that a fully rational agent is one whose state of mind is representable by a *set* of consistent, maximally opinionated attitudinal states or avatars. Now, on the view defended in the second part of the book, the degrees of belief and desire of a maximally opinionated rational agent are measured by a pair of probability and desirability functions defined on a Boolean algebra of prospects. So the associated view on non-opinionated agents can be rendered more precisely as:

Imprecise Bayesianism A rational agent's state of mind, or judgemental state, is representable by a structure $\mathcal{J} = \langle \Omega_A, S \rangle$, where Ω_A is the Boolean algebra of prospects recognised by the agent and $S = \{(P, V)\}$ is a set of pairs of probability and desirability functions defined on Ω_A.

The basic idea here is that the features common to the pairs of functions constituting the agent's judgemental state represent what is settled for her, the differences between them what is not. Thus, if the agent regards the probability of rain tomorrow to be no less than 0.5, then her judgemental state will contain only probability functions assigning values greater than or equal to 0.5 to the prospect of rain tomorrow. If she desires snow more than rain, then her state of mind will contain only desirability functions assigning higher value to snow than to rain. And so on.[5]

Although I will not attend to the second and third questions in this chapter, it will be useful to state up front the elements of the answers to them that seem to be common ground amongst most authors who accept Imprecise Bayesianism. Firstly, on the question of how an agent should revise her opinions in the light of experience, the natural rule for an Imprecise Bayesian to follow is for each of her avatars to revise their opinions by whatever form of conditionalisation is appropriate to her experience. Specifically, in the case in which experience takes the form of learning that some proposition is true:

Imprecise Conditionalisation A rational agent with prior state of mind S and whose learning experience is exhausted by learning that E is true should adopt posterior state of mind:

$$S_E = \{(P(\cdot|E), V(\cdot|E)) : (P, V) \in S\}$$

It is not difficult to see how this principle could be extended to the other kinds of experience and corresponding forms of conditionalisation studied in Chapter 10. But I will argue in the next chapter that the Imprecise Bayesian must also countenance quite different kinds of attitude change from those encoded in such revision principles.

On the issue of how less than fully opinionated agents should make choices, there is far less of a consensus. Indeed, the only principle commanding near-universal assent is a unanimity principle: that if one prospect is considered more desirable than another by every avatar of the agent then she should prefer it to the second.[6] But implicit in much of the literature is a commitment to a somewhat stronger claim, namely that when there is no unanimity amongst an agent's avatars it is permissible for her to choose *cautiously*, either by giving more weight to 'pessimistic' avatars (the ones assigning low desirability to options) or by rejecting options on which her avatars are in strong disagreement as to their desirability in favour of those on which they are not, or simply by sticking with a status quo option when there is no alternative that is unanimously judged better by her avatars. At this stage it is not possible to be more precise than this about what it means to act cautiously. But it is worth keeping this vague contention in mind because the claim that it is sometimes appropriate to act more cautiously than is permitted

5   The use of sets of probability measures to represent imprecision in agents' states of belief has a long history going back to at least John Maynard, Keynes, Bernard Koopman and Émile Borel (see Walley, 1991, for a history) and has advocated more recently by, amongst others, Walley (1991), Levi (1978, 1990), Jeffrey (1990/65), Kaplan (1998), Good (1983), Joyce (2010) and Nehring (2009b).
6   Even this principle is disputed, however, on the grounds that the unanimity can be spurious. See Mongin (2016).

by classical Bayesianism has been both an argument for Imprecise Bayesianism and a source of criticism of it.

Further discussion of these questions must await a later chapter. For now let's focus on the core claim of Imprecise Bayesianism, starting with a look at the various interpretations that can be given to the sets of measures formalism that it employs. There are three main ones in the literature – subjective incompleteness, measurement imprecision and objective indeterminacy – corresponding, respectively, to cases in which an agent fails to reach a judgement, cases in which the modeller's evidence underdetermines her representation of the agent's judgemental state, and cases in which there is no fact of the matter as to what judgement the agent has reached. While the second and third of these are important to descriptive theories, it is the first of these – subjective incompleteness – that matters most to the theory of rationality. But even within this broad domain there are a number of distinct sources or motivations for imprecision.

Boundedness The agent may not have made up her mind about a particular question either because she has not got around to thinking about it, because she does not have the cognitive resources to do so or because she has deliberately decided to suspend judgement. When the last of these is the case, this may be because the agent hasn't yet collected all the information available to her or because the issue is of insufficient importance to justify the expenditure of time and other deliberational resources required to reach a judgement. For instance, she may have decided that she will take a holiday, but not yet where she will take it, intending to do some reading about her options first.

Severe Uncertainty The agent may be unable to arrive at a precise judgement because she lacks the informational basis for doing so. Someone who has no information about the climate in the Okavango Delta, for instance, will have no basis for producing a probability for rain in the delta in the year 2061 on the 31st of July. Indeed, I doubt that most of us are even able to produce precise probabilities for banal events such as our great grandchildren using Apple computers. The same applies to desirability judgements. If you learn that there is a new restaurant in town but nothing further about it, you have no basis for assigning a desirability to it. Many authors argue that in these situations the agent is not merely in a state of uncertainty in which, though she doesn't know for sure whether something is true or not, she can assign a probability to it being so. Rather, she is in a state of severe uncertainty, in the sense that, such are the limits on what she knows and can learn, she has no non-arbitrary basis for assigning such a probability. Keynes, whose views were even more radical than those I am defending, put it thus:

> By 'uncertain' knowledge, let me explain, I do not mean merely to distinguish what is known for certain from what is only probable. The game of roulette is

not subject, in this sense, to uncertainty; ... The sense in which I am using the term is that in which the prospect of a European war is uncertain, or the price of copper .... About these matters there is no scientific basis on which to form any calculable probability whatever. We simply do not know.   (Keynes, 1937, p. 209–23)

Conflict or Disagreement   The agent may be in a state of unresolved conflict or disagreement which prevents her from reaching a judgement on an issue. One way this can arise is from conflicting epistemic or value commitments on the part of the agent (see, for instance, Nagel, 1979, and Levi, 1990). Jean-Paul Sartre told the story of a pupil of his who had to choose between going to England to join the Free French forces or staying near his mother and helping her to survive which nicely illustrates this.

> He found himself confronted by two very different modes of action; the one concrete, immediate, but directed towards only one individual; and the other an action addressed to an end infinitely greater, a national collectivity, but for that very reason ambiguous – and it might be frustrated on the way. At the same time, he was hesitating between two kinds of morality; on the one side the morality of sympathy, of personal devotion and, on the other side, a morality of wider scope but of more debatable validity. He had to choose between those two. What could help him to choose?
>
> (Sartre & Elkaïm-Sartre, 2007, p. 30)

Conflict can also arise because the agent is not psychologically unified, as may be the case in group decision making when the group lacks cohesion, or in intertemporal decision making when the agent's tastes change over time. Or it can stem from disagreement located outside the agent. For example, when an agent seeks to base her opinions on the predictions yielded by scientific models or on the judgements of experts that she consults, and competing models or different experts deliver different judgements on some issue, she may have no means of settling the question of which she should adopt.

These three cases lie on a spectrum running from those in which an agent has not got around to making up her mind, but has at her disposal the means to do so (e.g. she has all the necessary information, is able to make the correct inferences, has the time and energy to draw them, etc.), through those in which there are surmountable obstacles to her reaching a judgement (lack of attention, information or inferential tools) all the way to those cases in which she cannot make up her mind because of incompatible commitments, fundamental disagreements amongst sources or insurmountable obstacles to acquiring or processing relevant information (e.g. because of the complexity of the issue or intractability of the conflict).

When an agent is in any of these situations then I claim that she is not rationally required to make up her mind. So imprecision of attitude is permitted in a wide range of circumstances. Some authors want to go further than this, however, and argue that such situations, or at least some of them, rationally require the agent *not* to make up her mind.[7] I do not think this view has any more justification in general than the one it rejects (namely, that the agent must make up her mind). But both the motivation and the implications of this view depend very much on which of the three circumstances – boundedness, lack of information or disagreement – is applicable.

In cases of boundedness when an agent is unable to make up her mind because she lacks the cognitive resources to do so, then it follows from the dictum that 'ought implies can' that she is not required to. It is not true, however, that 'cannot implies must not'; indeed, the very question of whether she should not make up her mind doesn't seem to arise in circumstances when she cannot. In many situations, of course, it is not literally impossible for the agent to reach a judgement, but it would be costly for her to do so. When the costs of deriving an opinion are higher than the potential benefits of having one, then the principle of maximisation of expected benefit, applied to the alternatives of reaching an opinion or not, counsels that the agent should not do so. On the other hand, the benefits may outweigh the costs; and then rationality requires opinionation.

When the agent has conflicting epistemic or value commitments then she is unable to reach a single opinion of an issue without violating one of her commitments. There are two different cases to consider. When the conflict arises because the values are objectively incommensurable, or at least believed to be so by the agent, then it would seem not only reasonable for her to suspend judgement, but indeed wrong to resolve the disagreement. Isaac Levi (1997) takes this view, for instance, arguing that the only appropriate response to value conflict is suspension of opinion and that anything less than this would be to prematurely foreclose moral enquiry and debate. On the other hand, if the conflict derives from subjective incommensurability, and the agent believes there to be an underlying, but unknown, fact of the matter as to which value should be upheld, then the case for the permissibility of suspending judgements remains strong, but not for it being mandatory.

The case of severe uncertainty is the most disputed of the three, and so before we tackle it we need to look more carefully at the grounds for the claim that adoption of imprecise degrees of belief is the appropriate response to conditions of this kind.

11.4 SEVERE UNCERTAINTY

> There are two ways of looking at our duty in the matter of opinion ... *We must know the truth*; and *we must avoid error* – these are our first and great commandments as would-be knowers. [...] Believe truth! Shun error!
>
> (James, 1897, 17–18, emphasis in original)

---

7 See, for instance, Levi (1978) and Kaplan (1998).

What must we believe? At least all logical truths and whatever is entailed by the evidence we hold. So much is pretty much common ground. But beyond this there is serious disagreement amongst epistemologists both as to the extent to which belief is objectively constrained (by evidence, reason, etc.) and as to the correct way to balance William James' commandments. 'Equivocators' favour avoidance of error in judgement by counselling against the adoption of opinion beyond what is mandated by the evidence; 'dogmatists' urge that one should maximise truth by adopting the strongest opinions consistent with this evidence. When it is uncertain whether the dark and threatening clouds in the sky portend rain, the former might say that one should suspend belief on the question. Equivocation, the latter might say, is one way of getting wet.[8]

Bayesians refuse this binary choice between believing and not believing. In the face of uncertainty (they say) the question to ask is not whether to believe something, but to what degree one should. By assigning a probability to the prospect of rain one calibrates one's belief state, and hence one's decisions to the strength of the available evidence. For taking an umbrella to be a reasonable course of action it suffices that dark clouds are strong enough evidence for rain to counterbalance the inconvenience of carrying it.

Bayesianism has not been spared dispute, however. Although Classical Bayesians are united in their commitment to the view that agents should adopt precise degrees of belief in all propositions, subjectivists and objectivists disagree on the question of what precise degrees of belief to adopt and in particular on how *this* choice is constrained by the evidence. Subjectivists, as we have seen, allow that any consistent set of degrees of belief is rationally permitted, while objectivists forbid those that are not maximally non-committal and that extend further than is strictly mandated by the evidence. Imprecise Bayesians, on the other hand, refuse the classical Bayesians' demand for precision and allow, for one or more of the reasons sketched before, that a rational agent not make up her mind on some questions. We are thus presented with a cross-cutting dispute along two dimensions: Subjectivism–Objectivism and precision–imprecision. The four possible positions this yields are displayed in Table 11.1, along with a tentative ascription of them to some well-known figures.

I will argue for the position represented by the bottom-right cell of the table; in particular, for a Bayesianism that does not regard any level of precision as mandatory. But to get a grip on the precision–imprecision dispute, let

---

[8]   In contemporary epistemology this dispute has shifted from belief to belief change and the question of how cautious one should be in discarding currently held opinions and adopting new ones.

TABLE 11.1. *Types of Bayesian*

	Precise	*Imprecise*
Objective	Jaynes Williamson	Levi
Subjective	Savage De Finetti	Jeffrey

us assume for the moment that there is a rationally mandatory belief state to adopt and ask ourselves what it might be in each of the following three situations.

1. You have before you an urn containing an equal number of black and white balls and you are asked to say how likely you think it is that a ball drawn at random will be white.
2. Same as before, except you are not told what the proportions of white and black balls are.
3. Same as (2), except you are not told what ball colours the urn contains.

On the face of it, the uncertainty you face regarding the colour of the drawn ball is quite different in each of these three situations. In the first, you don't know for sure whether white will be drawn or not, but you do know the probability with which it will. In the second, you don't know what the probability of a white draw is, but you know what the possible probabilities for a white draw are: 0, 0.01, 0.02, ..., or 1. In the third, you don't even know this. To capture these differences I will refer to situations of Grade 1 uncertainty (or risk), Grade 2 uncertainty (or ambiguity) and Grade 3 (or severe) uncertainty.

Everyone in this debate would agree that in the first situation, that of risk, it is natural to adopt a probability of one-half for a draw of a white ball (though extreme subjectivists might still claim that such an ascription is not mandatory). But what about the second? Here Precisers, at least those of objectivist inclination, will want to apply the Principle of Indifference to the set of hypotheses concerning the composition of the urn, assigning equal probability to each and hence probability one-half to the draw of a white ball. Imprecisers too can appeal to a version of this principle. While Precisers say 'If you have no reason to ascribe a different probability to two prospects, then you should ascribe equal probability to them', Imprecisers say 'If you have no more evidential reason to adopt one probability judgement rather than another, then you should either adopt both or neither'. This latter application of the principle leads to ascription of the unit interval to the prospect of a draw of a while ball – to suspension of opinion rather than adoption of an even-handed one – an assignment that Imprecisers claim better reflects your state of uncertainty than the Precisers' assignment of equal probability.

The problem is a familiar one: different applications of the Principle of Indifference support different conclusions, without it being obvious as to which is the correct application. But the stand-off dissolves when we turn to the third situation. For in cases displaying Grade 3 uncertainty, it is no longer possible to determine degrees of belief by application of the Principle of Indifference in the manner required by the objective Bayesian. The problem here is that there is no salient partition of prospects on which to base the indifferences. And each choice of partition will give a different answer to the question of which probability to adopt. If, for instance, you take your propositions to be the proportion of balls that are white, then you will be led once again to a probability of one-half. On the other hand, if you take the basic propositions to concern the proportions of different colours, then you will be led to something less than one-half (exactly what will depend on how you individuate colours). In this case it seems much more reasonable to acknowledge that you have not the faintest idea what the probability of a white ball is and adopt a very imprecise probabilistic belief.

Even setting aside cases of Grade 3 uncertainty, objective Precisers face another difficult challenge deriving from the difference between situations of risk and uncertainty. According to them, the probability of a draw of a white ball is the same in both kinds of situation. But the uncertainty we face regarding the colour of the drawn ball seems very different in the two cases. So it would seem to follow that probability alone cannot be an adequate measure of uncertainty.[9]

There is a corresponding pragmatic version of the difficulty. Suppose that for both urns a bet can be purchased for $5 which pays $11 if a white ball is drawn from the urn. Suppose that you can purchase a bet on only one of the urns and assume linear utility for money. The Precise Bayesian (of objectivist inclination) should be indifferent between the bet on the first urn and that on the second because both have an expected value of $5.50. There is a good deal of evidence, however, that many people would prefer to bet on the first urn rather than the second, citing greater uncertainty about the outcome in the former than in the latter. This evidence is, on the face of it, very damaging to the equivocal Preciser, who must, it now seems, either declare such people irrational or find some other way to account for the difference between the two urns.

There are interesting responses that the Precise Bayesian can make to both of these challenges that we will look at in the last chapter. But on the face of it they provide some support for the Imprecise view. On the one hand, Imprecise Bayesians can capture the differences between situations of Grade 1

---

[9] There are strains here of Popper's Paradox of ideal evidence. We return to this in the final chapter.

uncertainty and Grade 2 uncertainty by the precision of the belief state they mandate. On the other hand, these differences in belief state allow Imprecise Bayesians to advocate different decision rules in the two situations. These strengths, it turns out, are also sources of other difficulties for Imprecise Bayesians, but we will have to defer discussion of this issue until later.

For the moment, let us return to the second dimension of disagreement amongst Bayesians. The dispute between Subjectivism and Objectivism has largely been conducted within the framework of Precise Bayesianism. But the issue is just as important for those who accept the argument for imprecision. Many Imprecise Bayesians accept the objectivist view that one should not adopt beliefs not mandated by the evidence but see its implication as being that one's degrees of belief should be no sharper than the evidence requires, rather than that they be sharp in some particular way. Joyce (2005), for instance, argues that since the information we receive is often incomplete, imprecise or equivocal, the correct response is to have opinions that are equally incomplete, imprecise or equivocal. Indeed it is widespread in the philosophical literature (especially amongst the critics of the Imprecise view) to assume that Imprecise Probabilism finds its support in some version of what Roger White calls the Chance Grounding Thesis, namely that

> only on the basis of known chances can one legitimately have sharp credences. Otherwise one's spread of credence should cover the range of chance hypotheses left open by the evidence.
>
> (White, 2009, p. 174)

This is a claim that a subjectivist Imprecise Bayesian should reject. A subjectivist should regard it as permissible not only to have imprecise beliefs but also to have sharp beliefs, even when the evidence does not fully determine what these beliefs should be. Most of the time we don't *know* what the relevant objective chances are and so we would be required to have maximally imprecise probabilities if the Chance Grounding Thesis were true. But, although I don't know what my chances are of suffering coronary heart disease in the next 10 years, on the basis of what I do know I can reasonably exclude chances above 30 per cent and below 1 per cent from my deliberations. Similarly, when faced with the second urn discussed above, it would not be irrational to judge that the symmetries built into the situation warrant the adoption of a precise degree of belief of one-half for a white ball being drawn (even if it is not mandatory to do so).

The idea that it is knowledge of chances that allows for precise probability assignments derives, I suspect, from the seductive but mistaken conception that probabilistic judgement aims at identification of the true chances. But the Imprecise Bayesian does not adopt a probability interval for some prospect because she believes the true probability to lie within it, any more than the Precise Bayesian adopts a particular probability because she believes it to be the true one. I might know that the coin that is about to be tossed is either two-headed or two-tailed, and hence that the chances of it landing heads are either one or zero, but nonetheless reasonably adopt a probability of one-half for heads (a 'false' probability). Conversely, I might not know for sure that the coin is not two-headed or two-tailed, but nonetheless adopt an imprecise probability that does not extend to the unit interval.

Such talk of adopting probabilities needs clarification. For the Precise Bayesian, it means taking them as a basis for choice aimed at maximisation of subjective expected benefit, as exemplified in a willingness to bet both for and against the truth of propositions for stakes reflecting these probabilities (canonically a bet paying $1 priced at the adopted probability). For the Imprecise Bayesian too, it means taking the adopted probabilities as the basis for choice. Roughly, we can think of it as specifying upper and lower bounds on the bets one would be willing to accept. But making the connection to action clearer must await the discussion of decision making under severe uncertainty.

In summary, we should reject both the dogma of rationally necessary precision and that of rationally necessary imprecision. In probabilistic judgement, as with categorical judgement, we must balance the Jamesian imperatives of seeking truth and avoiding error. The question of how precise to make one's judgement is something that the agent must answer for herself, weighing the epistemic advantages of keeping an open mind against the practical need to consider only a limited range of possibilities. The agent who seeks to avoid error will not adopt any opinions not required by the evidence; the agent who seeks truth will adopt as many opinions as she consistently can, given the evidence. There is no general requirement to give precedence to either aim: the reasonable person will assess what the costs and benefits of each are in the circumstances in which she finds herself.

11.5 RATIONAL INCOMPLETE PREFERENCE

In earlier chapters we sought foundations for our theory of maximally opinionated rational agency in a set of representation theorems establishing correspondences between the postulated properties of rational numerical belief and desire, on the one hand, and the rationality conditions on relational belief and preference, on the other. They showed, firstly, that if an agent has consistent degrees of belief and desire then the relational beliefs and preferences based on them are assured of satisfying these rationality conditions. And, secondly, that if an agent's relational beliefs and preferences satisfy the rationality conditions (along with some structural ones) then her degrees of belief are required to be probabilities and her degrees of preference to be desirabilities.

The correspondences in question rest on the assumption that the agent's relational attitudes (her comparative credences and preferences) and quantitative attitudes (her degrees of belief and desire) should cohere in such a way that the latter explains or rationalises the former. The same thought applies to the less opinionated agents whose states of mind are given by sets of probabilities and desirabilities. In this more general case, we say that an agent's state of mind explains or rationalises her preferences whenever every pair of probability and desirability functions making up her state of mind implies a ranking of the prospects that is consistent with these preferences. More formally, we say:

Rationalisation A state of mind $S = \{\langle P_j, V_j \rangle\}$ explains or rationalises the preference relation $\succsim$ iff, for all α, β in the domain of $\succsim$, it is the case that

$$\alpha \succsim \beta \Leftrightarrow \forall \langle P_j, V_j \rangle \in S,\ V_j(\alpha) \geq V_j(\beta)$$

Clearly, an agent's preferences could be rationalised by more than one state of mind, since any subset of a state of mind will rationalise the preferences rationalised by the state of mind itself. Only one rationalising state of mind contains *all* pairs of probability and desirability measures jointly consistent with her preferences, however. Such a state we call *maximal*.

The more complete an agent's preferences, the smaller the maximal set of pairs of functions rationalising them. But even complete preferences can be rationalised by more than one maximal state of mind, because the same preference ordering can be induced by different desirability functions. Where the differences in values assigned to prospects by two desirability functions are arbitrary – because, for instance, they are attributable to nothing more than different choices of scale – these functions may be regarded as substantially equivalent. Similarly, two states of mind may be regarded as substantially equivalent if each of the functions belonging to one is equivalent to a function belonging to the other, and vice versa.

The claim that different states of mind may be substantially equivalent is consistent with different views as to what makes for such equivalence. On the view adopted in this book, any differences in numerical measures of the agent's attitudes not derivable from features of her relational attitudes are without empirical content. One has an opinion on a prospect iff one is able to place it in an ordering of prospects, and substantial differences of opinion should be reflected in differences in this ordering. Hence, to establish what properties of a numerical representation of an agent's state of mind are real, one should look at whether differences in them give rise to differences in the properties of the orderings they determine.

The question that now arises is: under what conditions can an agent's preferences be rationalised by an imprecise state of mind – i.e. by a set of pairs of probability and desirability functions? The question has only recently received much attention; indeed, in decision theory general recognition of the implications of incomplete preferences has been slow in coming, despite seminal work by Bewley (2002) and Aumann (1962). (The main reason for this, I would speculate, is because, on the interpretation of preferences most favoured by economists, namely as actual or hypothetical choices, completeness is essentially built in.) In economics and statistics, the question has largely been explored within either the von Neumann and Morgenstern framework or the more general Anscombe–Aumann one (see, for instance, Seidenfeld *et al.*, 1995, and Karni & Vierø, 2013. Our task is to answer the question within the propositional framework used in this book.

11.5.1 Coherently Extendable Preferences

What does rationality require of the relational beliefs and preferences of agents who are not maximally opinionated? In the representation theorems for Precise Bayesianism given earlier on, we assumed that the credibility relation $\trianglerighteq$ was complete, transitive and ∨-Separable and that the preference relation $\succsim$ was complete, transitive and satisfied ∨-Betweenness and Coherence with respect to $\trianglerighteq$, along with various structural conditions. Evidently, the completeness axiom must now be dispensed with. But the background assumption of maximal opinionation is much more deeply embedded in these theorems than this. For example, if it is not irrational for an agent to fail to develop a preference, then Transitivity cannot be a general requirement of rationality. For Transitivity requires me to have a preference for α over γ whenever I have a preference for α over β and for β over γ, even if I haven't got around to thinking about α in comparison to γ and will never be confronted with a choice between them. To put it somewhat differently, Transitivity combines two requirements, namely that (i) if I have a preference for α over β and for β over γ then, if I have a preference between α and γ, I must prefer α to γ and (ii) I must have a preference between α and γ. The second is just the completeness requirement again and should once again be dispensed with. The upshot is that Transitivity should be weakened to something like (i).[10]

Similar considerations apply to the other rationality conditions (and indeed the structural ones too). But, instead of weakening each of them individually, I propose to characterise rational relational belief and desire by a single requirement, namely that an agent's relational attitudes can be coherently 'completed' – i.e. extended to a complete and rational credibility or preference relation, where 'rational' means satisfying the conditions for complete relational attitudes defended in the second part of the book. This means, in the case of credibility, satisfaction of Transitivity, Monotonicity and ∨-Separability and, in the case of preference, of Transitivity, ∨-Betweenness and Impartiality (or Coherence). So, if an incomplete relation is extendable to a complete one, then it must be possible for all the gaps in this relation to be filled in without the result being a relation that violates one of the rationality axioms.

---

[10] More exactly, it should be weakened to a requirement that the weak preference relation be *Suzumura consistent*. See Bossert & Suzumura (2010) for a definition and argument as to why it is the appropriate rationality condition for incomplete preferences.

To make this more precise, let R be a binary relation on some domain Ω. We say that another binary relation R' on Ω is

- an *extension* of R iff $(\alpha, \beta) \in R$ implies $(\alpha, \beta) \in R'$
- a *minimal extension* of R iff if $(\alpha, \beta) \in R$ and $(\beta, \alpha) \notin R$ then $(\alpha, \beta) \in R'$ and $(\beta, \alpha) \notin R'$
- a *coherent extension* of R iff R' is complete and transitive

Clearly, if R is itself complete and transitive then it has a coherent extension (namely itself). This extension is not typically unique, but it is the only one that is minimal.

A rational incomplete relation must at least have a coherent extension. But this is far from being sufficient, for a relational attitude can be inconsistent but still have a coherent extension. For example, suppose the agent's preferences are such that $\alpha \succsim \beta, \beta \succsim \gamma$ and $\gamma \succsim \alpha$, but not $\alpha \succsim \gamma$. Then she violates Transitivity. But her preferences can be coherently extended simply by adding the weak preference $\alpha \succsim \gamma$. In fact, *every* weak binary relation R has a coherent extension, namely the relation R^{sup} defined by, for all α and β in the domain of R, $(\alpha, \beta) \in R^{\text{sup}}$. The relation R^{sup} is obviously reflexive and complete, and, because, in effect, it regards all alternatives as indifferent to all others, it is transitive too. But R^{sup} is a superset of all relations defined on its domain. So every relation on this domain has R^{sup} as its coherent extension.

The requirement of coherent extendability is, it seems, entirely toothless when applied to weak relations (not so for strict relations). What has the necessary bite is the requirement that they have a *minimal* extension that is coherent. Let us say that a binary relation on domain Ω is *weakly consistent* iff it has a coherent minimal extension on Ω and that R is *strongly consistent* iff every minimal extension of R is coherent. The difference between the two, as we saw in our earlier informal treatment of these notions, is that weak consistency does not entail logical omniscience. For instance, someone with the preferences $\alpha \succsim \beta$ and $\beta \succsim \gamma$, but not $\alpha \succsim \gamma$, not only has incomplete preferences but does not satisfy Transitivity, which should be regarded in this context as the deduction rule appropriate for preferences. Strong consistency, on the other hand, does entail the transitivity of the binary relation. So whether or not one requires weak or strong consistency of preference depends on whether or not one thinks that rationality requires that the agent's attitudes be deductively closed.

11.5.2 Representation Theorems

Rationality requires of agents that their relational attitudes be at least weakly consistent. This implies that their relational beliefs regarding prospects must be minimally extendable to a weak credibility order and that their preferences must be minimally extendable to a weak preference order that coheres with the agent's beliefs. The state of mind of an agent whose relational beliefs and preferences are at least weakly consistent can be represented, given the usual technical conditions, by sets of probability and desirability functions on the domain of prospects. More formally, let $\Omega = \langle X, \models \rangle$ be a complete, atomless Boolean algebra of prospects and let $\unrhd$ and $\succsim$ be, respectively, a credibility and a preference relation on Ω. Consider:

Weak Axiom of Consistency (Belief) The relation $\unrhd$ has a minimal coherent extension on Ω that is a continuous credibility relation

Weak Axiom of Consistency (Preference) The relation $\succsim$ has a minimal coherent extension on Ω that is a continuous and impartial preference relation

Then, as is proved as Theorems A.24 and A.25 in the Appendix:

Imprecise Probabilism Let $\unrhd$ be a non-trivial credibility relation on Ω that satisfies the Weak Axiom of Consistency for belief; then there exists a maximal set of probability functions $S = \{P_1, ..., P_n\}$ that explains or rationalises $\unrhd$ in the sense that for all $\alpha, \beta \in \Omega$,

$$\alpha \unrhd \beta \Leftrightarrow \forall P_i \in S, \ P_i(\alpha) \geq P_i(\beta)$$

Imprecise Bayesianism Let $\succsim$ be a non-trivial preference relation on Ω that satisfies the Weak Axiom of Consistency for preference; then there exists a maximal set of pairs of probability and desirability functions $S = \{\langle P_i, V_i \rangle\}$ on Ω that explains or rationalises the preference relation $\succsim$ in the sense that for all $\alpha, \beta \in \Omega'$,

$$\alpha \succsim \beta \Leftrightarrow \forall V_i \in S, \ V_i(\alpha) \geq V_i(\beta)$$

These theorems are offered as vindications of the claim that the state of mind of a rational, non-opinionated agent can be represented by a set of pairs of probability and desirability functions. They thereby give foundations to Imprecise Probabilism and its more broad-ranging analogue, Imprecise Bayesianism. For they show that considerations of rational relational belief and preference imply that it is sufficient that an agent's quantitative beliefs and desires are of this form for them to stand in the right kind of explanatory and/or rationalising relations with respect to her relational attitudes.

A couple of comments of a more technical nature. Firstly, these representations are *unique*, in the sense that the maximal sets that are referred to in the theorems are unique. But, as a form of representation – in terms of sets of pairs of functions – it need not be. For instance, it may sometimes be possible to represent an agent's relational beliefs by probability intervals or upper and lower probabilities. But, to ensure the existence of such alternative representations, further conditions on relational attitudes may be required. And I doubt that any of them will be pure conditions of rationality.

Secondly, it is noteworthy how simple these theorems are, compared to those for the von Neumann–Morgenstern and Anscombe–Aumann frameworks referred to before. This stems essentially from the fact that in the framework adopted in this book beliefs and preferences are defined on the same objects (prospects or propositions). The consequence is a much cleaner and more compelling defence of Imprecise Bayesianism than can be provided in these other frameworks.

Thirdly, the two Weak Axioms of Consistency require only that the agent's relational beliefs and preferences, respectively, be coherently extendable to a credibility and preference order. But I argued in Part II of the book that rationality requires more of an agent's attitudes than this. For instance, when the set of prospects contains indicative conditionals then her preferences for them should have the Indicative Property and her preferences for conjunctions of orthogonal indicative conditionals should be $\rightarrow$-Separable. Beefing up the requirement of coherently extendability to include these additional rationality conditions will produce corresponding stronger restrictions on the probability and desirability pairs representing the agent's preferences.

Finally, it is notable that an imprecise probabilistic/Bayesian representation of the agent's relational attitudes is possible even when she is not logically omniscient. This happens because, in effect, the representation identifies the agent's state of mind with her consistent avatars rather than the full set of them (recall that a non-omniscient agent has strictly inconsistent avatars). There is clearly something unsatisfactory about this from a descriptive point of view because it leads to an inability at the level of numerical representation to distinguish agents with the same set of strictly consistent avatars, but different sets of inconsistent ones. And from a normative point of view, because the numerical representations could be said to over-rationalise the agent's relational state of mind by ignoring inconsistent avatars.

There are two ways of dealing with this difficulty. Firstly, one could simply replace the requirement of weak consistency on relational attitudes by a requirement of strong consistency and obtain corresponding representation theorems for Imprecise Probabilism and Imprecise Bayesian based on the following axioms:

Strong Axiom of Consistency (Belief) Every minimal coherent extension on Ω of the relation $\unrhd$ is a continuous credibility relation

Strong Axiom of Consistency (Preference) Every minimal coherent extension on Ω of the relation $\succsim$ is a continuous and impartial preference relation

Secondly, one could weaken the numerical representation obtained with the Weak Axioms of Consistency, by allowing that the set of functions representing a state of mind include numerical functions that are not strict probabilities or desirabilities. This has certain advantages, not least that it would allow for representations of inconsistent agents. But I will not explore this possibility further here.

11.6 CHALLENGES TO IMPRECISE BAYESIANISM

Let me turn to some of the challenges facing the Imprecise view and which will set the agenda for discussion in the next couple of chapters. They can be illustrated using variants of a single example. Suppose that 200 balls, 100 white and 100 black, are divided between two urns, respectively labelled A and B. The division is such that there are 100 balls in each but with any ratio of black to white balls. A fair coin is tossed and a ball drawn from urn A if it lands heads and from urn B if it lands tails. You will be given an opportunity to bet on a white ball being drawn; the bet costs $10 and you win $50 if the drawn ball is white. Just before being offered it, however, you are told that the coin landed heads. There are three problems now that arise for you if you handle your uncertainty in the manner (apparently) advocated by Imprecise Bayesianism.

1. *Dilation*: Consider your epistemic situation. On the face of it, the initial probability of drawing a white ball must be one-half, as the number of white and black balls is equal and the ball is drawn at random. But, once you are told that the coin landed heads, you know that the ball must have been drawn from urn A. Since you have no idea of the proportion of balls in urn A that are white, Imprecise Probabilism recommends that you now adopt the full range of possible probabilities, [0,1] (or at very least more than a single probability). But this seems odd. How can learning how the coin landed undermine your judgement about the probability for drawing a white ball? The way the coin lands doesn't seem evidentially relevant to what ball is drawn, given that there is no correlation between the two.

2. *Learning*: Suppose that you are offered the opportunity to sample the urns with replacement and that you do so, repeatedly drawing a white ball from urn A. Intuitively, this should increase your confidence in a white ball being drawn and hence the attractiveness of the bet on offer. But Imprecise Probabilism seems to disallow this. For, no matter how many times in a row your sample is white, the possibility that there is no more than i balls in the urn remains undefeated, for $i \in (0,1]$.

3. *Decision Making*: Suppose that you value small sums of money linearly so that $50 is worth five times as much to you as $10. Then, before learning how the coin has landed, the prospect of betting on the white ball being drawn should be very attractive to you: its expected monetary value is $15, after all. But after learning that the coin landed heads you find yourself unable to say whether the bet on white is a good one or not, for its expected monetary value now ranges from −$10 to $40, for

degrees of belief in the $[0, 1]$ interval. If you are cautious you might now reject the bet. But how could learning how the coin landed change your decision?

Dilation Let us start with dilation, a phenomenon recognised fairly early on by Walley (1991) and Seidenfeld & Wasserman (1993), but which has been much discussed more recently because White (2009) and others have used examples involving dilation to attack the cogency of Imprecise Probabilism. White's examples are more elaborate than the one presented here but are essentially of the same variety. They are very extensively discussed in Joyce (2010) and Pedersen & Wheeler (2014), so I will content myself with bringing out what I think to be the central lessons of the debate.

To help us understand what is going on here, we need to unpack our grounds for saying that the probability of drawing a white ball is one-half. The imprecise probabilist responds to her ignorance about the numbers of balls in the urn by allowing for the possibility of different views on the question. Each of her opinionated avatars can be identified with a hypothesis as to the number of white balls in urn A. Let $\{H_1, H_2, ..., H_{100}\}$ be a set of such hypotheses with H_i being the hypothesis that there are i white balls in A and let $S = \{P_1, P_2, ..., P_{100}\}$ be the state of mind of the imprecise probabilist, with each P_i being the avatar that believes that H_i is true. If H_i is true then the chance of drawing a white ball from urn A is $\frac{i}{100}$ and from urn B $\frac{100-i}{100}$ because any white balls not in A must be in B. So, for any avatar i, the probability of (W) drawing a white ball is

$$P_i(W) = P_i(W|A) \cdot P_i(A) + P_i(W|B) \cdot (1 - P_i(A))$$
$$= \left(\frac{i}{100} \cdot \frac{1}{2}\right) + \left(\frac{100-i}{100} \cdot \frac{1}{2}\right)$$
$$= \frac{1}{2}$$

In other words, although the agent's avatars disagree on the distribution of balls in each urn, they nonetheless agree on the probability of a white ball being drawn. This agreement is accidental, however: it rests entirely on the symmetries built into the situation and in particular on the agreed fairness of the coin toss. As soon as the symmetry is broken, the underlying disagreement reveals itself.

What makes dilation seem mysterious is that it appears that something evidentially irrelevant to a judgement is undermining it. But this is not in fact what is happening in these cases. On the contrary; for each avatar the result of the coin toss *is* evidentially relevant to the distribution of balls, and each avatar will draw a different inference from it. At the aggregate level this fact is masked by the initial spurious agreement on the probability of W. So dilation does not in itself pose a problem for Imprecise Bayesianism.

Learning The associated learning problem is more difficult to dismiss. The reason why the agent does not learn much from the sampling of the urns is that each of her avatars is dogmatic. Each avatar i fully believes hypothesis H_i and interprets the evidence yielded by the sampling through the prism of this hypothesis. Since the evidence of draws is consistent with H_i, conditioning on it does not disturb her belief in H_i (the exception being avatar 0; as the evidence has probability zero for her she cannot condition on it). One solution therefore might be to insist that the imprecise probabilist should not countenance dogmatism and instead admit only probability functions that are regular – i.e. that assign probability one to logical truths only. The agent's avatars would then be all the linear combinations of the P_i excluding those that give maximum weight to any of them, each representing a possible assignment of probability to the various hypotheses.[11] Each avatar will respond to a white ball sample by shifting probability from 'mainly black' hypotheses to 'mainly white' ones.

Unfortunately, this move creates a new problem. For, although each avatar is non-dogmatic about the number of white balls, no matter how many white ball samples are produced there will always be avatars that assign very low probability to a white ball being drawn. Indeed, for any probability of a white ball greater than one, no matter how small, some avatar will have that value as her degree of belief following the sampling. For, as 'each "extremist" finds her views tempered by the data, an even more radical extremist slides in from the wings to take her place' (Joyce, 2010, p. 9). So, even though the agent's avatars all shift probability in the same direction, the agent's belief state remains unchanged, and extremely non-committal, as a result of the sampling.

This problem seems to arise because the agent is too open-minded from the beginning; indeed, so open-minded that she cannot learn from experience. I have already argued that an agent is not required to countenance all probabilities consistent with the known chances. So, simply because she does not know the distribution of balls doesn't mean that she must countenance every possible probability for a white ball being drawn. Indeed, the fact that doing so would seem to prevent learning provides a good reason for adopting a more restricted set of probabilities. Precision must be permissible.

Even if this is true, however, it does not completely solve the problem, since there does not seem to be anything irrational about having a maximally open mind on the question of what ball will be drawn when one lacks clear evidence on the question. The force of the point is not so much that sharper prior beliefs are permitted but that it is open to the agent to sharpen her beliefs on the basis of the evidence she receives even when the evidence does not compel such a sharpening. (This is the essence of ampliative inferences: they sharpen one's beliefs more than is achieved by simply believing the evidence and drawing the deductively implications of this.) If this is right, our question becomes: how

---

[11] This is the solution proposed by Halpern (2003).

can the agent discard certain hypotheses about the distribution as she obtains information? Ordinary Bayesian conditionalisation does not allow it except in cases where the evidence refutes the hypothesis. So some other way of making up one's mind is needed. This issue will be addressed in the next chapter.

Decisions We are in no position to do justice to the difficulty posed for decision making by this example, as we have not as yet looked at what decision rules Imprecise Bayesianism might support. But it will help to steer future discussion if we get some sense of what is causing the difficulty. Prior to the coin toss, standard Bayesian decision theory is applicable (because of unanimity amongst the avatars on relevant prospects) and under the assumptions of the example, it prescribes acceptance of the bet on white. Dilation undermines the grounds for this application, because it leaves the avatars in disagreement. In the face of such disagreement, it does not seem unreasonable to be cautious and to forgo the option of betting. After all, one no longer knows what the chances of drawing a white ball are, so for all one knows it might be a very bad bet (some avatars say it is, some say that it is not). But if it is permissible to refuse a bet on white on these grounds then it is surely also permissible to refuse a bet on black with the same stake and prize. So if caution is reasonable then it is permissible to refuse both a bet on white and a bet on black. But refusing a bet on both is strictly worse than accepting both bets, for accepting both is guaranteed to leave one $30 better off. So it seems that acting cautiously is irrational.

   This presents Imprecise Bayesianism with a dilemma. If it permits refusal of both bets then it seems to allow irrational behaviour. But if it doesn't permit behaviour that is more cautious than that permitted by Classical Bayesianism then what, from a pragmatic point of view, is it contributing to the discussion? I will return to this problem in the chapter after the next.

12

Changing Your Mind

Agents can change their minds in a number of different ways. They can revise their opinions about one or more prospects, for instance as a result of acquiring new information or of deliberating on some question; they can form new opinions when they need to; and they can withdraw or suspend opinions currently held when confronted with undermining or conflicting evidence. They can also extend their attitudes to larger domains, or restrict them to smaller ones, as they become aware of new prospects, or forget about some.

Of these forms of attitude change, only one – revision – can be completely adequately studied within the Classical Bayesian models of belief and desire. The main obstacle to the study of the other forms is the assumption that agents are fully opinionated, an assumption that leaves no space open for cases in which they make up their minds, rather than simply change them. Nor does it allow for the treatment of cases in which an agent has reason to suspend judgement on some question or to retreat from a previously held attitude; for instance, when someone discovers that the evidence supporting one of her beliefs is unreliable (prompting retreat) or when two experts are found to disagree on some question (prompting suspension). Combined with the assumption of logical omniscience, it also leaves no space open for the treatment of deliberation, understood as a process by which an agent derives the view that she should adopt with regard to some prospect in her attention from her attitudes to other related prospects.

These limitations are not common to all theories of attitude change. In particular, the 'AGM' approach pioneered by Carlos Alchourrón, Peter Gärdenfors and David Makinson (see Gärdenfors, 1988), and extended by Sven Ove Hansson (1995) to preferences, in which attitudinal states are represented by sets of sentences satisfying conditions of logical consistency, but explicitly *not* required to contain every sentence or its negation, allows for a characterisation not only of belief revision but also of various other types of attitude change. My strategy will be to draw on some of the insights of this approach to build quantitative models of the various forms of attitude change. In subsequent sections, I carry over the insights gained from this exercise to the more complex quantitative models of belief and preference.

12.1 ATTITUDE CHANGE

Let us start by distinguishing five 'basic' types of attitude change.

1. **Attitude Revision** The process by which an agent changes her attitudes as a result perhaps of acquiring new information or of deliberating on some matter; a process that requires her to replace some previously held attitudes with the newly acquired ones. For instance, I might have wanted to take my holiday in Greece, but after reading about rioting in Athens I come to the view that Turkey would be better.
2. **Attitude Formation** The process by which an agent develops an attitude where she held none before, or develops an attitude that is more precise than that previously held. For instance, I might think I would like a holiday by the sea, initially have no view as to where I would most like to go but settle on a particular destination after reading several guidebooks.
3. **Attitude Withdrawal** The process by which an agent gives up an attitude, perhaps because her grounds for holding it are brought into doubt for some reason. For instance, someone might point out that the guidebook I was using to inform me about different seaside destinations was out of date. As a result, I withdraw my preference for the place I had settled on previously. Not having given much consideration to any alternative I go back to being in a state in which I don't know where I want to take my seaside holiday.
4. **Attitude Extension** The process by which an agent extends her attitudes to a superset of the prospects about which she currently entertains attitudes. For instance, suppose that I was previously unaware of the possibility of flying to the Greek islands without going through Athens, so that I have no initial attitude to this prospect. Once the possibility of doing so is brought to my attention, I bring it into consideration in my deliberations.
5. **Attitude Restriction:** The process by which an agent restricts her attitudes to some subset of those prospects about which she previously entertained attitudes. For instance, if I forget about the possibility of contracting malaria while on holiday, my deliberations about what medicines to take will be restricted to considerations not involving this prospect.

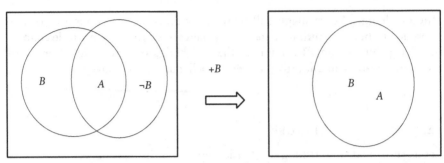

FIGURE 12.1. Attitude Formation

These types of attitude change are, clearly, quite closely connected. For instance, attitude withdrawal and formation are 'opposite' processes, while revision may be thought of as consisting of a bit of each. Equally, they can compose in a variety of ways to make up hybrid forms of attitude change. Extending one's attitudes to a new domain, for instance, is typically followed by attitude formation with respect to the prospects newly entertained (but not necessarily; one may simply bring them to attention without reaching a judgement).

In the qualitative model of belief developed in the preceding chapter, an agent's belief state was represented by a set of sets of propositions (the agent's avatars). The five types of attitude change just identified correspond to particular transitions between these belief states, displayed diagrammatically in Figures 12.1 to 12.4, driven by changes either in the number of avatars or the number of propositions in them. When new beliefs are formed the number of avatars of the agent is reduced (as in Figure 12.1); when they are withdrawn, the number increases (as in Figure 12.2). When beliefs are revised, the number of avatars does not change, but each is replaced by a new one consistent with the revision-inducing experience (as in Figure 12.3), a process that is equivalent to first expanding and then reducing the set of avatars. Belief extension and restriction correspond to the cases in which the number of propositions in the domains of the agent's avatars are increased or decreased (as in Figure 12.4).

The hypothesis that this informal modelling leads to is the following. All changes in attitudinal states can be represented as a sequence of four basic operations: reducing and increasing the number of avatars and reducing or increasing the number of propositions in the avatars. In the rest of this chapter, I will try to put some flesh on this claim.

12.2 CHANGING AN INCOMPLETE STATE OF MIND

To adapt this informal theory of attitude change to the model of states of mind as sets of probability and desirability functions, let's take as our

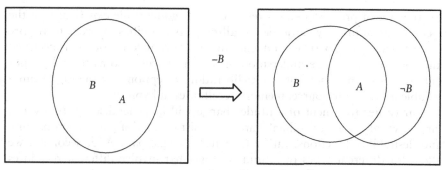

FIGURE 12.2. Attitude Withdrawal

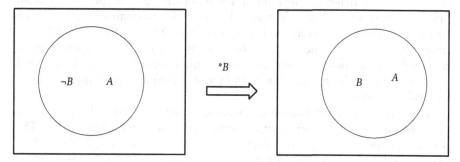

FIGURE 12.3. Attitude Revision

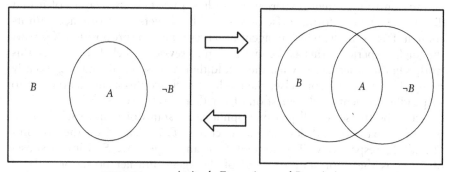

FIGURE 12.4. Attitude Extension and Restriction

starting point the family of conditioning rules for handling different kinds of experience developed in Chapter 10. Like the AGM theory, they are based on 'perturbation–propagation' models of learning, in which the change in an agent's state of mind is viewed as a two-step process. In the first stage, the agent changes one or more attitudes to a particular prospect – or, more generally, to some set of prospects – as a result of undergoing some experience. In the second, she adjusts her attitudes to all other possibilities in order to restore consistency. The experience inducing the initial change is not itself modelled and the notion is to be understood in the widest sense so as to include not only sensory experience but also deliberation, reception of a message from a reliable information source, memory recall or even hypnosis.

Our earlier treatment of attitude change will be generalised in two ways. Firstly, we allow an attitudinal state to consist of a *set* of pairs of probability and desirability functions, rather than just a single pair. And, secondly, we allow for different types of attitude change that involve different responses to constraints deriving from experience. For, as we recognised before, agents with imprecise attitudes don't change their mind only by revising them. Sometimes they make up their mind about something and come to hold an opinion where once they held none. Equally, they sometimes withdraw or suspend judgements, thereby giving up opinions. These actions are different from revision: in essence, while revision is a passage from one set of permissible judgements to another set, opinion formation and withdrawal involve, respectively, a reduction in the set of permissible judgements and an expansion of them.

As before, a rule for changing a state of mind is specified by a function that maps an experience and a prior attitudinal state to a posterior state. The notion of experience requires further refinement, however. The *content* of an experience can be specified as before – i.e. either intensionally, as a set of constraints, or extensionally, as the set of pairs of probability and desirability functions consistent with these constraints. But now we must recognise that, in addition to a content, experience also has a *valency*: the feature of it that directs the agent to form, revise or suspend her beliefs in accordance with its content. For example, when an agent observes that some proposition X is true, she might experience this as the instruction to revise her beliefs to include this truth. On the other hand, when she is told that X by someone, having recently been told $\neg X$ by someone else, she might experience this as an instruction to suspend judgement on the question of whether X is true or not.

Let Σ be the set of all possible opinionated states of mind, each being a pair of probability and desirability functions $\langle P, V \rangle$ defined on the domains Ω and Ω', respectively. The content E of any experience $\mathcal{E}$ is just a subset of Σ, as is the prior state of mind of the agent and her posterior state of mind following the change induced by $\mathcal{E}$, respectively denoted by S and S_E. The experience $\mathcal{E}$ itself is a function from subsets S of Σ to a member S_E of Σ. Hence $\mathcal{E}(S) = S_E$. What state of mind an experience induces depends on its valency, which can be thought of as an imperative to do something with the content of that experience (see van Fraassen, 1989). To denote the relevant valencies of experience, I will use the symbols $+$, $-$ and $*$ for attitude formation, suspension and revision, respectively. So $*\mathcal{E}$ corresponds to the command 'Revise attitudes in accordance with $\mathcal{E}$', $+\mathcal{E}$ to the command 'Form attitudes in accordance with $\mathcal{E}'$ and $-\mathcal{E}$ to the command 'Suspend attitudes to $\mathcal{E}'$. Similarly, we will denote the posterior states of mind obtained by revising, forming or withdrawing attitudes in response to an experience $\mathcal{E}$ by S_E^*, S_E^+ and S_E^-, respectively. Hence $*\mathcal{E}(S) = S_E^*$ and so on.

In thinking about responses to experiences that have both contents and valency, let us continue to be guided by considerations of minimal change and in particular the idea that if an experience is silent on some conditional probability or desirability then the agent's attitudinal state with respect to it should not change. As the focus of interest is on the different kinds of attitude change, rather than the different kinds of experience, I will restrict attention to one rather general kind of experience content, namely that represented by a distribution of probability and desirability across a particular partition of prospects. Recall that, in the case of fully opinionated states of mind, minimal attitude revision in response to experiences of this kind respects a Rigidity condition requiring preservation of conditional probabilities and desirabilities given the elements of the partition (see page 203).

To extend this analysis to the non-opinionated case, let $\mathcal{A} = \{\alpha_i\}$ be a partition of the Boolean algebra of propositions and $\Xi = \{E^k\}$ be a partition of the set of states of mind Σ in terms of opinions regarding the elements of $\mathcal{A}$. So each $E \in \Xi$ is a set of pairs of functions $\langle P, V \rangle \in \Sigma$ such that $P(\alpha_i) = a_i$ and $V(\alpha_i) = b_i$, for some real numbers $a_i > 0$ and b_i such that $\sum_i a_i = 1$ and $\sum_i a_i \cdot b_i = 0$. The set Ξ serves here as the set of contents of the experiences that we consider. As such experiences concern only attitudes to the partition elements and are silent on the requisite conditional attitudes to take, given their truth, considerations of minimal change require that:

Rigidity with respect to $\mathcal{A}$ For all $[E] \in \Xi$ and $\langle P, V \rangle \in S$, for all $\langle Q, W \rangle \in \mathcal{E}(\langle P, V \rangle)$:

$$Q(\cdot|\alpha_i) = P(\cdot|\alpha_i)$$
$$W(\cdot|\alpha_i) = V(\cdot|\alpha_i)$$

I will now propose three rules of attitude change, respectively appropriate as responses to the instruction to revise, form or suspend one's state of mind, that are minimal in the sense of satisfying this Rigidity requirement.

12.2.1 Attitude Revision

In Chapter 10 I argued that generalised conditioning was the correct rule of attitude revision for maximally opinionated agents when experience took the form of a redistribution of probability and desirability across a particular partition of prospects. Extending this rule of revision to the states of mind of non-opinionated agents is most naturally achieved by taking a posterior state of mind to be the *union* of the probability–desirability pairs obtained by generalised conditioning on the elements of the prior state of mind.

Recall that a maximally opinionated agent revises her state of mind by generalised conditioning on an experience $E \in \Xi$ just in case her new state of mind $\langle Q, W \rangle$ is related to her prior state of mind $\langle P, V \rangle$ by, for all prospects $\beta \in \Omega$ and for all $\alpha_i \in \mathcal{A}$ such that $P(\beta | \alpha_i) > 0$,

$$Q(\beta) = \sum_i P(\beta | \alpha_i) \cdot a_i$$

$$W(\beta) = \sum_i [V(\beta | \alpha_i) + b_i] \cdot Q(\alpha_i | \beta)$$

where the a_i and b_i are, respectively, the new probability and desirability values for the α_i that are yielded by experience.

Generalising to imprecise states of mind, we can say that new state of mind S_E^* is obtained by conservative opinion revision on the experience $E \in \Xi$ iff:

Conservative Revision $S_E^* = \{\langle Q, W \rangle : \langle P, V \rangle \in S\}$

where Q and W are the probability and desirability functions determined by generalised conditioning from P and V. It is evident that conservative revision on a partition $\mathcal{A}$ satisfies Rigidity with respect to it since each element in S_E^* is constructed in accordance with it. Note also that attitude revision of this kind takes an agent from an imprecise state of mind to another state of mind that is no less precise than before and, typically, no more either (the exception being when $a_i = 0$ for some $\alpha_i \in \mathcal{A}$).

12.2.2 Attitude Formation

The second kind of change we consider is that of attitude formation, when an agent makes up her mind on some question by adopting some constraint on her attitudes and rejecting as impermissible all attitudinal states that are incompatible with it. She might, for instance, adopt a definite probability for some event on the basis of statistical evidence or adopt a particular preference for a holiday destination after reading through brochures. In doing so she does not simply revise each of the opinions she initially regarded as permissible, she entirely dispenses with those that are ruled out by the adopted constraint. So attitude formation leads to a reduction in the number of the agent's avatars.

Formally, let E be any experience content that is consistent with the agent's prior state of mind – i.e. $S \cap E \neq \emptyset$. Let S_E^+ be her posterior state of mind after forming her opinions on the basis of E. Then we say that the agent changes her state of mind S by conservative opinion formation from E just in case:

Conservative Formation $S_E^+ = S \cap E$

In the case where the new constraint takes the form of an assignment of probabilities and desirabilities across a partition $\mathcal{A} = \{\alpha_i\}$, with each α_i being assigned probability a_i and desirability b_i, the posterior state S_E^+ is just the member of the prior S that satisfies this new constraint. It follows that

$$S_E^+ = \{\langle P, V\rangle \in S : P(\alpha_i) = a_i, V(\alpha_i) = b_i\}$$

Attitude change by conservative formation is clearly different from change by conservative revision, with S_E^+ typically being a strict subset of S_E^*. Indeed, only in the special case where the agent already accepts E, in the sense that $S \subseteq E$, will the two coincide. Nonetheless, conservative formation also preserves conditional probabilities and desirabilities in accordance with Rigidity, and in that sense also constitutes a form of minimal attitude change.

Attitude formation provides a formal mechanism for avoiding the learning problem discussed in the previous chapter. Recall that, to the question of how an imprecise probabilist who was initially open-minded about the distribution of balls in an urn could become less open-minded as she acquired information about the distribution, we answered that she could simply discard certain hypotheses regarding the distribution when they become sufficiently implausible. To do this would in effect be to revise her beliefs by conservative formation in response to constraints that she takes to be imposed by her observations; in particular, regarding which hypotheses about the distribution are no longer sustainable in the light of the evidence of samples. Of course, to present a formal possibility is not to provide an answer to the substantial question as to the conditions under which an agent should make up her mind in this way; for instance, as to how many white balls must be drawn before she can reasonably exclude extreme hypotheses such as H_1 (that there is only one white ball). We return to this issue in the final chapter.

12.2.3 Attitude Withdrawal

The last case to consider is one in which an agent wishes to suspend or withdraw judgement on some issue for one of the reasons mentioned before. In suspending judgement she *reduces* the constraints on her attitudes, so the state of mind that she achieves should be less opinionated than her original one – i.e. her new state of mind should be a superset of her initial one.

Before looking at belief withdrawal in abstract, let us look at a special case. Suppose that you are in state of mind S and that you learn the state of mind S' of someone whose judgement you respect and whom you believe to be well informed. What should you do? One thing you could do is simply hold on to your current opinions without modifying them at all, but this seems reasonable only when you regard the other person's opinions as completely uninformative about the truth. Another thing you might do is adopt all her opinions along with your own, so that your state of mind is the intersection of your and the other individual's initial states of mind, $S \cap S'$. This amounts to treating her opinions as constraints on yours, something that might be reasonable in special circumstances, but hopeless in situations when your opinions disagree (for then

$S \cap S' = \varnothing$). Thirdly, you could revise your opinions to some degree, perhaps adopting ones that are intermediate between yours and the other individual's. Fourthly, you could suspend judgement on the question of whose opinions are correct by adopting the union of your states of mind, $S \cup S'$. Finally, of course, you could aim for some combination of these, holding on to those opinions of your own that you regard as well founded, completely or partially deferring to those of the other where you believe them to have special information and suspending judgement in other cases.

Let's focus on the fourth of these possibilities. When is adopting the union of your and the others' opinion states the correct way to suspend judgement? Suspending judgement in the face of disagreement with someone will involve some weakening of one's opinion state, but how much of one's opinion must be withdrawn depends on the extent of the disagreement. Suppose this disagreement is centred on a particular partition $A = \{\alpha_i\}$, in the sense that you differ on the probabilities and desirabilities you ascribe to the α_i but not on your conditional probabilities and desirabilities given the α_i. (Disagreement can always be localised to some such partition in this way, though it might have to be a very fine one when the disagreement is severe.) Then $S \cup S'$ will just be the opinion state that minimally departs from your prior one but which does not preclude the other person's opinions. Any further weakening of your opinion would be unnecessary from the point of view of suspending judgement on the question of who is correct; anything less than this weakening would require ruling out the other's opinion on some proposition.

The point generalises to disagreement with any number of other individuals. If a disagreement between you and n of your peers can be localised to some particular partition A, and you wish to suspend judgement on who is correct, then the minimal departure from your current opinion state which fits this bill is the opinion state that is the union of yours and the others'. In the limit, if each $E \in \Xi$ is represented in this disagreement, then once you have suspended judgement on who is correct you will have no opinion at all on the α_i.

This then gives us the notion of opinion withdrawal that we were looking for. When you withdraw judgement on some issue you in effect grant the permissibility of different opinions on the issue from the one that you currently hold and so adopt the state of mind that is appropriate to cases of complete disagreement amongst you and your peers over the issue. In other words, you withdraw your opinions based the constraints imposed by the experience $E \in \Xi$, represented by a set of probability and desirability judgements on the α_i, by completely suspending judgement on the α_i.

Formally, let $\Xi = \{E^k\}$ be a partition of Σ in terms of opinions on the α_i and let us suppose that the agent's prior state of mind is given by $S \subseteq E \in \Xi$. Now, for any $E^k \in \Xi$, the state of mind $S^*_{\mathcal{E}^k}$ obtained by conservative revision on the partition A is just the state of mind of someone whose disagreement with the agent can be localised to A, with the agent holding opinion E and the other holding opinion E^k. For, by the definition of conservative revision, $S^*_{\mathcal{E}^k}$

differs from S only in terms of the probabilities and desirabilities assigned to the α_i and not on the conditional probabilities and desirabilities given the α_i. So we can say that the agent changes her mind by conservative withdrawal of her opinions $E \in \Xi$ just in case her new state of mind S_E^- satisfies:

Conservative Withdrawal $S_E^- := \bigcup_{E^k \in \Xi} \{S_E^*\}$

Opinion withdrawal, so defined, is clearly consistent with the Rigidity requirement: that the agent's conditional attitudes, given some prospect, should be unaffected by the suspension of attitude towards that prospect. (This just follows from the fact that conservative *revision* respects Rigidity.) In fact, it is the least opinionated state of mind satisfying this requirement, so that any state of mind achieved by a less complete withdrawal of opinion can be recovered by the addition of further constraints on opinion.

12.2.4 Connections*

The three operations of revision, formation and withdrawal are related in a variety of ways. In the previous subsection, opinion withdrawal was defined in terms of revision. But it is more natural perhaps to take withdrawal as the primitive notion and define revision in terms of it and opinion formation. For, intuitively, one can revise any set of opinions on some issue by first suspending one's opinion on the issue and then forming a new one. To do this, however, we need to provide a somewhat different characterisation of withdrawal.

Suppose that the agent is in prior state of mind $S = \{\langle P, V \rangle\}$ such that $P(\alpha_i) > 0$ for all elements α_i of the n-fold partition $\mathcal{A}$. Let $\mathcal{K} = \{k_j = \langle (a_1, b_1), ..., (a_i, b_i), ..., (a_n, b_n) \rangle\}$ be the set of n-tuples of pairs of real numbers such that $0 \leq a_i \leq 1$, $\sum_i a_i = 1$ and $\sum_i a_i \cdot b_i = 0$. Then we define for each $k \in \mathcal{K}$:

$$Q_k(\beta) = \sum_i a_i \cdot P(\beta | \alpha_i)$$

$$W_k(\beta) = \sum_i [V(\beta | \alpha_i) + b_i] \cdot Q_k(\alpha_i | \beta)$$

Then:

$$S_E^- = \{\langle Q_k, W_k \rangle : \langle P, V \rangle \in S, k \in \mathcal{K}\}$$

This definition of opinion withdrawal does not apply to partitions on which the agent's prior probability for one of the elements is zero. As in the case of opinion revision, this limitation could be overcome by working with measures of conditional attitudes that are defined for zero-probability prospects. Note that the state of mind achieved by opinion withdrawal so defined satisfies the Rigidity requirement in the sense that each pair $\langle Q_k, W_k \rangle$ assigns the same conditional probabilities and desirabilities, given the α_i, as the pair $\langle P, V \rangle$ from which they are derived.

We now have independent characterisations of conservative opinion withdrawal, formation and revision. So construed, these operations are related by the following identity. Suppose that the agent's prior state of mind is given by $S \subseteq E \in \Xi$ and let $E' \in \Xi$ be the constraints imposed by experience. Then, as is proved in (Bradley, 2007a, proposition 9):[1]

Decomposition $S^*_{E'} = (S^-_E)^+_{E'}$

It is natural to read Decomposition as a formal expression of the intuition that opinion revision can be reduced to a sequence of opinion withdrawal followed by opinion formation. It also follows immediately from it that opinion formation and withdrawal can be regarded as 'opposite' operations in the sense that withdrawing an opinion and then re-forming it returns one to one's original state of mind – i.e. for any $E \in \Xi$ such that $S \subseteq E$:

Recovery $S = (S^-_E)^+_E$

Opinion formation can therefore be defined in terms of opinion withdrawal, namely as the unique operation on states of mind that ensures satisfaction of Recovery. The situation is not symmetrical, however, since forming an opinion using some assignment of probability and desirability to a particular partition, and then withdrawing one's opinion on that partition, need not return one to one's original state of mind. Indeed, it will do so iff the original state of mind was completely non-opinionated with regard to the partition in question. In general, however, we have only that $S \subseteq (S^+_E)^-_\varepsilon$, and this does not suffice for a unique characterisation of opinion withdrawal in terms of opinion formation. So, from a formal perspective, it is natural to take withdrawal as the primitive operation and define formation and revision operations in terms of it.[2]

12.3 UNAWARENESS

The assumption that agents are aware of all relevant prospects, though commonly made, is as dubious normatively and descriptively as the assumption that they have opinions on all of those that they are aware of. Descriptively, because we must often make decisions without knowing what all our options are or what all the contingencies are that might affect their outcomes. Normatively, because it is not irrational for agents to be unaware of some

---

[1] A rather similar equivalence is central to the AGM theory, where it carries the title of the Levi Identity. By the Levi Identity, revising a belief set by some sentence A is equivalent to withdrawing $\neg A$ from the belief set and then expanding with A. This suggests that the withdrawal operation characterised here is a good candidate to serve as the counterpart to contraction in the AGM framework.

[2] Hansson (2009) shows that both revision and contraction can be united in a single operation, a Sheffer stroke for belief change.

prospects even if it is typically desirable that they are. In this section, therefore, we investigate rational agency in the context of less than full awareness.

Although we tend to speak as if someone is either aware of a prospect or not, there are at least three different senses or grades of unawareness that should be distinguished. Firstly, one can be unaware of a prospect when one has not encountered it or heard anything about it before, such as when one is unaware of the fact that there is a bus that goes to the town one wants to visit because one has never been there before. Secondly, one might have heard about something but have since forgotten about it or fail to recall it at a particular time because it slips one's attention. And, thirdly, one may deliberately exclude possibilities from consideration by 'blinding' oneself to them, by removing them or by having them removed from one's attention. This can be very difficult for individuals to achieve, but mechanisms for doing this are common in certain institutional settings. When selection panels consider applications for a job, for instance, they may not be allowed to take certain features of the applicants into consideration, such as their age or race, and so will take steps to prevent knowledge of these characteristics from being available. Similarly, jurors are screened to ensure unawareness of certain characteristics of the defendant which may bias their judgements and are required to avoid doing things that might result in them becoming aware of these characteristics (such as searching the web for background information).

What these situations of unawareness have in common is that certain contingencies or prospects are not available to the agent's consciousness at the time at which she is deliberating on some question. What makes them different is the cause of the unavailability: experiential lack, forgetfulness or deliberate blocking of access. These causal mechanisms and their opposites are also at work in changes of states of awareness. We typically become aware of something by encountering it or hearing about it (either accidentally or as a result of a deliberate search), but it can also come about through an act of recall. On the other hand, we typically lose awareness of something by forgetting about it, either temporarily or permanently, though there are more dramatic possibilities (such as brain damage). Forgetting is largely an unconscious process but we can of course do things to either delay or speed it up, both individually and collectively.

Our task now is to investigate what is required of an agent who is rational, but less than fully aware. This has three components: what constraints does rationality impose on less than fully aware states of mind? What form do rational changes of awareness take? And how should we make decisions in the light of the possibility of unawareness? I shall argue that, contrary to what might seem to be the case, the less than fully aware agent faces just the same rationality constraints as the fully aware one. On the other hand, the possibility of changes in the set of things of which she is aware means that entirely new forms of attitude change need to be examined and that decisions might have to draw on different considerations.

12.3.1 Rational Unawareness

An agent can form attitudes only towards prospects that she is aware of. But what should she do if she is conscious of the possibility (as she should be) that she may not be aware of all relevant prospects? The question seems to pose a threat to the account of rationality that was developed under the implicit assumption of full awareness. For what good does it do to ensure that our beliefs and desires are consistent, or even coherently extendable, when we believe that we have not taken everything into consideration? Why bother, for instance, to ensure that our degrees of belief over some set of alternatives sum up to one when we suspect that the set might not be exhaustive? In short, is it not the case that the standards of rationality so carefully built up over the previous chapters must be totally overhauled in the light of unawareness?

To address this two kinds of cases need to be considered. A simple example will suffice to illustrate each. Suppose you regard tomorrow's weather as relevant to your decision about what activity to undertake and are aware that it can be cloudy, rainy, snowy or sunny and also that it can be either cold or hot. This implies a set of eight prospects to consider: cold and cloudy, hot and cloudy, cold and rainy, and so on. Now suppose it is brought to your attention that humidity also affects how enjoyable the various activities are, a fact of which you had previously been unaware. In this case, you should refine your set of prospects so as to take the possible properties of being humid or not into account and then extend your attitudes to cover these more refined possibilities. In doing so, you need to make no changes to your attitudes to the prospects of which you were aware all along, for the need to develop more refined attitudes does not in itself imply that there is anything wrong with one's coarse-grained ones. Similarly, the possibility that one will have to refine one's attitudes some time in the future does not give one any reason to change one's current attitudes: they could be satisfactory just as they are. So this case poses no immediate threat to our picture of a rational state of mind.

The second kind of case is, on the face of it, more difficult to deal with. Suppose that it is brought to your attention that it could also be misty tomorrow, a possibility that had not occurred to you before, but which seems relevant to your decisions. In this case you need to add the prospects 'cold and misty' and 'hot and misty' to your set. Your new set is not a refinement of your old one as the new prospects are incompatible with any of those that you originally considered. This means that you cannot simply extend your attitudes to them without modifying your opinions about the latter. You must, for example, reduce the probability of at least one of the prospects in the initial set. (How this should be done will be discussed later.)

It is this second case that gives rise to unease. For if we are aware that we could face such a case then we know that at least some of our current opinions must be wrong. It would seem, therefore, that we should adjust our current attitudes to take into account this possibility. But how are we to do this?

Two strategies present themselves. Firstly, one could adopt non-probabilistic degrees of belief; for instance, degrees that don't sum to one on every partition or that can sum to a number greater than one. This would allow one to give positive probability to new possibilities as and when one becomes aware of them. Secondly, one could create a catch-all prospect – the prospect corresponding to 'all those possibilities that I have not thought of' – and give positive probability to it.

Neither strategy is appealing. For, given that we don't know anything about the prospects that we are potentially unaware of, on what basis are we to determine what non-probabilistic degrees of belief are appropriate to adopt, or what probability we should assign to the catch-all prospect? But if we recognise the lack of such a basis then we will be driven towards a state of radical non-opinionation, which does not seem to be much of an improvement over that of conscious unawareness.

This line of thinking is, in any case, based on a misunderstanding. What our current opinions reflect are the *relative* plausibility and desirability of the prospects that we are aware of. Although such relational judgements imply specific degrees of belief and desire relative to a given set of prospects, these degrees cannot be compared across different sets of prospects. If I am aware only that the weather can be cloudy or sunny then the judgement that it is as likely to be sunny as it is to be cloudy implies a degree of belief of one-half in the prospect of sunny weather. But when I become aware of the fact that misty weather is also possible, I will need to revise this degree of belief. Not because my judgement regarding the relative credibility of sunny and cloudy weather was wrong, but because relative to the larger set of prospects that I am now aware of my relational judgement implies different degrees of belief.

The upshot of this is that the possibility of unawareness does not imply the need for a radical revision of our account of rationality. The rationality constraints on an agent's relational beliefs and desires do not depend on the assumption that the agent is in a state of full awareness, but apply straightforwardly to her attitudes regarding those prospects that she is aware of. On the other hand, as we shall see, the possibility of unawareness has significant implications for attitude revision. But, before we turn to this topic, let us consider the question of how to model unawareness more formally.

12.3.2 Representations of Unawareness*

How should we represent the state of mind of an agent who lacks full awareness? Several proposals for modelling unawareness already exist, of which the most prominent are perhaps the awareness structures proposed by Ronald Fagin and Joseph Halpern (1987) and Halpern (2001), and the Generalised Standard Structures of Eddie Dekel, John Barton Lipman and Aldo Rustichini and Salvatore Modica and Aldo Rustichini (1999), both of which extend the standard models of epistemic game theory. The approach that I will take is much simpler, because it is designed to achieve much less. The essence of the proposal is to introduce a distinction between the background domain of prospects (the universal or modeller's domain) and the agent's subjective domain: the set of all prospects of which the agent is aware at a particular time. Her credibility and preference relations will be defined on the subjective domain only, but as her awareness changes so too will the domain of these relations. It is these dynamics that make them interesting.

Let $\Omega = \langle X, \models \rangle$ be a background Boolean algebra of propositions, which we can think of as the set of all possibilities, or the set of all possibilities that the modeller can conceive of. In contrast, let $\Omega_A = \langle A, \models \rangle$ with $A \subseteq X$ be the set of possibilities of which the agent is aware. Then the agent's state of mind can be represented by a structure $S = \langle \Omega, \Omega_A, \{\langle P, V \rangle\} \rangle$, where $\{\langle P, V \rangle\}$ is a set of pairs of probability and desirability functions on Ω_A. While the fact that the agent is less than fully opinionated is reflected in the fact that she countenances multiple sets of probability and desirability functions, the fact that she is less than fully aware is reflected in the fact that these functions are defined on a restricted algebra of prospects.

Recall that the basic rationality condition on an incomplete state of mind is that of weak consistency – i.e. that of the coherent minimal extendability of the agent's relational beliefs and preferences. To the set of prospects of which she is aware or to the entire background set? It doesn't matter, since if they are extendable to the former then they are to the latter. So the basic rationality requirement on agents, including those who are aware that they may be unaware of certain possibilities, is to have weakly consistent relational attitudes to those prospects that they are aware of.

This means that only a slight tweak is required in the representation of a Bayesian-rational agent to allow for less than full awareness, in the form of a relativisation of her attitudes to her subjective domain of awareness. More formally, let $\unrhd$ and $\succsim$ be, respectively, a credibility and a preference relation on Ω_A, the Boolean subalgebra of propositions of which the agent is aware. Then:

Bayesianism without Full Awareness Let $\succsim$ be a preference relation on Ω_A that satisfies the Weak Axiom of Consistency for preference. Then there exists a maximal set of pairs of probability and desirability functions $S = \{\langle P_i, V_i \rangle\}$ on Ω_A that explains or rationalises the preference relation $\succsim$ in the sense that, for all $\alpha, \beta \in \Omega'_A$:

$$\alpha \succsim \beta \Leftrightarrow \forall V_i \in S, \ V_i(\alpha) \geq V_i(\beta)$$

12.4 CHANGES IN STATES OF AWARENESS

Our state of awareness can, and often does, change, as we become aware of new possibilities and forget about others. For example, when new goods come onto the market, or you meet someone for the first time or you are told about a restaurant that has just opened in the neighbourhood, you become aware of certain possibilities for the first time. When your attention shifts to other things or your memory fades, you may lose awareness of them.[3] In reality, these kinds of changes may come in grades, with possibilities moving in or out of the centre of our attention rather quickly, but in and out of more general awareness much more slowly.

When an agent becomes aware of prospects of which she was previously unaware then the domain on which she has attitudes is extended. On the other hand, when she forgets about certain issues or deliberately suppresses them, then the domain over which she has attitudes is reduced. Our question is: how should an agent change her attitudes when the domain of her awareness changes in either of these ways? The question has received a small amount of attention within the standard framework of Savage's decision theory (see, in particular, Karni & Vierø, 2013). Here we will seek an answer to it within the framework of the book, drawing again on considerations of minimal change.

There are two basic types of domain expansion to consider. Firstly, the domain of an agent's attitudes can expand as a result of a refinement of the possibilities that she entertains. In this case, a reasonable conception of minimal change will dictate that she should not revise her attitudes on the coarser domain. For the mere fact that one's attitudes are defined on a domain that has proved to be too coarse does not give one any reason to change one's attitude to the coarse-grained prospects themselves. It follows that, in this case, attitude change should take the form of an extension of one's degrees of belief and desire to the 'new' prospects (those of which one has just gained awareness) in a manner consistent with leaving one's attitudes to the 'old' prospects unchanged. That is, one's new degrees of belief and desire should agree with one's old ones over their common domain.

Secondly, the domain of awareness can expand because the agent entertains one or more possibilities disjoint from those she entertained initially. In this case, her attitudes to the possibilities initially entertained must change. In particular, since both her old and new degrees of belief in the elements of any partition must sum to one, if a new element is added to the partition as a result of the agent's new awareness of it, and she does not consider it impossible, then she must diminish her degrees of belief in the other partition elements. Nonetheless, as we saw before, this does not entail that her relational attitudes to the 'old' prospects must change. Becoming aware of the possibility of misty weather does not affect the relative credibility of snowy and sunny weather, for example. So, by a principle of minimal change, they should not. In other words, the key to conservative attitude change in cases where we become aware of prospects that are inconsistent with those that we previously took into consideration is that we should extend our relational attitudes to the new set in such a way as to conserve all prior relational beliefs and preferences.

---

3   See Titelbaum (2012) for a discussion of the rationality constraints on forgetting.

12.4.1 Domain Extension*

Within the Bayesian framework, conservation of the agent's relational beliefs is ensured by the rigidity of her conditional probabilities. So we can conclude that conservative belief change requires the agent's new conditional probabilities, given the old domain, for any members of the old domain should equal her old unconditional probabilities for these members. A similar argument applies to her new degrees of desire, leading to the conclusion that when an agent becomes aware of new possibilities she should extend her current degrees of desire to the new domain in such a way as to ensure that her new conditional degrees of desire, given the old domain, should equal her old degrees of desire.

To make this more precise, let $\Omega = \langle X, \models \rangle$ be a Boolean algebra of prospects of which the agent is initially aware and let $\vee X$ be its upper bound: the disjunction of all prospects of which she is aware. Let $S = \{\langle P, V \rangle\}$ be the agent's initial state of mind, with the $\langle P, V \rangle$ being pairs of probability and desirability functions on Ω. Let E be some set of prospects not contained in X. Let Y be the closure of $X \cup E$ under the Boolean operations and $\Omega_E^\oplus = \langle Y, \models \rangle$, called the *extension* of Ω by the set of prospects E, be a Boolean algebra of prospects based on Y. Note that $\vee X$ belongs to Y. Then, for any $\langle P, V \rangle \in S$, a corresponding pair of probability and desirability functions $\langle P_E^\oplus, V_E^\oplus \rangle$ on $\Omega_E^\oplus$ is called a *rigid extension* of $\langle P, V \rangle$ to $\Omega_E^\oplus$ iff, for all $\alpha \in X$,

$$P_E^\oplus(\alpha \mid \vee X) = P(\alpha)$$
$$V_E^\oplus(\alpha \mid \vee X) = V(\alpha)$$

Finally, we can define $S_E^\oplus$, the extension of an agent's initial state of mind S to the domain $\Omega_E^\oplus$, as the set of rigid extensions of the agent's probabilities and desirabilities to the elements of $\Omega_E^\oplus$ – i.e.:

Domain Extension $S_E^\oplus := \{\langle P_E^\oplus, V_E^\oplus \rangle : \langle P_E^\oplus, V_E^\oplus \rangle$ is a rigid extension of a $\langle P, V \rangle \in S\}$

We can illustrate this with our earlier examples of domain refinements and expansions. Take X to be the closure of the set of mutually exclusive propositions {cloudy, rainy, snowy} under the Boolean operations and the agent's state of mind to be given by a single pair $\langle P, V \rangle$. First, suppose that E is a set of propositions {hot, cold} that refines X. Then the extension of X by E will be the Boolean closure of the set {cloudy and hot, cloudy and cold, rainy and hot, ..., snowy and cold} and any rigid extension $\langle P_E^\oplus, V_E^\oplus \rangle$ of $\langle P, V \rangle$ will agree with $\langle P, V \rangle$ in the assignment of probabilities and desirabilities to the elements of X – i.e. $P_E^\oplus(\text{cloudy}) = P(\text{cloudy})$, $V_E^\oplus(\text{cloudy}) = V(\text{cloudy})$, and so on.

Second, suppose that E just contains the proposition {misty} which is disjoint from the elements of X. Then the extension of X by E will be the Boolean closure of the set {cloudy, rainy, snowy, misty} and any rigid extension $\langle P_E^\oplus, V_E^\oplus \rangle$ of $\langle P, V \rangle$ will agree with $\langle P, V \rangle$ in the assignment of probabilities and desirabilities to the elements of X conditional on it being either cloudy, rainy or snowy – i.e. $P_E^\oplus(\text{cloudy}|\text{cloudy or rainy or snowy}) = P(\text{cloudy})$, $V_E^\oplus(\text{cloudy}|\text{cloudy or rainy or snowy}) = V(\text{cloudy})$, and so on.

12.4.2 Domain Restriction*

We can treat domain contraction in much the same way as domain extension. When the domain is coarsened, the principle of minimal change once again dictates that the agent's attitudes on the coarser domain should not change. The only difference is that the coarser domain is now the new domain, not the old one. Similarly, when the domain is restricted because the agent loses awareness of a possibility that is disjoint from those that she continues to entertain, she should adjust her degrees of belief and desire only as much as is needed in order to renormalise them. Jointly, this implies that she should conserve her old conditional probabilities and desirabilities for elements of the new domain, given the new restricted set of atomic possibilities.

As before, let $\Omega = \langle X, \models \rangle$ be a Boolean algebra of prospects of which the agent is initially aware, $S = \{\langle P, V \rangle\}$ be the agent's initial state of mind and R be some set of prospects contained in X. Let Z be the largest Boolean subset of X not containing R and $\Omega_R^\ominus = \langle Z, \models \rangle$, called the *restriction* of Ω by $R = \{R_i\}$, be a Boolean algebra of prospects based on Z. Let $\vee Z$ be its upper bound. Then, for any $\langle P, V \rangle \in S$, a corresponding pair of probability and desirability functions $\langle P_R^\ominus, V_R^\ominus \rangle$ on $\Omega_E^\ominus$ is called a *rigid restriction* of $\langle P, V \rangle$ to $\Omega_R^\ominus$ iff, for all $X \in \Omega$,

$$P_R^\ominus(X) = P(X | \vee Z)$$
$$V_R^\ominus(X) = V(X | \vee Z)$$

Finally, we can define $S_E^\ominus$, the restriction of the agent's initial state of mind S to the new domain $\Omega_E^\ominus$, as the set of rigid restrictions of the agent's probabilities and desirabilities – i.e.:

Domain Restriction $S_E^\ominus := \{\langle P_E^\ominus, V_E^\ominus \rangle : \langle P_E^\ominus, V_E^\ominus \rangle$ is a rigid restriction of $\langle P, V \rangle \in S\}$

Note that $S_E^\ominus \subseteq S \subseteq S_E^\oplus$. Furthermore, domain extension and restriction are 'opposite' operations in much the same way as attitude formation and withdrawal are. For, if an agent first becomes aware of new prospects and extends her attitudes rigidly to her new domain of awareness and then subsequently loses awareness of these same prospects, she will return to her original state of mind iff she rigidly restricts her attitudes to the old domain. Formally:

Domain Recovery $S = (S_E^\oplus)_E^\ominus$

The opposite is not true, however, as an agent who rigidly restricts her attitudes and then rigidly extends them back to the old domain will typically find herself in a less precise state of mind than before – i.e. $S \subseteq (S_E^\ominus)_E^\oplus$. So, while it is possible to define the domain restriction in terms of domain extensions as the unique operation satisfying Domain Recovery, it is not possible to do it the other way round. If we seek parsimony in our basic attitudinal change operations then we must take as our primitives the operations of attitude withdrawal and domain extension, use Recovery (see 252) to define attitude formation and domain restriction and then derive any more complex revisions as sequences of these four operations.

12.4.3 The Problem of New Hypotheses

Clark Glymour (1980) presents Bayesianism with the following problem. Suppose a new scientific hypothesis (e.g. General Relativity) is proposed which is able to explain some known but hitherto unexplained fact (e.g. the advance of the perihelion of Mercury). Intuitively, the new hypothesis's ability to explain this fact should afford it some support. Traditionally, Bayesians conceive of confirmation in terms of the probability-raising effect on a hypothesis of learning some fact. But by Bayes' theorem, the posterior probability of any hypothesis given evidence of probability one is just its prior probability. So the confirmatory effect of known or 'old' evidence apparently cannot be explained by Bayesians.

As John Earman (1992) and James Joyce (1999) observe, Glymour's challenge raises at least three distinct, if related, issues: what probability to attach to new hypotheses; how to account for the confirmatory effect of learning that an 'old' hypothesis explains some known evidence; and how to measure the confirmation relation between hypothesis and evidence in a manner which is independent of the probability of the evidence. Plausible solutions to the second and third issues are to be found in the literature, but it is to the first that the formal work of the previous sections contributes. Let me illustrate its implications by drawing on an example of Sylvia Wenmackers and Jan-Willem Romeijn (2016).

A food safety inspector is testing dishes at a restaurant for salmonella. She interprets a positive result on a food test as certain evidence that a dish has salmonella and considers the prior probability of infection in a kitchen fully implementing safety regulations to be 0.01 and of infection in one which is not to be 0.2. She orders and tests two dishes, finding them to be both contaminated, leading her to conclude that it is very probable that the restaurant is in violation of regulatory standards. To be certain, however, she orders three more dishes, all of which test positive for salmonella. Since this is 'too good to be true' she now wonders whether the test kits have been contaminated, reinterpreting the 'old evidence' from the first tests in the light of this hypothesis.

Let V be the hypothesis that the kitchen has violated regulatory standards, O the hypothesis that regulatory standards have been met but that one of the other salmonella causes is present and C the hypothesis that the tests are contaminated. Let E_i be the proposition that the ith dish tests positive for infection with salmonella. The inspector's initial domain of awareness is $\{V, O\} \times \{E_1, \neg E_1\} ... \times \{E_5, \neg E_5\}$. Suppose that initially $P(V) = 0.05$, $P(O) = 0.5$, $P(E_i|V) = 0.1$ and $P(E_i|O) = 0.01$. After testing, her odds on the two hypotheses are given by

$$\frac{P(V|E_1)}{P(O|E_1)} = \frac{P(E_1|V) \cdot P(V)}{P(E_1|O) \cdot P(O)} = 1$$

$$\frac{P(V|E_1, E_2)}{P(O|E_1, E_2)} = \frac{P(E_1, E_2|V) \cdot P(V)}{P(E_1, E_2|O) \cdot P(O)} = 10$$

Since she considers V and O to be the only possible explanations for the test results, the two tests drive $P(V)$ to $\frac{10}{11}$ – i.e. to near certainty of a regulatory violation. But the positive results in the third test lead instead to a domain extension by C, the new hypothesis, yielding a new domain $\{V, O, C\} \times \{E_1, \neg E_1\} ... \times \{E_5, \neg E_5\}$. If the inspector modifies her probabilities by rigid extension, she must assign some probability to C and modify her probabilities for V and O, without changing their ratios. Suppose she adopts new probabilities $P_C^{\oplus}(C) = 0.01$, so that $P_C^{\oplus}(V) = 0.05 \times 0.99 \approx 0.05$. Then:

$$\frac{P_C^{\oplus}(C|E_1, E_2, E_3)}{P_C^{\oplus}(V|E_1, E_2, E_3)} = \frac{P_C^{\oplus}(E_1, E_2, E_3|C) \cdot P_C^{\oplus}(C)}{P_C^{\oplus}(E_1, E_2, E_3|V) \cdot P_C^{\oplus}(V)} = \frac{1}{0.05}$$

So the inspector concludes that it is 20 times more likely that the test results are caused by contamination then by violations of regulatory standards. She decides not to close down the restaurant.

13

Decision Making under Ambiguity

13.1 INTRODUCTION

Agents must make decisions in situations characterised by uncertainty that differs both in kind and in severity. Differs in kind because they face not only factual uncertainty about the state of the world but also option uncertainty about what the consequences would be of performing one or other of the actions available to them, evaluative uncertainty about the desirability of these possible consequences and modal uncertainty about the space of relevant contingencies. Differs in severity because the quality, amount and coherence of the information that the agent has about relevant prospects can vary to a considerable degree. Mainstream Bayesian decision theory recognises some distinctions in severity (between risk and uncertainty, for instance) but measures all the different kinds of uncertainty in the same way, namely by means of a probability function defined on the set of possible states of the world.

In the first two parts of the book, I argued that such reduction of all uncertainty to factual uncertainty is not always possible or useful, and offered an alternative theory that was applicable even when it was not. Firstly, the probability measure of factual uncertainty was complemented with a desirability measure of evaluative uncertainty that explicitly incorporated dependence on both beliefs about the facts and belief-independent judgements of value, and which could be revised as these beliefs and evaluations changed. Secondly, option uncertainty was captured by a suppositional probability on prospects conditional on an intervention of some kind. These fed into a decision rule prescribing choice, from the set of available options, of the ones that maximise expected desirability gain, relative to the status quo, on the supposition of its performance. When options can be formulated as

Savage-style acts this decision rule coincides with that of maximising subjective expected utility.

This broadly Bayesian decision theory shares with its mainstream cousins the assumption that agents come to decision problems equipped with a complete set of probability and desirability judgements. It is, in other words, a decision theory suitable for a maximally opinionated agent. This implies that, if decision makers want to use such a Bayesian decision theory as a guide to their choices, then they need to reach precise judgements on at least all contingencies relevant to the decision problem they face. This view has come under considerable criticism of late, with many philosophers and economists arguing that in situations of severe uncertainty and/or irresolvable disagreement it is neither possible nor desirable for the decision maker to make precise judgements about all decision relevant contingencies nor for her to make decisions in the manner prescribed by Bayesian decision theory.

This chapter will be devoted to the examination of this contention and its implications. My starting point will be the framework of Imprecise Bayesianism developed in the previous two chapters in which an agent's uncertainty is captured by sets of pairs of probability and desirability functions and which in effect encodes the permissibility of incomplete judgement. I will start by asking how a rational but not maximally opinionated agent might reach decisions, surveying some of the many proposals that have been made in this regard. I will then focus on the question of what role considerations of caution can play in resolving decision problems, asking whether the forms of cautious decision making that are frequently observed are rational and whether they involve violations of Bayesian norms. This will give me the opportunity to take up the last of the three challenges to Imprecise Bayesianism posed in Chapter 11.

In the final chapter I will turn to a second criticism of Bayesian theory, namely that probability does not suffice to measure all the factual uncertainty that agents face. Unlike the first, this criticism extends to the Imprecise Bayesianism defended in the previous chapters. Elaboration of it will lead us to consideration of the role of confidence in judgement and to a proposal as to how confidence judgements can help to resolve some of the problems facing the Imprecise Bayesian.

13.2 REACHING A JUDGEMENT

How should a decision maker choose amongst the courses of action available to her when she lacks precise probabilities and desirabilities for the contingencies relevant to her decision? There are, broadly speaking, two possible responses to this question. Firstly, the decision maker can try and make up her mind to the degree needed to apply Bayesian decision theory, by settling on the required precise desirability and probability judgements. And, secondly, she can make use of a different decision rule from that of maximisation of subjective expected utility; one that is much less demanding in terms of the judgemental precision it requires. In the next two sections I will put some flesh on these alternatives, without trying to settle immediately the question of which is the best route to take. Indeed, later on I will argue that different responses are applicable under different circumstances.

Until quite recently the accepted solution to the problem of decision making under severe uncertainty and/or disagreement was a version of the first response. Classical Bayesians argued that we must turn to subjective judgement for the probability and desirability values required to implement the rule of subjective expected utility maximisation, pointing out that Leonard Savage had shown that considerations of rationality require that decisions are made *as if* they maximise the decision maker's subjective expectation of benefit relative to her degrees of belief and her preferences. So precise judgements (whether explicit or not) are mandatory, on pain of irrationality, irrespective of the circumstances in which the decision is made.[1]

The Classical Bayesian view is most compelling when the source of the less than maximal opinionation on the part of the agent derives from the considerations previously collected together under the label of 'Boundedness' (see Section 11.3). If, for instance, her preferences are incomplete simply because she has not given the matter much thought, then it is perfectly reasonable to expect her to put in the effort required to reach a judgement. But as a general prescription it faces two obvious challenges. The first is that under conditions of severe empirical and evaluative uncertainty and/or disagreement the decision maker may find it very difficult to arrive at a precise subjective judgement about all relevant factors. Indeed, if she is uncertain about the state space itself, she may find it impossible to do so in a non-arbitrary way. More generally, she may reasonably regard each of a range of different probability and utility judgements as equally justifiable.

The second challenge derives from the famous Ellsberg Paradox and experiments based on it, which show that under conditions of severe uncertainty many apparently rational agents seem not to conform to the dictates of Savage's theory. Those who regard this empirical evidence as normatively significant argue that it undermines the pragmatic case for expected utility maximisation and reveals a need for alternative decision rules in conditions of severe uncertainty. In reply, Bayesians argue that this experimental evidence does little to undermine the normative appeal of their theory and argue that we have no choice but to 'bite the bullet' and do the best we can to come up with reasonable probability and utility judgements

---

[1] As Binmore (2008) notes, Savage himself was more cautious and acknowledged that his argument was suited only to circumstances in which you could 'look before you leap' because all contingencies have been foreseen, but this qualification was largely ignored by those who followed him.

by resolving the uncertainty and/or disagreement one way or another.[2] John Broome, for instance, argues:

> The lack of firm probabilities is not a reason to give up expected value theory. You might despair and adopt some other way of coping with uncertainty... That would be a mistake. Stick with expected value theory, since it is very well founded, and do your best with probabilities and values.
>
> (Broome, 2012, p. 129)

Such biting of the bullet need not involve a renunciation of Imprecise Bayesianism. Contexts of enquiry impose different requirements from contexts of decision, and it is reasonable to hold the view that the judgements that an agent takes as the basis for action may be more precise than those she forms on the basis of the evidence she holds. In enquiry it is the Jamesian imperative to avoid error and to keep an open mind that takes precedence; in decision making it is the Jamesian imperative to seek truth and to form an opinion that does. So long as the precise opinions formed for the purposes of making a decision can subsequently be suspended if there is opportunity for further enquiry, there is no reason to fear spells of pragmatic dogmatism (hence the importance to Imprecise Bayesianism of the rule of opinion withdrawal described in the previous chapter).

Clearly, though, this imperative to form opinions when they are called for needs to be backed up with some advice on how this might be done when the decision maker lacks information about relevant contingencies or is divided in her evaluations. If she has the time, she can seek further information or deliberate further in the hope that this will help to settle matters. This would normally require postponement of the decision, a possibility examined in the next section, so I shall set it aside for the moment. If postponement is ruled out there are two (not mutually exclusive) strategies she can pursue: she can try and identify the 'best' opinion by applying additional (non-evidential) considerations; or she can try and form an aggregate of the permissible opinions that is, in some sense, the best compromise between them. Let's look at some examples of each.

13.2.1 Picking the 'Best'

We have already encountered a salient version of the first strategy. When the evidence does not fully discriminate between various hypotheses, objective Bayesians look to the Principle of Indifference to determine a unique probability assignment. Recall that application of this principle to a set $O = \{O_i\}$ of mutually exclusive and exhaustive outcomes leads, in the absence of any information distinguishing these outcomes, to an assignment of equal probability to each. But what about when we have partial information about these outcomes? Then, it is argued, we should pick a probability

---

[2] See, for instance, Nabil Al-Najjar and Jonathan Weinstein (2009).

assignment consistent with this information that departs minimally from an equal assignment to all outcomes. On a natural metric for minimal departure, this yields the rule of Maximum Entropy (MaxEnt).[3] As its name suggests, this rule picks the member P of the set of probability functions consistent with the evidence the agent holds which maximises entropy relative to the set of outcomes – i.e. it minimises the measure

$$H(P) = \sum_i P(O_i) \cdot \log(P(O_i))$$

By departing minimally from the equal assignment, we avoid giving any more probability to outcomes than the evidence requires us to. And, in that sense, we thereby adopt the most equivocal set of degrees of belief that we can, given the evidence we hold. But what reasons do we have to be equivocal in this sense? Jon Williamson (2007b) argues that the adoption of equivocal degrees of belief leads, on average, to more cautious decision making and hence that caution furnishes pragmatic grounds for MaxEnt. Suppose I don't know the bias on a coin that is to be tossed and must bet on how it lands. An epistemic 'equivocator' will bet cautiously in the sense of refusing bets that pay $1 on heads that cost more than 50c. In contrast, the epistemically 'reckless', who adopts a probability of one for heads, will accept a bet at any price up to $1 for the bet. On the other hand, however, while Reckless will not sell such a bet at any price, Equivocator will willingly sell at 50c.

Is there anything that we can say in favour of or against one or other of these betting decisions? Nothing much at all of substance, it would seem. We don't know how the coin will land; indeed, we don't even know what its chance of landing heads is. So, we not only don't know who will do better as a result of her choices, we can't even say who can be expected to do better. If Reckless pays $1 and loses then she loses big; bigger anyway than Equivocator who pays only 50c. So equivocation can help to minimise losses. But it also minimises gains. For Equivocator will forgo opportunities to bet and some of these will pay out for Reckless. Only if losses matter more to the agent than gains is there anything to be said in favour of caution. But then the attraction of equivocation cannot be a purely epistemic matter, nor can it be completely general. So the claim that equivocation is objectively required or rationally mandatory seems completely without foundation.

When the dust settles we are left with little more than the original thought behind the Principle of Indifference: that absence of evidential reasons for differential probability assignments is a reason to make equal ones. That this reason is not itself an evidential reason should not be held against MaxEnt since our problem is precisely to pick a probability when evidential reasons give out. And at least it does provide the subjectivist Impreciser with a way to settle on a precise opinion for the purposes of making a decision.

3   See Williamson (2007a), Landes & Williamson (2013), Paris (2006) and Jaynes (1968, 2003).

There are dangers here for the subjectivist, however, and she should consume with moderation. For one thing, the Principle of Indifference is notoriously sensitive to the choice of description of outcomes, yielding different prescriptions depending on the level of refinement of the problem; a difficulty that carries over to the MaxEnt rule. And, for another, application of MaxEnt can conflict with Bayesian conditionalisation (see Williamson, 2011). Somewhat ironically, however, neither need be too much of problem for the subjectivist seeking only to apply MaxEnt for decision purposes. For in many decision-making contexts, and in particular those characterised by what I previously called Grade 2 uncertainty, what is needed is a probability assignment for the decision problem *as it is formulated*. A subjectivist wishing to avail herself of MaxEnt will represent her decision problem in a way that she judges is appropriate for application of the rule – i.e. where the symmetries that she takes to be present in the situation she finds herself in are captured in the descriptions of the outcomes. Furthermore, in using MaxEnt to determine a probability for the decision at hand she need not commit herself to holding onto these probabilities in the future and can perfectly well opt to suspend opinion again after the decision has been made and implemented. So conflict with Bayesian norms of belief revision can be avoided.

13.2.2 Aggregating

The second strategy that can be pursued is to reach an opinion by aggregating all those opinions that the agent regards as permissible. In essence, the idea is to exploit an analogy between a group agent and an individual agent with multiple avatars in disagreement on the question of what opinion to adopt, and then to draw on the large literature in social choice theory to provide rules for fashioning precise aggregate judgements.[4] Two classes of aggregation rules are particularly salient in this literature. Voting rules select the opinion with the greatest support by counting the number of voters (in this context, avatars) endorsing it, perhaps weighting them on the basis of other considerations, such as how competent they are or how affected they are by the decision. Averaging rules, such as 'splitting the difference' and linear averaging, on the other hand, select opinions that are the best compromises between the individual ones. The Principle of Indifference makes an appearance here too. When there is no reason for favouring one opinion over the others in virtue of who holds it or its

---

4 Methods for aggregating different kinds of opinion have been extensively studied. The problem of aggregating probabilities has received most attention from statisticians (see Genest & Zidek, 1986, for a survey), while Social Choice theory has mainly focused on the problem aggregating preferences and/or utilities (see Sen, 1970, for a classic discussion), but there is some work on the joint aggregation of probabilities and utilities (see, for instance, Mongin, 1995, and Bradley, 2005a). Finally the theory of judgement aggregation, developed by Christian List, Franz Dietrich and others, tackles the problem of aggregation in a very general way (see List & Puppe, 2009, for a recent survey).

content, the principle dictates treating each equally: rules such as majority rule and equal weighted averaging respect this dictum. But the literature recognises a great variety of contexts in which considerations favour either particular opinion holders or particular propositions and provides aggregation rules appropriate to them.

Some perspective on the scope and limits of such rules can be gained by looking in more detail at one of the most widely endorsed proposals, namely that an agent should form an aggregate of a set of probability judgements by taking a weighted average of them. Formally, given a set $C = \{P_i\}$ of probability functions defined on a common domain of propositions, a linear average, P_0, of these probabilities is obtained by setting, for some set of corresponding weights $\{w_i\}$ such that $w_i > 0$ and $\sum_i w_i = 1$:[5]

Linear Averaging $P_0 = \sum_i w_i \cdot P_i$

There are a variety of possible interpretations of the probabilities and the weights on them occurring in this formula. One rather salient one, particularly relevant to the context of uncertainty, treats the P_i as the various candidate hypotheses as to the true probabilities or objective chances of the prospects in their common domain and the weights as second-order probabilities on an extended domain containing chance propositions. So interpreted, Linear Averaging is simply an implication of the Principal Principle. And the proposal it supports, namely to adopt the expected chances of prospects as one's aggregate degrees of belief for them, amounts to what might be called 'second-order Probabilism', since it enriches the standard probabilistic framework in a way which allows for rationality constraints rooted in beliefs about objective probabilities.

The obvious problem with this proposal in this context is that it is hard to see how agents who lack the information necessary to form first-order probability judgements would nonetheless be able to form second-order ones, in particular to assign probabilities to hypotheses about what the chances are. Indeed, second-order Probabilism is more demanding cognitively than the simple first-order version, which requires no recognition on the part of the agent of objective chances. So this interpretation doesn't usefully apply in contexts of severe uncertainty insofar as we are interested in providing agents with guidance as to how to deliberate about their uncertainty.

A second interpretation, more appropriate to contexts of disagreement, treats the P_i occurring in Linear Averaging as the probabilistic judgements of different experts and the weights as some measure of either the experts' competence or reliability of the confidence the agent has in them (which is to be distinguished from confidence in a belief or a probability judgement).

---

5 The stipulation of strictly positive weights reflects the assumption that no probability function not deserving of a positive weight should be in the permissible set in the first place.

Lehrer & Wagner (1981), for instance, promote this as a method for forming opinions in cases of disagreement amongst experts, including ones in which the 'experts' are simply the agent's epistemic peers. A variant treats the P_i as the outputs of different models or lines of enquiry, with the weights once again measuring either the reliability of the methods they employ or the confidence that the agent has in them (see, for instance, Gärdenfors, 1988). On both, the aggregate probability P_0 can be construed as a confidence-weighted average of the probability judgements that the agent regards as worthy of consideration.

Contrary, however, to the claim of Lehrer (1976, 1983) that linear averaging is the uniquely rational way of forming one's beliefs in the face of disagreement amongst experts, it has a number of significant weaknesses. I'll mention two here. The first is that this rule is insensitive to whether the opinions expressed by different individuals on the same proposition are independent or not. But compare a situation in which two scientists conduct separate experiments to try and settle some question with one in which they conduct a single experiment together. Suppose that in both cases the scientists report that as a result of their experiments they consider X to be highly probable. In the former case, we would want to raise our own probability for X quite considerably because of the convergence of their expert testimony. In the latter case too we would want to raise our probability for X, but less so, because their joint testimony in favour of X is based on the same information. To revise once in the light of the testimony of the first scientist and then again in the light of that of the second would in effect be to update twice on the same evidence, akin to an individual scientist conditioning twice on the same experimental result.

A second problem is that the confidence weights that this rule places on the different experts or models are proposition-*independent*. Consider a simple case in which we consult two scientists with different domains of expertise, one being an oceanographer and the other a meteorologist. It would be natural to have more confidence in what the former says about sea temperature but more confidence in what the latter says about cloud formation. But Linear Averaging requires us to assign the same weight to the probabilistic judgements of the oceanographer and the meteorologist on the second question as we do to the first. So it asks us to apply confidence considerations in an unsatisfactory way.

Could we not avoid the problem by employing proposition-*dependent* confidence weights? Unfortunately not, for if we do so then we will land up with incoherent degrees of belief (see Bradley, 2007b). Consider the following example. Suppose that you know that Anne has observed that A is true while Bob has observed that B is true. Suppose, furthermore, that they report the probabilistic degrees of belief across the partition $\pi = \{AB, A\neg B, \neg AB, \neg A\neg B\}$, as displayed in Table 13.1. Now Anne and Bob's observations make them maximally reliable, respectively, on the question of whether or not A is true and whether or not B is true (supposing absence of observational error). So we should simply adopt their reported beliefs as our own, leaving us with degree

TABLE 13.1. *Linear Averaging*

	AB	A¬B	¬AB	¬A¬B
Anne	0.1	0.9	0	0
Bob	0.1	0	0.9	0
Linear average	0.1	0.9*a*	0.9*b*	0

of belief 1 in *AB* and 0 in all the other propositions. But this goes against the recommendations of Linear Averaging, which, irrespective of the weights *a* and *b* assigned to Anne and Bob, will yield a probability of 0.1 for the proposition *AB* and 0 for ¬*A*¬*B*.

We can put the problem slightly differently. If you form your beliefs by averaging Ann and Bob's opinions on the four element partition π using weights *a* and *b* for Anne and Bob, respectively, then your degrees of belief on the partitions {*A*, ¬*A*} and {*B*, ¬*B*} must, on pain of probabilistic incoherence, be linear averages of Anne's and Bob's degrees of belief obtained by applying weights *a* and *b*. But Anne and Bob have different competencies over these partitions, so these weights cannot be adequate to both.

These problems are not peculiar to linear averaging; any of the usual averaging rules found in the literature will face similar ones. Indeed, the root of the problem, it seems to me, lies in the way that confidence considerations are applied by such rules, namely as weights on experts or models rather than on the probabilistic judgements that are supported by what these experts or models say. But it may be that we are simply asking too much from these techniques by making them live up to epistemic standards appropriate to opinion formation rules that are designed to be sensitive to the evidence. For our problem is precisely how to form an opinion when such evidence gives out. In which case, the rationale for the adoption of these techniques may simply lie in the fact that they deliver a consistent solution to this problem.

On this line of reasoning, it might be reasonable to seek pragmatic grounds for the assignment of weights to experts. But letting pragmatic factors shape belief formation risks putting the cart before the horse, as we usually want our decisions to be guided by our beliefs rather than the other way around. Arguably, therefore, the appropriate point to apply pragmatic considerations is at the moment of choice, rather than during attitude formation.[6] So let's turn to the second strategy, of leaving attitudes imprecise when circumstances do not warrant greater precision and applying an alternative decision rule.

---

[6] There are grey areas, I think. In group decision making it is often impossible to reach a decision on what to do without getting agreement on the reasons that ground that choice. In which case, pragmatic considerations are likely to slip into the phase of attitude formation.

13.3 ALTERNATIVE DECISION RULES

A decision maker who is unable or unwilling to form precise probability and desirability judgements on all prospects relevant to the decision problem she faces cannot, of course, choose in accordance with expected utility maximisation. But she might instead apply a different decision rule, one that is tailored to her state of severe uncertainty or conflict. A great many different proposals for such rules exist in the literature, involving more or less radical departures from Bayesian theory and varying in the informational demands they make. Our focus will be on rules that take as inputs the set of expected utilities associated with an act that characterise the decision maker's uncertainty, organising them in terms of the additional considerations they appeal to in order to settle matters.

More formally, let us suppose that the agent is in decision situation $\mathcal{D} = \langle O, S \rangle$, with $O = \{f, g, ..., h\}$ the set of actions available to her and $S = \langle \Omega, \unrhd, \succsim \rangle$ her judgemental state with, as before, $\unrhd$ and $\succsim$, respectively, being her credibility and preference relations on the Boolean algebra of prospects $\Omega = \langle X, \models \rangle$. It will not matter to the discussion here whether we think of actions in the manner of Savage, as functions from a set of states (a partition of X in terms of the features of the world that are causally independent of the actions) to consequences (a partition of X in terms of the features of the world that matter to the agent), or in terms of the partitioning indicative conditionals that pick them out.

Let $\mathcal{J} = \langle \Omega, \mathcal{A} = \{A_i\} \rangle$ be the set of avatars of the agent determined by the relations $\unrhd$ and $\succsim$, with each avatar being a pair of probability and desirability functions that jointly represents them. For any action f, let $\mathbb{E}_i(f)$ be the expected utility of f according to avatar i – i.e. its expected utility calculated by applying the pair of probability and desirability functions constituting that avatar.

The problem the agent faces is to settle on a choice of action on the basis of the set of expected utilities associated with each of her options and any other considerations that she can apply. One principle of choice commands universal assent: that, if all the agent's avatars assign higher expected utility to one action than another, then the latter should never be chosen when the former is available. More formally, if C is a choice function on decision situations $\mathcal{D} = \langle O, S \rangle$, then:

Unanimity If $f, g \in O$ and for all $A_i \in \mathcal{A}$, $\mathbb{E}_i(f) \geq \mathbb{E}_i(g)$, then

$$g \in C(\mathcal{D}) \Longrightarrow f \in C(\mathcal{D})$$

Unanimity can be strengthened a bit by adding a second clause to the effect that if, in addition, for some $A_{i^*} \in \mathcal{A}$, $\mathbb{E}_i(f) > \mathbb{E}_i(g)$, then $g \notin C(\mathcal{D})$. But even so strengthened the Unanimity principle is unlikely to resolve the agent's decision problem in a significant number of contexts. So further considerations will have to be brought to bear on the problem in order to resolve it.

There are four such considerations that are particularly salient: *caution, confidence, robustness* and *flexibility*. To study how they can be used to settle decisions, let $\geq$ be a ranking of options in terms of their choice-worthiness and which determines what the agent can permissibly choose. Formally, if C is a choice function on decision situations $\mathcal{D} = \langle O, S \rangle$, then for all $f, g \in O$

Choice-Worthiness Ranking $f \geq g \Leftrightarrow \left[g \in C(\mathcal{D}) \Rightarrow f \in C(\mathcal{D}) \right]$.

Unanimity implies that $\geq$ contains the intersection of the avatars' preference orderings. Let us now look at how additional considerations can be used to complete it.

13.3.1 Caution

When a decision maker regards a range of probabilities and/or desirabilities as reasonable, she may wish to be cautious in her decisions by giving more weight to the 'downside' risks – the possible negative consequences of a choice of action – and less to the 'upside' chances. Someone who is cautious in this sense will tend to hedge against risks by choosing actions with less variance in their expected outcomes. Hedging seems particularly compelling when the costs and benefits of an action in each state of the world accrue differently to different individuals, for in this case reducing the variance can serve the goal of treating individuals more equally. But, in general, it has the advantage of assuring the agent that her expected losses will not exceed some amount.

A salient decision rule encoding such caution is the maximin-EU (MMEU) rule, which recommends picking the action with the greatest minimum expected utility. In its usual formulation the expected utilities are determined relative to a fixed single utility and a set of probabilities, but it is very naturally generalised to the case in which both utilities and probabilities are imprecise. This more general MMEU rule says

MMEU $f \geq g \Leftrightarrow \min[\mathbb{E}_i(f)] \geq \min[\mathbb{E}_i(g)]$

MMEU and near variants have been advocated by a number of philosophers and economists, including Levi (1990), Peter Gärdenfors and Nils-Eric Sahlin (1982) and Itzhak Gilboa and David Schmeidler (1989), and the latter have provided an elegant representation theorem for it. Arguably, however, the rule is much too cautious, paying no attention at all to the full spread of possible expected utilities, though the force of this criticism somewhat depends on how the range of probabilities and associated expected utilities is determined in any given decision situation.

These problems can be avoided to some extent by adopting one of the rules for decision making under ambiguity, which draw on further information about the set of expected utilities determined by the agent's imprecise beliefs. Daniel Ellsberg (1961), for instance, proposes maximising a weighted average of the minimum and mean expected utility, where the relative weights on the minimum and mean can be thought of as either reflecting the decision maker's pessimism or her degree of caution. This rule yields much the same prescriptions as maximisation of a weighted average of the maximum and minimum expected utility (often called the α-MEU or Hurwicz rule). Formally, this latter rule dictates that, for some $\alpha \in [0,1]$, canonically taken to be a measure of the agent's degree of pessimism or caution and assumed to be greater than 0.5:

α-**MEU** $f \geq g \Leftrightarrow$

$$\alpha \min_i[\mathbb{E}_i(f)] + (1 - \alpha) \max_i[\mathbb{E}_i(f)] \geq \alpha \min_i[\mathbb{E}_i(g)] + (1 - \alpha) \max_i[\mathbb{E}_i(g)]$$

The α-MEU rule has been defended by Binmore (2008) and axiomatically characterised by Paolo Ghirardato, Fabio Maccheroni and Massimo Marinacci, 2004. Like Ellsberg's proposal, it generalises MMEU by allowing decision makers with the same imprecise beliefs to differ in the degree of caution that they display in their choices.

A question that all such rules must address is the specification of the set of probabilities that the expected utilities are based on. When the evidence does not determine a single probability then the Bayesian insistence on a single probability seems too extreme. But, if all probabilities *consistent* with the evidence are included, then it is likely to determine very wide probability intervals for decision-relevant contigencies. In the case of the MMEU rule, with $\alpha > 0.5$, this will tend to lead to very cautious decision making; in all cases the extremes of the probability intervals have considerable influence on the choice of action. A natural thought is that the set should determine intervals that are sufficiently broad that the decision maker is confident that the 'true' probabilities lie within them or that they contain all reasonable values. For instance, if the source of these probabilities is the opinions of others, the decision maker does not need to consider every possible opinion consistent with the evidence, only those that they have some confidence in. But how confident do they need to be? We return to this question later, once we have discussed the notion of confidence in more detail.

13.3.2 Confidence

A second set of alternative rules draws on considerations of confidence and/or reliability. The thought here is that, even if you do not know what the 'true' expected utility of an action is, you can be more or less confident about the various candidate estimates. For instance, when the estimates derive from different models or experts, the decision maker may regard some models as better corroborated by available evidence than others or some experts as more reliable than others in their judgements. In these cases it is reasonable, *ceteris paribus*, to favour actions of which you are more confident that they will have beneficial consequences. One way of doing this is to weight each of the expected utilities associated with an action in accordance with how confident you are about the judgements supporting them and then choose the action with the maximum confidence-weighted expected utility. Formally, given a set of weights $\{\alpha_i\}$ such that $\alpha_i > 0$ and $\sum_i \alpha_i = 1$, this rule counsels choice in accordance with

CWEU $f \geq g \Leftrightarrow \sum_i \alpha_i.\mathbb{E}_i(f) \geq \sum_i \alpha_i.\mathbb{E}_i(g)$

Note the 'kinship' of CWEU with Linear Averaging, the rule for forming aggregate probability judgements that we looked at in the previous section. Indeed, when the agent has complete preferences for consequences so that her avatars disagree fundamentally only in their beliefs, then the ranking over acts induced by CWEU is just the same as that induced by expected utility maximisation relative to a linear average of the avatars' probabilities.[7]

There is room here too for different interpretations of the confidence weights occurring in the CWEU equation. In some cases they can be construed as measures of the reliability of the expert or model that are their source. In other cases they can be construed as second-order probabilities; for instance, the probability that the expectation $\mathbb{E}_i$ is the best one to use in evaluating the action. In this case, CWEU becomes a form of 'second-order' Bayesianism according to which the value of action is determined by the subjective expectation of its 'true' expected utility. As such it does little to address the problem of decision making under severe uncertainty, since it returns us to the problem of how to form precise second-order beliefs. If combined with one of the techniques for forming a judgement described in the previous section, some progress can be made, however. For instance, if there are no grounds for greater confidence in any one of the expected utility judgements than another, appeal might be made to the Principle of Indifference to motivate the assignment of equal confidence weights to the agent's avatars. In this case, CWEU reduces to a well-known rule for decision making under conditions of ignorance: maximisation of mean expected utility. Formally:

MaxMean $f \geq g \Leftrightarrow \sum_{i=1}^{n} \frac{\mathbb{E}_i(f)}{n} \geq \sum_{i=1}^{n} \frac{\mathbb{E}_i(g)}{n}$

Such second-order Bayesianism – and, indeed, simple maximisation of CWEU under any interpretation of the confidence weights – leaves no room for the kind of caution considered before. But a close variant of it, the 'smooth

---

7   There are hidden dangers here, though. When the agent's avatars disagree in both their beliefs and their desires, then the linear average of their expected utilities may not cohere with the agent's incomplete preferences. For discussions of these aggregation problems, see Mongin (1995) and Bradley (2005a)

ambiguity' model of Peter Klibanoff, Massimo Marinacci and Sujoy Mukerji (2005), allows for an aversion to wide spreads of expected utilities, by valuing actions in terms of a linear average of a concave transformation of their expected utilities, rather than in terms of the expected utilities themselves, where this transformation reflects the agent's degree of aversion to the spread. Formally, let $\phi : \Re \to \Re$ be such a concave mapping on the real numbers. Then, according to the smooth ambiguity model of Klibanoff, Marinacci & Mukerji,

SAM $f \geq g \Leftrightarrow \sum_i \alpha_i . \phi \left(\mathbb{E}_i(f) \right) \geq \sum_i \alpha_i . \phi \left(\mathbb{E}_i(g) \right)$

A second model, due to Alain Chateauneuf and José Faro (2009), combines consideration of confidence and caution in a quite different way. They postulate a confidence threshold for determining the set of probabilities relative to which the decision maker applies the maximin-EU decision rule. In doing so they partially resolve the problem of the determination of the set C of priors, though they do not say anything about what level of confidence should be required. But considerations of confidence can be used even when precise confidence weights cannot be provided, though they then need to be supplemented with other considerations (such as caution). Peter Gärdenfors and Nils-Eric Sahlin (1982), for instance, suggest simply excluding from consideration any estimates that fall below a reliability threshold and then picking cautiously from the remainder. Similarly, Brian Hill (2013) uses an ordinal measure of confidence that allows for stake-sensitive thresholds that can be combined with other considerations. We will return to Hill's model later.

13.3.3 Robustness

A third consideration that can be appealed to is the robustness of the decision rationale.[8] The basic thought here is that the decision maker should work out which dimensions of uncertainty make the most difference to the outcomes of her decisions and then choose actions that do sufficiently well for a reasonable range of values on these dimensions. Actions chosen on this basis will usually be 'regret-free' in the sense that, even if they do not always turn out to be optimal, they are likely not to turn out to have been a bad choice.

What counts as a reasonable range of values? Most approaches that appeal to robustness assume that a best estimate or preferred model is available and then consider small deviations away from this estimate or small changes in the model parameter values (see, for instance, Ben-Haim, 2006, and Hansen & Sargent, 2001). A robust action is one that can be expected to have beneficial consequences relative not just to the best estimates of the values

---

[8] See especially Nehring (2009a), who uses the criterion of the robustness of a decision rationale in a more far-reaching way than examined here.

of relevant variables but also to a class of estimates that deviate from the 'best' one to some degree. The wider the class in question, the more robust the action. When the option that maximises expected utility, relative to the best estimate, is robust in this sense, then one gets an extra reason to choose it. But sometimes the expected utility-maximising option may be less robust than alternatives that are nonetheless satisfactory in terms of their expected utility. Then some trade-off between the two considerations, expected utility and robustness, must be made in order to resolve the question of what to choose.

13.3.4 Flexibility

The final consideration that can be appealed to is flexibility. In some contexts, an option that is available to decision makers is to delay all or part of the decision until more information is available or some of the disagreement is resolved through deliberation. The basic motive for delaying a decision is to maintain the ability to respond flexibly to contingencies that arise. Suppose, for instance, that a choice must be made between building a cheap, but low, sea wall or a high, but expensive, one, and that the relative desirability of these two courses of action depends on unknown factors, such as the extent to which sea levels will rise. In this case it would be sensible to consider building a low wall first but leave open the possibility of raising it in the future. If this can be done at no additional cost, then it is clearly the best option: at worst, no new information is acquired by the time the low wall is completed and you are in much the same situation as you started; at best, you are able to make the optimal choice at the later time. Typically, of course, flexibility comes at a cost and some judgement must be made as to whether the cost is worth bearing (and this decision may be no easier to make than the initial one). So the extent of the benefit that can be extracted by pursuing this strategy will depend on the possibility of keeping these costs down by breaking the original decision problem down into relatively autonomous, subsidiary decision problems that can be settled sequentially.

A preference for flexibility is reasonable even under conditions of normal uncertainty, when the decision maker has precise probabilities for the future contingencies that determine how beneficial a course of action is. Indeed, there are well-established models of dynamic decision making that exploit this fact (see Kreps & Porteus, 1978, and Arrow & Fisher, 1974, for instance). But the central principle at stake here, namely that, *ceteris paribus*, you should prefer actions that leave more options open to those that restrict them, can do even more important work in conditions of severe uncertainty. For, if you are unable to determine all the consequences of your possible actions or if you are unable to predict what value you will attach to these consequences at their time of realisation, then you have a strong incentive to avoid making irreversible

TABLE 13.2. *The Ellsberg Paradox*

	Red	Black	Yellow
L_1	$100	$0	$0
L_2	$0	$100	$0
L_3	$100	$100	$0
L_4	$0	$100	$100

commitments too early on. Finally, when you find yourself in a situation of conscious unawareness, when you are unsure about true state space and believe that you may not be aware of all relevant contingencies, the ability to respond flexibly to changes in circumstances becomes crucial.

13.4 CAUTIOUS DECISION MAKING

Drawing on additional considerations such as caution, confidence, robustness or flexibility yields a wide range of alternative decision rules to that of maximisation of expected utility, each with pretensions of being appropriate in at least some circumstances of severe uncertainty and/or disagreement. Exhaustive examination of each of these considerations and associated proposed rules is beyond the scope of this book and I will focus on the role of just two of them. This rest of this chapter will be devoted to an assessment of the role of caution in decision making, with particular attention to the question of the rationality of cautious attitudes. The next chapter will look at the role of confidence.

Much of the current debate about what decision rules are appropriate to conditions of severe uncertainty has been driven by the desire to explain the pattern of preferences frequently observed in the Ellsberg Paradox (exhibited again, for convenience, in Table 13.2) and attributed by Ellsberg to an attitude that has come to be called ambiguity aversion. Behaviour consistent with the kind of aversion to ambiguity postulated by Ellsberg has been established in numerous experiments involving set-ups similar to his.[9] And his characterisation of it, as a type of cautionary attitude that leads decision makers to prefer actions with known, or less uncertain, chances of reaching their goals, to those with more uncertain chances, has also found widespread acceptance. On the other hand, there is considerable disagreement about how to model ambiguity aversion and on the question of whether it is rational; in particular, whether it is compatible with the Bayesian theory of rationality.

---

[9] See Wakker (2010) and Trautmann & Van De Kuilen (2016) for a review of the literature.

Many decision theorists take the Ellsberg Paradox as evidence that agents do not have precise probabilities for draws of the black or yellow balls from the 'ambiguous' urn and then use this fact together with one of the decision rules for agents with imprecise probabilistic beliefs previously introduced to explain the Ellsberg preferences. The Maximin EU model, which prescribes choice of the alternative that maximises the minimum expected utility, affords one such explanation. If we set the utility of \$100 to 1, for instance, then the minimum expected utility of L_1 is $\frac{1}{3}$, of L_2 is 0, of L_3 is $\frac{1}{3}$ and of L_4 is $\frac{2}{3}$. So agents who employ MMEU will have the characteristic Ellsberg preferences: $L_4 \succ L_3$ and $L_1 \succ L_2$. They are not by any means the only ones, though. Agents who employ the α-MEU[10] – or, indeed, *any* other of the cautious rules that were presented in the previous section – can also display these preferences and, more generally, some form of ambiguity aversion.

Although these models differ in various ways, they all take it as given that the Ellsberg preferences and, more generally, ambiguity aversion are inconsistent not only with Savage's Sure-Thing Principle but also more generally with the view that individuals base their decisions on precise probabilities for the contingencies upon which the consequences of their choice depend. I have already argued (in Chapter 9) that this conclusion cannot be drawn from the Ellsberg Paradox and that the Ellsberg preferences are perfectly compatible both with Precise Probabilism and with expected utility maximisation, provided that agents do not value chances linearly. In this section I will develop this discussion by addressing three questions in more detail. (1) What kind of an attitude is ambiguity aversion? (2) Is ambiguity aversion rational? (3) Is ambiguity aversion consistent with Bayesian principles?

The first step to addressing these questions is to define ambiguity aversion more precisely. To do so, I will follow the characterisation of it given by Schmeidler (1989) and Itzhak Gilboa and David Schmeidler (1989) as a preference for hedging or randomisation, using a propositional variant of the framework developed by Anscombe & Aumann (1963) in which most of the recent debate on decision making under ambiguity has been conducted. In the Anscombe and Aumann (A–A) framework, actions are represented as functions from the set of states of the world to von Neumann and Morgenstern lotteries. So an action is something that determines for each state of the world an objective chance distribution over the set of prizes. Formally, let $S = \{S_1, S_2, ..., S_m\}$ be the set of states of the world, which we continue to consider to be a partition of the set of prospects in terms of all combinations of features of the world that are causally independent of the agent's actions but relevant to determination of the outcome of choosing a particular one. As before, let the $\bigwedge_{i=1}^{n}(Ch(X_i) = x_i)$ be lottery propositions: conjunctions of the propositions $Ch(X_1) = x_1$, $Ch(X_2) = x_2$, ..., and $Ch(X_n) = x_n$ that specify the chances of obtaining each of the 'prizes' represented by the propositions X_i.

---

[10]   See definition in Section 13.3.1.

Let Δ be the set of all such lottery propositions. As lottery propositions serve as consequences in the A–A framework, for them an action is just a function from $\mathcal{S}$ to Δ. Let $\mathcal{F}$ be the set of all such A–A actions and $\mathcal{H}$ be the subset of them consisting of actions with constant lottery consequences. For any $f, g \in \mathcal{F}$ and $\alpha \in [0, 1]$, let the α-mixture of f and g, denoted by $\alpha f + (1 - \alpha)g$, be an act whose consequence in each state of the world, S, is defined by

$$(\alpha f + (1 - \alpha)g)(S) = \alpha f(S) + (1 - \alpha)g(S)$$

In other words, an α-mixture of f and g is an act whose consequences lie between those of f and g in each state of the world. Note that $\alpha f + (1 - \alpha)g$ itself belongs to $\mathcal{F}$. It follows that $\mathcal{F}$ is closed under mixing.

Let $\succsim$ be a complete and transitive preference relation on $\mathcal{F}$. Anscombe and Aumann make two contentious assumptions about $\succsim$. Firstly, they assume that preferences for lotteries are state-independent. This assumption is no more plausible as a general principle in this framework than in Savage's, but since it is not at the centre of the questions we need to consider let's just accept it for convenience. Secondly, they assume that preferences satisfy a strong separability condition on mixtures of acts which says that, if any two acts are mixed with a third one, then this does not affect the preference ranking of them. Formally, for all $f, g, h \in \mathcal{F}$ and $\lambda \in [0, 1]$:

A–A Independence $\lambda f + (1 - \lambda)h \succsim \lambda g + (1 - \lambda)h \Leftrightarrow f \succsim g$

A–A Independence is a *very* powerful axiom. Not only does it imply both Savage's Sure-Thing principle and the von Neumann and Morgenstern Independence axiom, but in fact it implies much more than the conjunction of them. For example, suppose that you must choose between betting for or against a coin landing heads that is known to be either two-headed or two-tailed (the ambiguous options) or betting on a coin landing heads that is known to be fair (the risky option). These three options are represented in Table 13.3, respectively as the bets H, T and F, with the cell entries specifying the chances of winning the bet in each of the unknown states of the world. The bet H (T) on the ambiguous coin landing heads (tails) yields a chance one of the prize in the event that the coin is two-headed (two-tailed) and a chance zero of the prize in the event that it is two-tailed (two-headed), while the bet F on the fair coin yields a one-half chance of the prize in each event.

Now observe that bet F is a 50:50 mixture of bets H and T because the chances of winning associated with F in each state of the world are an equal-weighted average of those associated with H and T. So A–A Independence requires that, if the agent is indifferent between bets H and T, then she should also be indifferent between bets H and F. This feature of her preferences, that they be unaffected by the spread of the chances, is not required by the Sure-Thing Principle, which imposes no constraints at all on how consequences should be valued. Nor can vN–M's Independence axiom be

TABLE 13.3. *Hedging Your Bets*

	Heads bias	Tails bias
H	1	0
T	0	1
F	0.5	0.5

applied here, since the chance-consequences appear in different states. So A–A Independence must impose constraints that exceed those jointly imposed by these other two conditions.

We are now finally in a position to characterise the kind of cautionary attitude we are interested in. A–A Independence requires indifference to the spread of chances; when an agent satisfies it, we will say that she is *ambiguity-neutral*. In contrast, if she prefers g to both f and h when indifferent between the latter two, we will say that she is *ambiguity-averse*. Formally, following Schmeidler (1989):

Ambiguity Aversion If $f \approx g$ then $\alpha f + (1 - \alpha)g \succsim f$

So ambiguity aversion on this account is a cautionary attitude canonically revealed in a preference for hedging against spreads of chances. My task now is to assess its significance for our understanding of rational decision making in conditions of severe uncertainty, looking first at competing models of ambiguity aversion and then at the question of whether it is rational.

13.5 MODELS OF AMBIGUITY AVERSION[*]

Ambiguity aversion plays a crucial role in the characterisation of the rules of cautious decision making presented in the previous section; most notably in the representation theorem of Itzhak Gilboa and David Schmeidler (1989) for the MMEU rule. Gilboa and Schmeidler adopt all the Anscombe and Aumann axioms, with the exception of A–A Independence, which they replace with a weaker version – called C-Independence – which restricts the separability requirement to mixtures with constant acts. More precisely, it requires that, for all acts $f, g \in \mathcal{F}$ and constant acts $\bar{h} \in \mathcal{H}$ and for all $\lambda \in [0, 1]$

C-Independence $\lambda f + (1 - \lambda)\bar{h} \succsim \lambda g + (1 - \lambda)\bar{h} \Leftrightarrow f \succsim g$

What they then prove is that an agent who satisfies these axioms plus ambiguity aversion can be represented as making choices in accordance with the MMEU rule relative to a set of probability functions, $C = \{P_1, ..., P_n\}$, and a utility function U on consequences – i.e. they can be represented as agents with imprecise degrees of belief C who maximise minimum expected utility relative to C in their choice of action.

Gilboa and Schmeidler's result has been used to support two different claims: an explanatory one and a normative one. The first is the claim that the hypothesis that agents maximise the minimum expected utility in their choices provides the best explanation of ambiguity-averse behaviour. The second is the claim that rationality requires agents with imprecise degrees of belief to maximise the minimum expected utility relative to these beliefs. An evaluation of these claims depends on two distinct issues. Firstly, whether the axioms adopted by Gilboa and Schmeidler are both necessary and sufficient for rationality (an issue of most relevance to the normative claim). And, secondly, whether the numerical functions that represent these preferences are indeed measures of the agent's degrees of belief and desire; in particular, whether C truly measures her imprecise degrees of belief. For their result establishes only that agents whose choices conform with their axioms are maximising minimum expected utility relative to *some* set of probabilities. Those with behaviourist inclinations will dismiss the second question on the grounds that the agent's 'true' beliefs are simply those that are revealed in her choices or, more radically, that her beliefs are nothing other than constructions out of her behaviour. But, even if we grant this, what grounds do we have for basing the construction on the MMEU rule? The fact that this construction works is not sufficient to establish the explanatory claim. It must be shown that there is no other way of constructing an agent's degrees of belief and desire from her preferences that affords an equally adequate explanation for them.

In fact, however, there are. The most salient one for our purposes is the smooth ambiguity model of Peter Klibanoff, Massimo Marinacci and Sujoy Mukerji 2005 (hereafter KMM), which provides a rival characterisation of ambiguity-averse choice. The smooth ambiguity model too is backed up with a representation theorem for acts with lottery consequences that shows that, if an agent's preferences over such acts satisfy Savage's axioms and her preferences over constant acts (those that can be identified with the constant lottery they determine in each state of the world) satisfy the vN–M axioms, then her preferences can be represented by a probability P defined on the extended Boolean algebra of prospects, and a utility defined on lottery consequences, such that for all acts f and g

$$f \geq g \Leftrightarrow \sum_i \phi(\mathbb{E}_i(f)) \cdot P(S_i) \geq \sum_i \phi(\mathbb{E}_i(g)) \cdot P(S_i) \qquad (13.1)$$

where $\phi : \Re \to \Re$ is a mapping on the real numbers representing the agent's attitudes to ambiguity and $\mathbb{E}_i(f)$ is the vN–M expected utility of f in state S_i. Crucially, if her preferences also satisfy ambiguity aversion then φ will be a concave transformation of the expected utilities.

On KMM's smooth ambiguity model the value of an act is an expectation based on (subjective) probabilities for (objective) probabilities – the chances of the goods at stake – with the notable feature that the two tiers of probability are not reducible to a single expectation for these goods. This model generalises the one I proposed in Chapter 9, which applied only to decision problems (such as the Ellsberg Paradox) in which only one kind of good was at stake. In this case, the vN–M expected utility of a consequence reduces to the desirability of the chance of this single good and so the chances themselves serve as a vN–M index for preferences over the chances of such goods (because the greater the chances, the better). In the KMM model, the transformation is applied to the vN–M expected utilities without reference to the types of good determining them. This seems to be reasonable when the goods are just different quantities of money, but not when the chances concern very different types of goods, such as health and money, with regard to which agents may reasonably have rather different degrees of ambiguity aversion. In these cases, the desirabilities of the chances of the different goods must be determined first and only then aggregated.

My main disagreement with KMM, however, concerns the interpretation of the parameter φ representing the agent's uncertainty attitudes to chances. KMM take vN–M expected utility to be an appropriate measure of the desirability of chances (more generally, of lotteries) and ϕ as a transformation induced by the kind of epistemic attitude that Ellsberg postulates: a dislike of a lack of information that distorts the subjective probabilistic weighting of outcomes. I view $\phi(\mathbb{E}_i(\cdot))$ itself as the correct measure of the desirability of the lotteries, with ϕ a pragmatic attitude to uncertainty about chances that is encoded in the concavity of the utility function for chances. So interpreted, the smooth ambiguity model provides an explanation for ambiguity aversion that is perfectly consistent with Bayesian norms of rationality.

The upshot is that there is more than one model of ambiguity aversion on the table. I don't think we are in a position to definitively settle the question of which model is the most adequate descriptively and normatively, but I do think it's clear that MMEU is not it. Although it explains the ambiguity-averse pair of preferences $L_4 \succ L_3$ and $L_1 \succ L_2$ in the Ellsberg set-up (Table 13.2), it also implies that agents facing this decision problem will be indifferent between L_3 and L_1. Since L_3 weakly dominates L_1, this implication is both descriptively implausible and normatively unsatisfactory. The α-MEU rule does not have this implication but, like all rules that take into account only the maximum and minimum expected utilities of each action, does not discriminate between actions with quite different intermediate possibilities. We can illustrate this using the Ellsberg set-up once again, but dropping the assumption that the proportion of red balls is given, so that the situation is one of complete ignorance about the chances of drawing a ball of any particular colour. In this case, L_2 and L_4 have the same maximum and minimum expected utilities (respectively, the utility of \$0 and of \$100) and so must be regarded indifferently under the α-MEU rule. L_4 weakly dominates L_2, however, and I doubt anyone would choose L_2 in these circumstances. Nor should they.

In contrast, the smooth ambiguity model does, it seems to me, offer a plausible account of rational choice in the kinds of contexts exemplified by the Ellsberg Paradox, in which the state space is given in the description of the problem and in which the symmetries in the problem make application of the Principle of Indifference natural. In other contexts, those previously labelled as Grade 3 uncertainty in contrast to Grade 2 ambiguity, in which such symmetries are absent, the smooth ambiguity model is less compelling. For in such cases we have little to guide us in assigning subjective probabilities to the chances and no reason to trust them. In the next chapter I will defend a different model for such cases.

13.6 THE RATIONALITY OF AMBIGUITY AVERSION

Ambiguity aversion leads to decision making that is relatively cautious in the sense that ambiguity-averse agents will prefer actions with narrower spreads in the chances of outcomes. There is little doubt that agents do display caution of this kind, but is it rational? A number of philosophers and economists have argued that it is not (see, for instance, Adam Elga, 2010, and Nabil Al-Najjar and Jonathan Weinstein, 2009), others that it is (see, for instance, Levi, 1990, and Gilboa, 2009). In this section I will defend the latter position, but also stress the costs ambiguity aversion can impose on the agent.

Let's start with risk aversion by way of comparison. An agent is canonically said to be risk-neutral with respect to some divisible good if she is indifferent between a fixed amount of the good and a lottery which yields the same expected amount of it, but risk-averse (-loving) if she prefers the former (latter). For instance, someone who is risk-averse with respect to money will prefer \$50 to a lottery with a 50:50 chance of paying out either \$100 or nothing. More generally, risk aversion with respect to a divisible good G is manifested in a preference for a fixed quantity g of the good to a lottery with an expected return of g.

In a similar fashion, we can define uncertainty attitudes to goods. Consider Table 13.4, displaying prospects L1, L2 and L3, which have monetary consequences that depend on the truth or falsity of some event E. Suppose that an agent is indifferent between L2 and L3, thereby revealing that she regards E as likely to be true as not (assuming that the monetary consequences have desirabilities that are independent of E). Then we can say that she is neutral to the uncertainty regarding the monetary consequences if she is also indifferent between L1 and L2 and between L1 and L3; uncertainty-averse regarding money if she prefers L1 to both L2 and L3; and uncertainty-loving regarding money if she prefers both L2 and L3 to L1. These attitudes reflect, in essence, different desirability functions for quantities of money.

Now consider someone who is uncertainty-averse with respect to the *chances* of receiving some good (divisible or otherwise). She will prefer acts which yield constant chances of getting the good over acts with the same

TABLE 13.4. *Uncertainty*
Aversion regarding Money

	E	¬E
L1	$50	$50
L2	0	$100
L3	$100	0

TABLE 13.5. *Hedging Chances*

	RBB	RBY	RYY
B_0	$\frac{2}{3}$	$\frac{1}{3}$	0
B_1	$\frac{1}{3}$	$\frac{1}{3}$	$\frac{1}{3}$
B_2	0	$\frac{1}{3}$	$\frac{2}{3}$
—	—	—	—
B_3	$\frac{1}{3}$	$\frac{2}{3}$	1
B_4	$\frac{2}{3}$	$\frac{2}{3}$	$\frac{2}{3}$
B_5	1	$\frac{2}{3}$	$\frac{1}{3}$

expected chances when the chances vary by state of the world. Consider Table 13.5, for instance, which adds two acts (B_0 and B_5) to the simplified Ellsberg set-up presented before. Someone who regards the distributions RBB and RYY as equiprobable will be indifferent between B_2 and B_0 and between B_3 and B_5. If, furthermore, she is neutral with regard to the chances of monetary gain that are the outcomes of these acts, she will regard B_1 as equally good as both B_2 and B_0 and B_4 as equally good as both B_3 and B_5. But if she is uncertainty-averse with respect to these chances she will prefer B_1 over the other two because the desirability difference between a chance of one-third of the $100 and no chance of it exceeds that between a chance of two-thirds and a chance of one-third. Similarly, she will prefer B_4 to both B_3 and B_5.

Uncertainty attitudes to goods and uncertainty attitudes to the chances of these goods are logically independent. One could be uncertainty-neutral with regard to money, but uncertainty-averse with respect to the chances of obtaining it. Or just the other way around. But in one crucial respect they are similar: on the face of it, there is nothing particularly rational or irrational about having one uncertainty attitude rather than another. We certainly do, as a matter of fact, care about the chances of outcomes as well as the outcomes themselves. There is a difference, we tend to think, between having no lottery ticket at all and having a lottery ticket which is not, in fact, a winner. And between succeeding at a task when the chance of doing so was low and

TABLE 13.6. *Cost of Hedging*

Options	Composition of urn A				
	H_0	H_1	H_2	...	H_{100}
A	0	0.01	0.02	...	1
B	1	0.99	0.98	...	0
C	0.5	0.5	0.5	...	0.5

succeeding at it when the chance of doing so was very high. (This is just what the Chapter 9 examples of Ann the mountaineer and Bob the surly sibling revealed.)

Now, on my account, ambiguity aversion, as characterised by a preference for hedging, can be rationalised by aversion to uncertainty about chances. So the caution that it induces is perfectly rational from the perspective of the moderate Humeanism informing this book (which passes no judgement on the content of desires, only on their consistency). On the other hand, there *is* a price to be paid for such an aversion to spreads of chances. For an agent who prefers to hedge her chances will be willing to pay for such an opportunity even when it does not improve her expected gains. To see this, consider a close variant of the example that was presented as a challenge to Imprecise Bayesianism in Chapter 11.

Suppose 200 balls, 100 white and 100 black, are divided between two urns, respectively labelled A and B, with any ratio of the colours in the urns. You will be given an opportunity to bet on a white ball being drawn; the bet costs $10 and you win $50 if the drawn ball is white. There are three options you must choose amongst: A, a bet a white ball drawn from urn A; B, a bet on on a white ball drawn from urn B; and C, a bet on a white ball drawn from either urn A or urn B depending on the result of a toss of a fair coin. These options are displayed in Table 13.6, with the cell entries recording the chances of winning a prize for each hypothesis as to the composition of urn A for each of the options.

Table 13.6 makes evident the fact that option C offers the opportunity to hedge on the chances yielded by options A and B. So an ambiguity-averse agent should be willing to pay some amount of money to secure it in preference to options A and B. Indeed, if she is very averse to ambiguity she will be willing to buy bet C but not A or B. She is now open to exploitation, however, for someone could sell her bet C, toss a coin and then, once it has landed, reveal which urn the ball will be drawn from. At this point the agent in effect holds either bet A or bet B. Should she not be willing to sell these bets back for some further amount, leaving her out of pocket and back where she started?

Not necessarily. The fact that she would not buy either bet at a particular price does not mean that she is willing to sell them at that price. For her position as buyer or seller is rather different. As a buyer she trades a fixed amount of money for the chance of a gain. As a seller she acquires a fixed amount for the chance of a loss. If her attitudes to chances of gains and losses are different then she will not view sales and purchases symmetrically.[11] And so she might refuse to sell the bet back. Nonetheless, it does not seem irrational to view chances of gains and losses in the same way. In which case, an ambiguity-averse agent will be vulnerable to exploitation of this kind.

Does this make ambiguity aversion irrational? I don't think so. The way to think about it is in analogy with someone whose preferences change in a predictable way. Suppose that on Monday I find the prospect of going to the opera on Saturday sufficiently attractive that I am willing to purchase a ticket at £200. By Friday, however, I am feeling tired and the prospect of a late night seems unappealing. I now wish that I had not bought the ticket and would be willing to sell it back for less than £200. This fact about me makes me vulnerable, for someone who could anticipate my attitude changes could 'exploit' me by first selling me the ticket and then buying it back, leaving me out of pocket. (Of course, if I predict that I will end up selling the ticket back, I should not purchase it in the first place. So too the ambiguity-averse agent should not purchase bet C if she knows that the coin toss will be revealed to her, since she can anticipate the effect of the information about how the coin has landed on her attitudes. Sophisticated agents will not allow themselves to be exploited.)

There is no doubt that to have preferences that can be exploited in this way can be detrimental. If I could change them, that would be to my advantage, but, in reality, transforming preferences can be expensive – often prohibitively so. In any case, instability of preference is not irrationality; nor, more generally, is vulnerability to exploitation a sure sign of it. So the possibility of exploitation does not in itself show that ambiguity aversion is irrational. The ambiguity-averse agent's preferences imposes costs on her that an ambiguity-neutral agent does not face. But, like agents with expensive tastes, ambiguity-averse agents simply have to do the best they can with the preferences they are endowed with.

13.7 CONCLUDING REMARKS

I set out in this chapter to assess how an Imprecise Bayesian might make decisions in situations of severe uncertainty in which she lacks precise probability and/or desirability values for some of the prospects relevant to the evaluation of her options. Two strategies were proposed: furnishing the required probabilities and desirabilities by reaching a subjective judgement on the basis of whatever information she holds plus various non-evidential

---

[11] There is some evidence that people do view chances of gains and losses differently. See Wakker (2010) for the evidence and Bradley (2016) for a discussion.

considerations; and adopting a decision rule that does not require precise probabilities and desirabilities as inputs. The question that now needs to be addressed more explicitly is whether pursuit of either of these strategies risks confrontation with any of the core Bayesian rationality principles and, if so, what lesson is to be drawn from this fact. The question is a rather complex one, and evidently there are not only many variants of the two strategies to consider but also many nuances to the issue of what it would mean to abandon rather than modify the Bayesian approach. But there are nonetheless some preliminary conclusions to be drawn from the discussion.

Pursuit of the 'making up one's mind' strategy is of course required by Classical Bayesianism in order that expected utility maximisation be applicable. But it is also compatible with the kind of subjectivist Imprecise Bayesianism that I argued for in the previous chapter, which permits precision beyond what is required by the evidence. The question remains, however, whether there are principled ways of doing so that are consistent with Bayesian principles. The outlook in this regard is quite mixed. On the positive side, although I don't think much of the claim that MaxEnt is the uniquely rational way of forming precise probabilities, it does seem to me to offer the subjectivist a useful tool for arriving at a judgement in the kind of circumstance that I called Grade 2 uncertainty, when the state space and its symmetries are given by the description of the problem. (The Ellsberg Paradox is a prime example of such a circumstance; hence the importance of explaining how an agent might employ the Principle of Indifference to assign probabilities to relevant chance hypotheses and still act cautiously.) The flip side of this is that in situations of Grade 3 uncertainty, when the state space is not given, the grounds for reasonable application of MaxEnt are lacking, as it is no longer possible to identify symmetries in a non-arbitrary way. In such circumstances we might look to other considerations, such as confidence, to evaluate possible judgements. But the most commonly proposed version of this, linear averaging by the application of confidence weights on probabilities, turned out to face a number of grave difficulties. Whatever its pragmatic merits it does not provide a method for reaching a judgement fully consistent with Bayesian norms. Nonetheless, there is something right about this approach, and, in the next chapter, I will offer an alternative method for applying considerations of confidence to judgement.

In contrast to the first strategy, the second has typically been interpreted as one of providing *rival* decision rules to the Bayesian ones and hence as calling into question the descriptive and/or normative validity of Bayesian decision theory. I see things somewhat differently. There is an important difference between proposing additional considerations to those characteristics of Bayesianism and applying them in ways which require violation of the Bayesian norms of rationality. Since completeness is not a requirement of rationality, the core Bayesian rationality conditions on preference, together with the broad requirement of preference-based choice, do not completely determine what choices an agent should make. So a Bayesian must apply additional considerations – such as caution, confidence, flexibility and robustness – to bridge the gap between preference and choice. But doing so should not require violations of the rationality requirements on incomplete preferences.

Tension between the two arises only when such considerations are allowed to trump the Bayesian norms. For example, if caution is applied in the manner encoded in the MMEU decision rule, then the preferences over actions it supports may violate the Sure-Thing Principle. On the whole, however, I think this counts against applying caution in this way. Nor does the permissibility of ambiguity aversion require that we do so, as caution of this kind is perfectly consistent with a broad-minded Bayesianism that countenances a variety of attitudes to chances. It is true that ambiguity aversion is not consistent with the combination of Bayesianism and the treatment of risk given by the vN–M theory, but I have already argued that the latter is too restrictive. So this does not provide compelling grounds for allowing violations of Bayesian norms for incomplete preferences.

All this provides no more than a partial answer to the question of how an Imprecise Bayesian should make decisions. For the kinds of situations exemplified by the Ellsberg Paradox display only Grade 2 uncertainty, and the fact that we are able to provide a coherent decision theory for them does not mean that we have a proper handle on situations of Grade 3 or severe uncertainty. Our task in the final chapter will be to address this problem.

14

Confidence

14.1 THE PARADOX OF IDEAL EVIDENCE

Two kinds of criticism have been lodged against the classical Bayesian view that agents should capture the empirical uncertainty they face by a single probability function on the set of factual prospects. So far we have concentrated on the objection that this is too demanding and that, for a variety of reasons, agents may be unable to determine precise probabilities for all prospects. We now turn to the second objection; that a probability assignment (whether precise or not) does not suffice to capture all relevant aspects of the agent's uncertainty.

Recall the example I presented in Chapter 11 to illustrate the difference between three grades of factual uncertainty, in which we have before us an urn containing black and white balls and you are asked to say how likely you think it is that a ball drawn at random will be white. In the 'risk' variant of the example, you are told that half the balls are black and half white; in the 'ambiguity' variant you are not told the proportions of black and white balls but you do know that they exhaust the colour possibilities, while in the 'severe uncertainty' variant you are told neither. Objective Bayesians are agreed that the answer of 'one-half' is the only reasonable one in the first situation, but disagree over the second and third, with Precisers continuing to advocate the answer of 'one-half' and Imprecisers counselling a wide probability interval; perhaps as wide as $[0, 1]$.[1]

---

[1] A subjective Bayesian, I argued, should regard both answers as permissable in the ambiguity case. The discussion in the previous chapter has now addressed one objection to this contention, namely that adoption of precise probabilities precludes cautious decision making of the kind often observed in the Ellsberg set-up. For a Bayesian agent who is averse to uncertainty about the chances and who assigns equal probability to each hypothesis about the proportion of balls of each colour will display precisely this kind of caution in her choices.

This example will serve equally well to bring out the second objection to the Bayesian view. Suppose now that you are in a situation of Grade 2 uncertainty, but are given the opportunity to sample the urn with replacement. Suppose that after 1,000 trials you have drawn 498 white balls and 502 black. How would the advice of the two kinds of objective Bayesian change as a result of the evidence? Not much at all, it would seem. For the Preciser, one-half remains the ascription that maximises entropy. Similarly, since the evidence rules out only extreme ball distributions, the set of probability functions consistent with it is the open interval $(0,1)$. So the objective Impreciser's judgemental state should remain almost as imprecise as before.

Popper (1959) dubbed the implication of invariance in the Precise Bayesian's probabilities 'the paradox of ideal evidence'. He took it to cast doubt on the claim that probability was an adequate measure of the evidential support that we have for a hypothesis, arguing that the fact that the probability of drawing a white ball conditional on all the evidence garnered was the same as its prior probability showed that this evidence was irrelevant to its probability (though obviously not, Popper thought, to the question of whether it should be adopted). The essential point had been made earlier by Charles Sanders Peirce, and indeed granted by John Maynard Keynes (one of Popper's targets) when he observed that the weight of evidence for a claim was distinct from the degree to which it supported it:

> As the relevant evidence at our disposal increases, the magnitude of the probability of the argument may either decrease or increase, according as the new knowledge strengthens the unfavourable or the favourable evidence; but something seems to have increased in either case – we have a more substantial basis upon which to rest our conclusion. I express this by saying that an accession of new evidence increases the weight of an argument. New evidence will sometimes decrease the probability of an argument, but it will always increase its 'weight'.
>
> (Keynes, 1973/1921, p. 78)

Acquiring evidence can thus have two distinct effects on our judgemental state regarding some prospect. On the one hand, it can push the probability of the prospect in one direction or another; on the other, it can strengthen the basis for any particular probability assignment. These effects are independent, Keynes argued, for the weight of evidence regarding some prospect can grow without affecting the balance of probabilities for and against its truth, just as one can add weights to both sides of a balance scale without causing it to tip in either direction.

Probabilists have several responses to Popper's challenge.[2] The first is to observe that, although the probability of a white ball does not change, the

---

[2] See Howson & Urbach (2006), Joyce (2005) and Jeffrey (2004) for more detailed versions of these arguments.

probabilities of other prospects do; in particular, those of the hypotheses as to the composition of the urn. For simplicity, suppose the urn contains just 10 balls and for any number i in $[0,10]$ let H_i be the hypothesis that the urn contains i white balls. Suppose that the agent initially assigns a probability of 0.1 to each, reflecting her ignorance as to the composition of the urn. Now, as she samples the urn, her probabilities for these hypotheses will change. Her evidence E is the outcome of a sequence of n draws containing k white balls and $n - k$ black. Now by Bayes' theorem her posterior probability for any hypothesis H_i given this evidence E is

$$P(H_i|E) = 0.1 \cdot \frac{P(E|H_i)}{P(E)}$$

Assuming sampling with replacement, the probability of the sequence under any hypothesis is given by the relevant binomial distribution. So:

$$P(E|H_i) = \frac{n!}{k!(n-k)!} \cdot \left(\frac{i}{10}\right)^k \cdot \left(\frac{10-i}{10}\right)^{n-k}$$

Consider H_5, for example, the hypothesis that the urn contains five white balls. After a sequence of draws that produces five heads and five tails, her probability for H_5 will rise to approximately 0.28 while her probability for H_1 will fall to below 0.002. After ten heads and ten tails her probability for the former will stand at approximately 0.37 while that for the latter will be approaching zero. So the evidential relevance of the sampling *is* reflected in a change in the agent's overall state of judgement as represented by her probability function, contrary to Popper's claim. This is clearly exhibited in Figure 14.1, which plots the posteriors for all ten hypotheses (identified by the probability they assign to a draw of a white ball) after the draws of five heads and five tails and after draws of ten heads and ten tails. Note how the probability increasingly concentrates on the 'middling' hypotheses as the weight of evidence increases.

Something else changes with increased sampling: the fixity or resilience of the agent's opinion (see (Skyrms, 1977)). Initially her degrees of belief in a draw of a white ball are very sensitive to the results of the sampling. For example, if she draws three white balls in a row she will dramatically revise her probability for drawing a white ball to 0.837. In contrast, if she draws three white balls in a row after 20 samples with an equal number of black and white balls, her probability for white will rise only to 0.57. So, although the sampling of equal numbers of white and black balls does not change her unconditional probability for the draw of the white ball, it does change her conditional probabilities for it, given the possible future sequences of draws. Or, to put it differently, it affects how quickly she will learn from a sequence of draws because the weight of evidence acts as a drag on belief revision. For those Bayesians who distrust talk of objective chances, it is this effect of the sampling that characterises the increase in the weight of evidence.

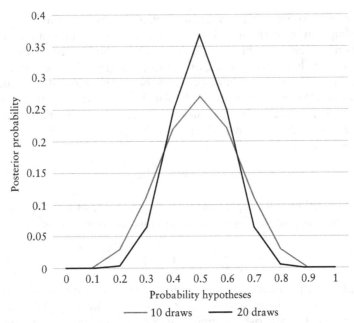

FIGURE 14.1. Posterior Probabilities of Hypotheses

On the face of it, the Imprecise Probabilist is much more vulnerable to Popper's argument. While the Precise Probabilist is quite right not to change her probability for a draw of a white ball despite the growth in the weight of the evidence, it seems that the Imprecise Probabilist *should* revise her belief state. For the evidence makes it ever more likely that the true proportion of white balls lies near to 50 per cent, a fact that should be reflected in a narrowing of the agent's probability interval for a draw of white. At the limit, the sampling should change the situation from one of ambiguity (Grade 2 uncertainty) to risk (the more benign Grade 1 uncertainty). But in the risk case, Imprecisers generally agree with Precisers that the best probability assignment for a white draw is one-half. So objective Imprecisers who accept that no precisification occurs must explain this discontinuity.

Subjectivist Imprecisers, I argued before, do not accept that states of belief must be maximally equivocal with respect to the evidence in the sense of containing every probability function consistent with it. So they will be anything but troubled by the argument that sampling should normally lead to a more precise belief state. The problem for them is to explain how this can happen, as no amount of conditionalisation on the part of her avatars will do this for the agent. The problem is just the one we noted before, that Imprecise Probabilists have trouble learning. In my response to it, I argued that we must recognise that probabilistic judgement can become more precise as a result of acquiring more evidence and that the process by which this occurs – belief formation – is quite different from the kind of probabilistic belief revision exemplified by conditionalisation. But I have yet to say very much about the substance of such a process or the judgements underpinning it.

14.2 CONFIDENCE-BASED LEARNING

> To express the proper state of belief, not one number but two are requisite, the first depending on the inferred probability, the second on the amount of knowledge on which that probability is based.
>
> (Weiss & Hartshorne, 1932, p. 421)

Peirce took the paradox of ideal evidence to show that probability was inadequate as a representation of our uncertainty and, further, that a second number was required to measure the weight of evidence supporting the probability ascription. Precise Probabilists have, on the whole, seen little merit in Peirce's claim, arguing that information about weight of evidence is already contained in the agent's full probability function; either in her subjective probabilities for the chances or, lacking these, by her conditional probabilities given possible evidence streams. But Peirce's insight gains more traction once we allow for imprecise probabilities. This is best approached by shifting attention from measurement of the weight of evidence to the attitudes that it supports; in particular the different levels of confidence that an agent can have in her probability judgements. Regarding the latter, the Precise Probabilist can say: 'As more balls are drawn from the urn, my confidence in the judgement of probability of one-half for a white ball grows.' The Imprecise Probabilist should be able to say the same, adding that her confidence in the extreme hypotheses diminishes as well. But more than this is possible too: she should be able to report how her confidence in the different probability intervals around the point of probability one-half shifts with the draws.

Compare the probability intervals $[0,1]$, $[0,25,0.75]$ and $[0.45,0.55]$. Initially, while the agent is fully confident in the judgement that the probability of a draw of a white ball lies in the first interval, she is much less so in the judgement that it lies in the second, narrower interval and even less that it lies in the third. The way she orders these probability judgements in terms of her confidence in them should not change as a result of the sampling, but the magnitudes of her confidence in them might. Even limited sampling yielding 50 per cent white balls could drive her confidence in the second judgement very close to the first. With sufficient evidence her confidence in all three will be very close to each other.

If this is right then the earlier claim that the agent's probabilistic beliefs become more precise as the weight of evidence increases will need modification. In this urn example, what happens is that, for a given level of confidence, the precision of the probability judgement achieving that level of confidence increases with increasing weight of evidence. So initially she has high confidence only in the unit interval. But with enough sampling she comes to have high confidence in the second, and eventually the third. *So precisification of judgement occurs relative to a fixed confidence threshold.* That is, when we say that an agent's judgements become more precise as a result of sampling, we mean that more precise judgements meet a relevant confidence threshold after sampling than before.

Suppose, for example, that there are just three confidence levels: high, medium and low. The effects of the sampling on the agent's confidence are illustrated in Figure 14.2. The upper diagram represents the situation before sampling and shows the agent to have low confidence in any interval more precise than [0.25, 0.75], medium confidence in any broader interval than that and high confidence only in the interval [0.05, 0.95]. The middle and lower diagrams represent her confidence after ten and 100 draws of the urn, which yield approximately half white balls. They show that, with increased sampling, the smallest interval meeting any particular confidence threshold narrows. For instance, after ten draws she has high confidence in any interval wider than [0.25, 0.75]. And after 100 draws she has high confidence in any interval wider than [0.375, 0.625]. Likewise, the confidence she has in the smaller intervals around 0.5 change. The interval judgement [0.3, 0.7] is initially held with low confidence, but after ten draws the agent's confidence in it is medium and, after 100, high. The upshot is that, if considerations of confidence work in the way I am suggesting, we should not conceive of an agent as simply having a particular set of imprecise probabilistic beliefs (at a point in time) but of having a set of them *at a particular level of confidence* (at a point in time).

This observation helps us understand how belief formation as previously conceived might occur. Fix a particular confidence level. Then, as the weight of evidence increases, more and more precise judgements will make it over the threshold for adoption at that level. This will look much like just the abandonment of extreme hypotheses. For instance, suppose that in the urn case a medium level of confidence is required by the agent for adoption of imprecise probabilistic beliefs. Then her initial probabilistic belief state can be identified with the most precise judgement held at that level – i.e. with the interval [0.25, 0.75]. After ten draws it can be identified with the interval [0.4, 0.6], and so on. So the evidence generated by the ten draws does lead to the elimination, *at that level of confidence*, of the agent's avatars occupying the intervals [0.25, 0.4] and [0.6, 0.75] even though it does not change the balance of probability for and against a draw of a white ball. (Belief formation might also occur because of a shift in the confidence requirements from one context to another, but I will not explore this possibility further here.)

The acquisition of evidence need not lead to a sharpening of belief; it can also lead to a suspension of opinion. For instance, suppose my confidence in probabilistic estimates based on a set of experiments is undermined by the discovery that the equipment that I have been using was not properly calibrated

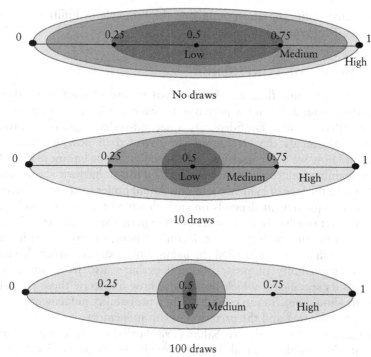

FIGURE 14.2. Confidence in Intervals

or is malfunctioning. Then I might have reason to regard these estimates as no better supported than a range of others, leaving me in a state in which the set of probabilistic judgements in which I have confidence is much less precise than before. This seems to be what is happening in cases of dilation. In these cases, the evidence makes our opinion less precise, by undermining our confidence in more precise judgements.

Confidence, it seems, steers belief formation and suspension. Plausibly, it also steers preference formation and decision making. For instance, suppose that an uncertainty-averse agent is deciding whether or not to accept bets on the draw of a white ball. Initially she may be unwilling to accept any bet that does not pay out substantially more than what she must stake on it, since she has no idea of what its expected value is. But, as her confidence that a draw of a white is as likely as that of a black grows, she will become willing to accept bets that pay out less. In the limit, she may accept a bet whose expected value, calculated using a probability of one-half for white, is only fractionally above the stake because she is confident that this is the correct probability to use. So her behaviour over time may approximate everything from what one might expect of someone using MMEU as her choice rule to what would be expected of an expected utility maximiser applying the Principle of Indifference to the ball colours. More on this in due course.

14.3 A MODEL OF CONFIDENCE

It is time to put some flesh on the informal notion of confidence that has been bandied around. From the previous discussion at least two central points can be distilled. Firstly, confidence is a second-order cognitive attitude to first-order judgements that is canonically revealed in how we learn from evidence and how we make decisions. In particular, if one has more confidence in one probability judgement than another then this judgement should change more slowly in the face of evidence and one should prefer to act on it. Secondly, confidence in a judgement depends on such things as the amount and quality of evidence that one has in its favour, one's experience in making judgements in this domain, and perhaps other factors. These aspects of evidence are distinct from what might be called the balance of evidence, which determines the relative probability of prospects (and perhaps also from what (Joyce, 2005), calls its specificity, which determines how imprecise these probability judgements should be). So too confidence in a probability judgement is distinct from both the content and the precision of this judgement.

Peter Gärdenfors and Nils-Eric Sahlin (1982) call the feature of a probability judgement that is determined by the weight of evidence its 'epistemic reliability', and measure it by a real-valued function on individual probability functions. From our point of view there are several problems with their approach. Firstly, to require of agents that they assign a measure of epistemic reliability to every permissible probability function is to make even greater informational and cognitive demands on them than ordinary Precise Probabilism. And, secondly, if they are able to assign such epistemic reliabilities, could they then not be used to determine a precise probability measure upon which expected utilities can be based, by simply taking an average of the permissible probabilities weighted by their epistemic reliabilities? But, then what would distinguish the decisions of the agent who makes second-order judgements of epistemic reliability from those of a classical Bayesian? Thirdly, evidence does not typically bear on probability functions, which are defined on entire algebras of prospects, but on probability claims regarding some subclass of them. Gathering a lot of information about rainfall patterns might raise our confidence in the rainfall predictions generated by our weather models, for instance, but have little implication for the predictions they make about sea levels. So further idealisation is required to turn confidence judgements regarding probability claims into judgements regarding probability functions.

For these reasons, I will treat confidence as a comparative relation (rather than as a numerical measure) defined on judgements (rather than on probability functions). Two types of judgement can serve as objects of confidence: categorical judgements, such as the judgement that it will rain tomorrow; and probability judgements, such as the judgement that the probability of rain tomorrow is one-half. The contents of the former are just the prospects or propositions that we have been working with throughout the book. The content of the latter will be modelled here by sets of probability functions. For instance, the content of the judgement that the probability of rain tomorrow is one-half will be represented by the set of probability functions (over propositions) assigning probability one-half to the prospect of rain tomorrow. Following Jeffrey (1992), I will call such probability propositions 'probasitions'.

One can have more or less confidence in both categorical and probability judgements. I can, for instance, be more confident that it will be hot tomorrow than that it will be precisely 32 degrees Celsius. And my confidence that it will be more than 30 degrees Celsius will be somewhere in between. Is this different in any way from saying that my degree of belief in the proposition that it will be hot tomorrow is greater than my degree of belief that it will be precisely 32 degrees Celsius? Yes, cognitively they are quite different. The objects of the confidence judgements are the weather judgements whereas the objects of the degrees of belief are the weather facts. So, while making probability judgements requires only awareness of the world and the different possible properties it might have, making confidence judgements requires a form of self-consciousness. The agent must be able to take themselves and their own judgements as objects of evaluation. They must be able to stand aside from the immediacy of their engagement with the external world and assess the nature of that engagement.

Nonetheless, first-order probability and second-order confidence are tightly bound together. I should regard α more probable than β iff I am more confident in the categorical judgement that α is the case than that β is. So it is true that introducing a confidence relation on categorical judgements seems to add little. The same will hold when there are probability facts (the chances) that are objects of my probability judgements: my confidence in any categorical judgement as to what the chances are should match my probability judgements for them. On the other hand, I can deny the existence of chance facts concerning tomorrow's weather, so have no degrees of belief regarding them, but still have different levels of confidence in my probability judgements about the weather. It follows that considerations of confidence regarding probability judgements can apply even when there are no probability facts. So the model we now introduce more formally genuinely increases the explanatory scope of our account, as well as having the advantage of being compatible with expressivist views about probability.

14.3.1 Confidence Relations*

Let $\Omega = \langle X, \models \rangle$ be a Boolean algebra of propositions, $\Pi = \{p_i\}$ be the set of all possible probability functions on Ω and $\Delta(\Pi)$ be the set of all subsets of Π.

Members of $\Delta(\Pi)$ will play a dual role here: as both the possible imprecise belief states of the agent and as probasitions – i.e. propositions about the probability of truth of the propositions in Ω. For instance, the probasition $\{p_i \in \Pi : p(X) = 0.5\}$ asserts that X has a probability of one-half, while the probasition $\{p_i \in \Pi : 0.25 \leq p(X) \leq 0.5\}$ asserts that the probability of X lies between one-quarter and one-half.

Let $\mathrel{\dot{\trianglerighteq}}$ be a reflexive and transitive binary relation on $\Delta(\Pi) = \{\pi_1, \pi_2, ...\}$, the set of probasitions. Intuitively, $\mathrel{\dot{\trianglerighteq}}$ captures the agent's relative confidence in the various probasitions about the state of the world, with $\pi_1 \mathrel{\dot{\trianglerighteq}} \pi_2$ meaning that she is at least as confident in the probasition expressed by π_1 as that expressed by π_2. I'll write $\pi_1 \mathrel{\dot{\vartriangleright}} \pi_2$ when $\pi_1 \mathrel{\dot{\trianglerighteq}} \pi_2$ but not $\pi_2 \mathrel{\dot{\trianglerighteq}} \pi_1$, and $\pi_1 \mathrel{\dot{\equiv}} \pi_2$ when both $\pi_1 \mathrel{\dot{\trianglerighteq}} \pi_2$ and $\pi_2 \mathrel{\dot{\trianglerighteq}} \pi_1$.

What are the properties of the confidence relation $\mathrel{\dot{\trianglerighteq}}$? Let's assume that it is non-trivial (i.e. $\Pi \mathrel{\dot{\vartriangleright}} \varnothing$) and monotonic (i.e. that $\pi_1 \mathrel{\dot{\trianglerighteq}} \pi_2$ whenever $\pi_2 \subseteq \pi_1$), but not necessarily complete, since we want to allow for the fact that the decision maker may have no grounds for a confidence judgement on some probasitions. Monotonicity is important, however, because it captures the idea that one should have more confidence in less precise propositions; a minimal condition on a notion of confidence.

Although not essential to what follows, I am inclined to add that the confidence relation should be $\vee$-separable – i.e. that, if $\pi_1 \cap \pi_2 = \pi_1 \cap \pi_3 = \varnothing$, then $\pi_2 \mathrel{\dot{\trianglerighteq}} \pi_3 \Leftrightarrow \pi_2 \cup \pi_1 \mathrel{\dot{\trianglerighteq}} \pi_3 \cup \pi_1$. This condition is clearly necessary if confidence in categorical judgement is to cohere with probabilistic judgement in the way that I claimed it should. This would make the confidence relation, formally speaking, a type of cognitive relation as defined in Chapter 5. From which it follows that, when the confidence relation is both complete and continuous over a complete, atomless Boolean algebra of judgements, it will determine a (second-order) probability function on this set, measuring the agent's degrees of confidence in her own first-order judgements. (The existence of just such a probability function was postulated in some of the rules for confidence-based belief formation and decision making that we canvassed in the previous chapter.)

A complete confidence relation partitions the space of probasitions into classes of judgements that the agent holds with equal confidence. But even when the confidence relation is incomplete it may partition the space of probasitions in a number of ways. Let us call a partition $\{\pi_0, ..., \pi_n\}$ of Π that is strictly ordered by $\mathrel{\dot{\trianglerighteq}}$ a *confidence partition*. A confidence partition for our urn example is illustrated in Figure 14.3, where the sets $\pi_0, ..., \pi_4$ are probasitions regarding the drawing of a white ball and the area occupied by each probasition corresponds to the number of probability measures making it true. For example, π_0 might be the probasition that the probability of drawing white is between 0.4 and 0.6, π_1 that it is either between 0.25 and 0.4 or between 0.6 and 0.75, π_2 that it is either between 0.1 and 0.25 or between 0.75 and 0.9 and π_4 that it is either less than 0.1 or greater than 0.9. The

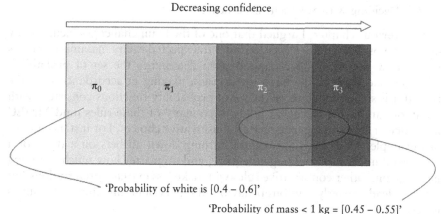

FIGURE 14.3. Confidence Partition

decision maker is represented by this figure as having a complete confidence relation over these (imprecise) probations for the colour of the ball drawn, but not for probations concerning other matters, such as the probability of the ball having a mass of less than 1 kilogram (which is a subset of Π that cuts across this confidence partition).

Of course, the agent's confidence relation may determine more than one partition of the space of probability functions. The agent may, for instance, be able to confidence-order probations about ball masses. And probations concerning both the colour and mass of the balls. In such cases, it is useful to know when information about the agent's confidence judgements contained in one such partition implies those contained in another. To this end, let us say that one confidence partition $C = \{\pi_0, ..., \pi_n\}$ of Π *refines* another confidence partition $D = \{\varpi_0, ..., \varpi_m\}$ iff, for every $\pi_i \in C$, there exists an $\varpi_j \in D$ such that $\pi_i \subseteq \varpi_j$. A *finest* confidence partition is simply a confidence partition such that there exists no other confidence partition that refines it. Intuitively, its elements are the strongest probations that the decision maker is able to rank in terms of her confidence in them.

In some cases a unique finest confidence partition will exist. This will be the case if the confidence relation is complete, for instance. But a unique finest confidence partition can exist even when it is not. Suppose that $C^* = \{\pi_0, ..., \pi_n\}$ is such a finest confidence partition of Π. If the maximal element, π_0, of C^* is a singleton $\{p^*\}$ then we call p^* the *best* probability. When $n = 2$, we call C^* a binary confidence partition: intuitively it is one that simply divides probations into those in which the agent has confidence and those in which she does not. As we shall now see, such a partition underpins many models of decision making under uncertainty.

14.3.2 Deciding with Sufficient Confidence

In the previous chapter, I argued that one of the main challenges facing many of the proposed alternative decision rules to expected utility maximisation was to say how the agent's decision situation determines the set of probability functions she takes as the basis for decision making in accordance with the rule. If it is simply taken to be the set of probability functions consistent with the information that the agent holds, then many of these rules (and MMEU in particular) prescribe unreasonably conservative choices. For instance in our urn example, MMEU would counsel turning down all bets on a draw of a white ball, no matter how large the prize, and no matter how much sampling has gone on. Other conservative rules will make less extreme prescriptions but still may lead to overly cautious behaviour if the set of permissible probabilities is very wide.

A natural solution now presents itself: take the set of probabilities on which an agent should base her decision to be those in which she has most confidence – i.e. the maximal element of her finest confidence partition. In the special case where a best probability exists, this would lead back to an instance of Classical Bayesianism, in the form of maximisation of expected utility calculated relative to the probability assignment in which one has the most confidence. In the more general case, it would allow for sensible application of any of the various decision models using imprecise degrees of belief presented in the previous chapter, especially in combination with our understanding of how considerations of confidence drive belief formation. Take the application of MMEU to the urn case, for example. On the proposed account of the relation between confidence and belief adoption, MMEU will initially prescribe relatively cautious decision making: refusal of bets with anything less than overwhelming favourable odds. But, as the evidence of sampling drives up her confidence in more precise judgements, the rule will begin to prescribe acceptance of a wider range of bets. At the limit, its prescriptions will be indistinguishable from that of maximisation of expected utility.

Compelling though this account may be, I do not think that it is entirely correct. For an agent should base her decisions not on the judgements in which she has *most* confidence but on those in which she has *sufficient* confidence, given what is at stake in the decision. Consider the decision problem illustrated by Table 14.1, requiring choice between acts which yield consequences dependent on whether a white ball is drawn or not and having the displayed utilities. (You can think of them as the utilities that an agent who is risk-neutral with respect to money gets from the corresponding monetary amounts.) Whether these acts have positive expected utility or not depends, of course, on how probable it is that a white ball is drawn. Now compare situations in which you are confident that the probability is close to one-half to those in which you are confident only that the probability lies in the interval $[0.25, 0.75]$ or those in which you are confident only that it lies in the unit

TABLE 14.1. *Decision Stakes*

	White	Black
f	1	0
g	10	−8
h	100	−96

interval. It is quite plausible that you would make different choices in these situations: perhaps h in the first, g in the second and f in the third. In the first you can be confident that h has the greatest expected benefit of the three, but in the other two this is no longer the case. In contrast, in the third situation the possibility of a severe expected loss can no longer be discounted. This is a reason to play it safe and choose f.

The insight that the level of confidence that we require in order to act can reasonably depend on what is at stake is central to the decision model proposed by Brian Hill (2013). Hill allows for confidence considerations to influence the decision by determining the range of probabilities on which the decision is based. The idea is that each decision situation will determine a confidence threshold for decision making based on what is at stake in that decision. When the stakes are low the decision maker may not need to have a great deal of confidence in a probability measure in order to base her decision upon it. When the stakes are high, however, the decision maker will need a correspondingly high degree of confidence in her probabilistic information in order to make her choices.

For determining a confidence-sensitive decision rule, Hill draws on a purely ordinal notion of confidence, requiring only that the set of probability measures forms a nested family centred on the set of measures in which the decision maker has most confidence. This structure is illustrated in Figure 14.4, where the rings are sets of probability measures, with the inner ring (the centre of the nested family) being assigned the lowest confidence level and each enclosing ring a higher confidence level than the one it encloses. A confidence level assignment to a set of measures can be thought as expressing the confidence that the decision maker has in any probasitions made true by all the probability measures contained in that ring. Since outer rings support less precise probasitions they will attract more confidence than the probasitions supported by the inner ones.

Given a confidence partition $\pi = \{\pi_1, ..., \pi_n\}$ it is straightforward to construct a nested family of sets $\{L_1..., L_n\}$. Put π_1 at the centre of the family by setting $L_1 = \pi_1$ and then let $L_i = L_{i-1} \cup \pi_i$. It then follows immediately that $L_i \subseteq L_{i+1}$ and hence, in virtue of the monotonicity of $\triangleright$, that $L_i \triangleright L_{i+1}$. So the set $\{L_1..., L_n\}$ constitutes a linearly ordered, nested family of sets of probability

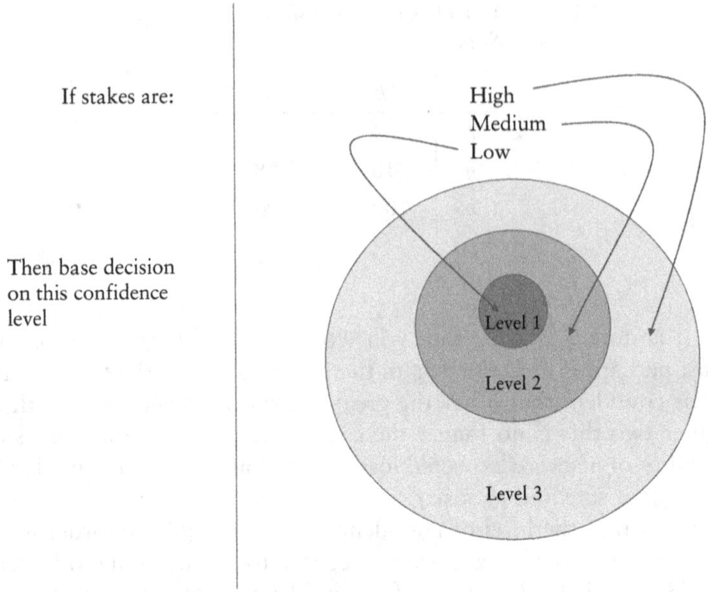

If stakes are:

High
Medium
Low

Then base decision
on this confidence
level

Level 1

Level 2

Level 3

FIGURE 14.4. Confidence Levels

measures. Our treatment of the confidence relation thus provides a natural foundation for the application of Hill's model.[3]

The decision stakes determine the confidence threshold for decision making by determining which level set is chosen as the basis for choice – intuitively, the smallest set of probability measures achieving this threshold. Formally, what is required thus is both a measure of the stakes associated with a decision problem and what Hill calls a 'cautiousness coefficient': something that determines, for each stake, a corresponding confidence threshold. Once the set of measures has been picked out in this way then a decision can be made. If the set is a singleton then ordinary expected utility maximisation is applicable. But if a set of probabilities is picked out then the decision maker can make use of one of the rules for cautious decision making under ambiguity discussed earlier, such as MMEU, α-MEU or the smooth ambiguity rule.

For example, consider Table 14.1 again and suppose that the agent is not comparing these acts with each other but, for each act, deciding whether or not to perform it. It is clear that, when she is confident in the assignment of probability one-half to the draw of a white ball, she should perform each of them. But, in the second situation, the minimum expected utility of the three

---

3 Hill treats confidence in a slightly different way from me, because he thinks one can only sensibly talk of confidence in those judgements that one accepts. But this difference does not matter here.

acts are respectively 0.25, −3.5 and −39, so only the first will be performed if she employs the MMEU rule.

As should be evident, what Hill provides is more of a schema for confidence-based decision rules than a specific one. Firstly, different notions of stakes and different accounts of cautiousness will determine different confidence thresholds. And, secondly, he leaves open the question of what decision rule to apply in conditions of ambiguity. But these details are less important than the fact that the schema allows for the application of confidence judgements in an intuitively compelling way and allows agents to handle the severe uncertainty that they face.

14.4 CONCLUDING REMARKS

I have sought, in this last part of the book, to give an account of the characteristics of bounded, but rational, agency: that of agents who face complex environments equipped with limited cognitive resources. I addressed three main questions: what does rationality require of the states of minds of bounded agents? How should such agents change their opinions as a result of experience? How should they make decisions?

To the first question, an unequivocal answer is on offer: rationality requires that agent's attitudes be coherently extendable. More precisely, their relational beliefs and preferences should have a coherent minimal extension on the full algebra of prospects, a requirement that implies neither that agents be aware of all prospects nor that they hold opinions on all that they are aware of. This rationality condition gives foundations to Imprecise Bayesianism, the view that a rational agent's state of mind is representable by a set of pairs of probability and desirability functions on the domain of her awareness.

Imprecise Bayesian agents, I argued, are permitted not only to revise their attitudes by a more general version of Bayesian conditioning but also to make them more or less precise, by reaching judgements or suspending them. Although rationality cannot dictate what an agent should take from her experience, it can discipline how such 'lessons' are absorbed by identifying the minimal modifications to the agent's judgements required to maintain consistency. Such conditions were captured in the rules of attitude revision, formation and withdrawal presented in Chapter 12.

The third question has proved the most difficult to address. On the one hand, there are a great variety of considerations and corresponding decision rules that an agent might apply in conditions in which she is unable to reach a precise judgement on all relevant contigencies. On the other, how sensible the rules are depends on the type of decision situation she finds herself in. I argued that, in situations of Grade 2 uncertainty, it is reasonable to be guided by the symmetries of the decision problem in reaching a precise judgement in order that standard expected utility maximisation could be applied. But, in situations of Grade 3 uncertainty, such techniques are not applicable and the agent must work with the fact that more than one opinion deserves consideration. It is in this context that considerations of confidence become salient as a possible means of evaluating 'competing' opinions.

In this chapter, I have tried to develop an account of confidence that might do justice to this potential. The exercise has confirmed, it seems to me, the important role that confidence can and does play in judgement and decision making. This means that the model of rationality presented by Imprecise Bayesianism needs to be filled out, at least when applied to agents capable of making second-order judgements about their own attitudes. In such cases, a rational agent can give a nuanced representation of the uncertainty she faces by grading potential judgements in accordance not only with whether the evidence favours them or not but also with how much evidence favours them and of what quality. Equipped with more nuanced representation, she can calibrate her decision making to what is at stake, so that high-stakes decision problems can be dealt with more cautiously than low-stakes ones.

This contention raises as many questions as it answers. But a book must end somewhere. I choose here.

Appendix: Proofs

Lemma A.1. *The axioms of probability and desirability imply:*

1. $P(\alpha) \cdot V(\alpha) = -P(\neg\alpha) \cdot V(\neg\alpha)$
2. $P(\alpha)[(V(\alpha) - V(\neg\alpha)] = -V(\neg\alpha)$
3. *If* $P(\alpha) = 1$ *then* $V(\alpha) = 0$
4. *If* $V(\alpha) = V(\neg\alpha)$ *then* $V(\alpha) = 0$
5. *If* $P(\alpha) \neq 1$ *then:*

$$\text{(a) } P(\alpha) = \frac{-V(\alpha)}{V(\alpha) - V(\neg\alpha)}$$

$$\text{(b) } \frac{P(\alpha)}{P(\neg\alpha)} = -\frac{V(\neg\alpha)}{V(\alpha)}$$

6. *If* $\{\alpha_i\}_{i=1}^{n}$ *is a partition of prospects then* $\sum_{i=1}^{n} V(\alpha_i) \cdot P(\alpha_i) = 0$

Proof. (1) By V2, $V(\top) = V(\alpha \vee \neg\alpha) = V(\alpha) \cdot P(\alpha) + V(\neg\alpha) \cdot P(\neg\alpha) = 0$ by V1. Hence $P(\alpha) \cdot V(\alpha) = -P(\neg\alpha) \cdot V(\neg\alpha)$. (2) and (3) By P1, it follows from 1. that $P(\alpha) \cdot V(\alpha) = (P(\alpha) - 1) \cdot V(\neg\alpha)$ and hence that $P(\alpha)(V(\alpha) - V(\neg\alpha)) = -V(\neg\alpha)$ and that if $P(\alpha) = 1$ then $V(\alpha) = 0$. (4) By 2. $P(\neg\alpha)(V(\alpha) - V(\neg\alpha)) = V(\alpha)$. Hence if $V(\alpha) = V(\neg\alpha)$ then $P(\neg\alpha) = 0$ or $V(\alpha) = 0$. But if $P(\neg\alpha) = 0$ then $P(\alpha) = 1$ so, by 3., $V(\alpha) = 0$. (5) Suppose that $V(\alpha) \neq 0$. Then, by 3. and 4., $P(\alpha) \neq 1$ and $V(\alpha) \neq V(\neg\alpha)$. Hence, $P(\neg\alpha) \neq 0$ and $V(\alpha) - V(\neg\alpha) \neq 0$. So, by 1., (b) $\frac{P(\alpha)}{P(\neg\alpha)} = -\frac{V(\neg\alpha)}{V(\alpha)}$. And, by 2., (a) $P(\alpha) = \frac{-V(\alpha)}{V(\alpha) - V(\neg\alpha)}$. Finally, let $\{\alpha_i\}_{i=1}^{n}$ be a partition of prospects. Then $\bigvee_{i=1}^{n} \{\alpha_i\} = \top$ and so, by V1, $V(\bigvee_{i=1}^{n}\{\alpha_i\}) = 0$. But, by V2, $V(\bigvee_{i=1}^{n}\{\alpha_i\}) = V(\bigvee_{i=1}^{n-1}\{\alpha_i\}) \cdot P(\bigvee_{i=1}^{n-1}\{\alpha_i\}) + V(\alpha_n) \cdot P(\alpha_n) = [V(\bigvee_{i=1}^{n-2}\{\alpha_i\}) \cdot P(\bigvee_{i=1}^{n-2}\{\alpha_i\}) + V(\alpha_{n-1}) \cdot P(\alpha_{n-1})] + V(\alpha_n) \cdot P(\alpha_n)... = \sum_{i=1}^{n} V(\alpha_i) \cdot P(\alpha_i)$. Hence $\sum_{i=1}^{n} V(\alpha_i) \cdot P(\alpha_i) = 0$. ∎

Lemma A.2. *Given* P^*1 *and* P^*2:

1. $P_\alpha^*(\neg\alpha\beta) = 0$

2. $P_\alpha^*(\alpha\beta) = P_\alpha^*(\beta)$
3. $P_\alpha^*(\beta|\alpha) = P_\alpha^*(\beta)$

Proof. By P*1, $P_\alpha^*(\beta) = P_\alpha^*(\alpha\beta) + P_\alpha^*(\neg\alpha\beta)$, and, by P*2, $P_\alpha^*(\alpha) = 1$, and so $P_\alpha^*(\neg\alpha) = 0$. So (1) $P_\alpha^*(\neg\alpha\beta) = P_\alpha^*(\beta|\neg\alpha) \cdot P_\alpha^*(\neg\alpha) = 0$. Hence (2) $P_\alpha^*(\beta) = P_\alpha^*(\alpha\beta)$. Similarly, $P_\alpha^*(\alpha\beta) = P_\alpha^*(\beta|\alpha) \cdot P_\alpha^*(\alpha)$. So, by 2., it follows that (3) $P_\alpha^*(\beta|\alpha) = P_\alpha^*(\beta)$. ∎

Theorem A.3. *Given P*1 and P*2:*

1. *Supposition Averaging implies Regularity.*
2. *Regularity implies P*3 Anchoring.*

Proof. (1) Since $P_\alpha^*(\alpha\beta) \geq P_{\neg\alpha}^*(\alpha\beta) = 0$ by Lemma A.2(1), it follows from Supposition Averaging that $P_\alpha^*(\alpha\beta) \geq P(\alpha\beta)$ and hence by Lemma A.2(2) that

$$P_\alpha^*(\beta) \geq P(\alpha\beta)$$

(2) Regularity implies that $P_\top^*(\beta) \geq P(\beta)$ and that $P_\top^*(\neg\beta) \geq P(\neg\beta)$. But $P_\top^*(\beta) + P_\top^*(\neg\beta) = 1 = P(\beta) + P(\neg\beta)$. So $P_\top^*(\beta) = P(\beta)$. ∎

Theorem A.4. *If both β and $\neg\beta$ are desirabilistically independent of α, relative to ⊛, then they are probabilistically independent of α, relative to ⊛.*

Proof. By Lemma A.1(5b)

$$\frac{P(\beta)}{P(\neg\beta)} = \frac{-V(\neg\beta)}{V(\beta)}; \frac{P_\alpha^*(\beta)}{P_\alpha^*(\neg\beta)} = \frac{-V_\alpha^*(\neg\beta)}{V_\alpha^*(\beta)}$$

Hence

$$\frac{V_\alpha^*(\beta)}{V_\alpha^*(\neg\beta)} = \frac{V(\beta)}{V(\neg\beta)} \Longleftrightarrow \frac{P_\alpha^*(\beta)}{P_\alpha^*(\neg\beta)} = \frac{P(\beta)}{P(\neg\beta)}$$

But

$$\frac{P_\alpha^*(\beta)}{P_\alpha^*(\neg\beta)} = \frac{P(\beta)}{P(\neg\beta)} \Longleftrightarrow P_\alpha^*(\beta) = P(\beta)$$

Now if both β and $\neg\beta$ are desirabilistically independent of α, relative to ⊛, then by definition $\frac{V_\alpha^*(\beta)}{V_\alpha^*(\neg\beta)} = \frac{V(\beta)}{V(\neg\beta)}$. It follows that $P_\alpha^*(\beta) = P(\beta)$ and, hence, that $P_\alpha^*(\neg\beta) = P(\neg\beta)$. So β and $\neg\beta$ are probabilistically independent of α. ∎

Theorem A.5. *Let $\Omega = \langle X, \models \rangle$ be an atomless Boolean algebra of prospects and $J = \langle \Omega, \trianglerighteq, \succsim \rangle$ be a Joyce structure. Let ⊛ be a supposition function both on $\mathcal{R}^\trianglerighteq$ and $\mathcal{R}^\succsim$ that satisfies Coherent Supposition. Let $A := \{\gamma \in X : \gamma \models \alpha\}$. Then:*

1. *(Existence) There exists a suppositional probability $P(\cdot, \cdot)$ on $X \times X'$, that numerically represents ⊛ on $\trianglerighteq$ and a suppositional desirability function $V(\cdot, \cdot)$ on $X' \times X'$, that numerically represents ⊛ on $\succsim$.*

2. *(Uniqueness)* $P(\cdot,\cdot)$ *is unique.* $V'(\cdot,\cdot)$ *is another such suppositional desirability iff, for all* $\alpha \in X'$, *there exists* $a > 0$ *such that, for all* $\beta \in A$, $V'(\beta,\alpha) = aV(\beta,\alpha)$.

Proof. By Joyce's theorem (Theorem 5.6), there exists a unique probability function P on X that numerically represents the credibility relation $\unrhd$ and a desirability function V on X', unique up to linear transformation, that numerically represents the preference relation $\succsim$. It also follows that there exists, for all $\alpha \in X'$, a unique probability function P_α^* on A that numerically represents $\unrhd_\alpha^*$ and a desirability function V_α^* on A', unique up to linear transformation, that numerically represents $\succsim_\alpha^*$. Furthermore, since α is the supremum of the algebra Ω_a, $P_\alpha^*(\alpha) = 1$ and $V_\alpha^*(\alpha) = 0$, in accordance with P*2 and V*2. Finally, in virtue of the fact that $\otimes(\unrhd,\top) = \unrhd$ and $\otimes(\succsim,\top) = \succsim$, it follows that $P_\top^*(\alpha) = P(\alpha)$ and $V_\top^*(\alpha) = V(\alpha)$ in accordance with P*3 and V*3. Now set

$$P(\beta,\alpha) = P_\alpha^*(\beta)$$

$$V(\beta,\alpha) = V_\alpha^*(\beta)$$

It is evident that $P(\beta,\alpha) \geq P(\gamma,\alpha) \Leftrightarrow P_\alpha^*(\beta) \geq P_\alpha^*(\gamma)$ and that $V(\beta,\alpha) \geq V(\gamma,\alpha) \Leftrightarrow V_\alpha^*(\beta) \geq V_\alpha^*(\gamma)$. Hence $P(\cdot,\alpha)$ and $V(\cdot,\alpha)$, respectively, represent $\unrhd_\alpha^*$ and $\succsim_\alpha^*$. It follows that $P(\cdot,\cdot)$ and $V(\cdot,\cdot)$ represent $\circledast$ *on* $\unrhd$ *and* $\succsim$. The uniqueness of $P(\cdot,\cdot)$ follows from that of the P_α^*. Now define $V'(\cdot,\cdot)$ by $V'(\beta,\top) = V(\beta,\top)$ and $V'(\beta,\alpha) = aV(\beta,\alpha)$ for all $\alpha \in X'$ and $a > 0$. Now

$$V'(\beta,\alpha) \geq V'(\gamma,\alpha)$$

$$\Leftrightarrow a \cdot V(\beta,\alpha) \geq a \cdot V(\gamma,\alpha)$$

$$\Leftrightarrow V(\beta,\alpha) \geq V(\gamma,\alpha)$$

$$\Leftrightarrow \beta \succsim_\alpha^* \gamma$$

So $V'(\cdot,\alpha)$ represents the $\succsim_\alpha^*$. It is obvious that $V'(\cdot,\cdot)$ is a suppositional desirability: $V'(\cdot,\alpha)$ is a linear transform of $V(\cdot,\alpha)$, so it is a desirability function on $\succsim_\alpha^*$; $V'(\alpha,\alpha) = aV(\cdot,\alpha) = 0$ in accordance with Certainty; and $V'(\alpha,\top) = V(\alpha,\top) = V(\alpha)$ in accordance with Anchoring. Conversely, by the uniqueness of the V_α^* on X', any suppositional desirability $V'(\cdot,\cdot)$ that represents the $\circledast$ *on* $\succsim$ must be such that, for each $\alpha \in X$, there exists $a > 0$ such that, for all $\beta \in A$, $V'(\beta,\alpha) = aV(\beta,\alpha)$. ∎

Corollary A.6. *(Representation of Evidential Conditional Attitudes)*

1. *If the* $\unrhd_\alpha^*$ *satisfy Suppositional Rigidity then* $P(\cdot|\cdot)$ *uniquely represents* $\circledast$ *on* $\unrhd$ *and* Ω.
2. *If the* $\succsim_\alpha^*$ *also satisfy Suppositional Rigidity then* $V(\cdot,\alpha) = a \cdot V(\cdot|\alpha)$, *for some* $a > 0$, *and* $V(\cdot|\cdot)$ *represents* $\circledast$ *on* $\succsim$ *and* Ω, *uniquely up to positive linear transformation.*

Proof. (1) By definition, $P(\beta|\alpha) \geq P(\gamma|\alpha) \Leftrightarrow P(\alpha\beta) \geq P(\alpha\gamma)$ and $P(\alpha\beta) \geq P(\alpha\gamma) \Leftrightarrow \alpha\beta \unrhd \alpha\gamma$. But, by Suppositional Rigidity, $\beta \unrhd_\alpha^* \gamma \Leftrightarrow \alpha\beta \unrhd \alpha\gamma$. Hence $\beta \unrhd_\alpha^* \gamma \Leftrightarrow P(\beta|\alpha) \geq P(\gamma|\alpha)$. So $P(\cdot|\alpha)$ numerically represents $\unrhd_\alpha^*$ and $P(\cdot|\cdot)$ represents $\circledast$ on $\unrhd$ and Ω. By Theorem A.5(2), $P(\cdot|\cdot)$ is a unique such representation. (2) By definition, $V(\beta|\alpha) \geq V(\gamma|\alpha) \Leftrightarrow V(\alpha\beta) \geq V(\alpha\gamma)$ and $V(\alpha\beta) \geq V(\alpha\gamma) \Leftrightarrow \alpha\beta \succsim \alpha\gamma$. But, by Suppositional Rigidity, $\beta \succsim_\alpha^* \gamma \Leftrightarrow \alpha\beta \succsim \alpha\gamma$. Hence $\beta \succsim_\alpha^* \gamma \Leftrightarrow V(\beta|\alpha) \geq V(\gamma|\alpha)$. So $V(\cdot|\alpha)$ numerically represents $\succsim_\alpha^*$ and $V(\cdot|\cdot)$ represents $\circledast$ on $\succsim$ and Ω. Let $V'(\cdot|\cdot) := aV(\cdot|\cdot)$ for some $a > 0$. Then

$$V'(\beta|\alpha) \geq V'(\gamma|\alpha)$$
$$\Leftrightarrow V'(\beta\alpha) - V'(\alpha) \geq V'(\gamma\alpha) - V'(\alpha)$$
$$\Leftrightarrow aV(\beta\alpha) - aV(\alpha) \geq aV(\gamma\alpha) - aV(\alpha)$$
$$\Leftrightarrow aV(\beta\alpha) - aV(\alpha) \geq aV(\gamma\alpha) - aV(\alpha)$$
$$\Leftrightarrow V(\beta|\alpha) \geq V(\gamma|\alpha)$$

So $V'(\cdot|\cdot)$ represents $\circledast$ on $\succsim$ and Ω as well. Now let $V'(\cdot|\cdot)$ be a conditional desirability that represents $\circledast$ on $\succsim$ and Ω. By Theorem A.5(2), for all $\alpha \in X'$, $V'(\cdot|\alpha) = aV(\cdot|\alpha)$ for some $a > 0$ and, in particular, $V'(\cdot) = V'(\cdot|\top) = kV(\cdot|\top) = kV(\cdot)$ for some $k > 0$. Now, by the definition of conditional desirability,

$$V'(\beta|\alpha) = V'(\alpha\beta) - V'(\alpha) = kV(\alpha\beta) - kV(\alpha)$$

But:

$$V'(\beta|\alpha) = aV(\cdot|\alpha) = aV(\alpha\beta) - aV(\alpha)$$

So $a = k$. So $V'(\cdot|\cdot) = kV(\cdot|\cdot)$. Hence $V(\cdot|\cdot)$ is unique up to linear transformation. ∎

Lemma A.7.

1. *Dominance implies Suppositional Rigidity*
2. *Suppositional Rigidity and ∨-Separability jointly imply Dominance*

Proof. (1) By Dominance, since $\neg\alpha$ is inconsistent with $\alpha\beta$ and $\alpha\gamma$, if $\alpha\beta R_\alpha^* \alpha\gamma$ then $\alpha\beta R\alpha\gamma$, and if $\alpha\gamma R_\alpha^* \alpha\beta$ then $\alpha\gamma R\alpha\beta$. So $\alpha\beta R_\alpha^* \alpha\gamma \Leftrightarrow \alpha\beta R\alpha\gamma$. (2) Suppose Suppositional Rigidity and that $\forall\alpha_i \in \{\alpha_i\}$, $\beta R_{\alpha_i}^* \gamma$. Then, $\forall\alpha_i \in \{\alpha_i\}$, $\alpha_i\beta R\alpha_i\gamma$. Then by ∨-Separability, $\vee_i(\alpha_i\beta)R\vee_i(\alpha_i\gamma)$. Hence $\beta R\gamma$. ∎

Lemma A.8. *Given the axioms of supposition, Variable Averaging implies that:*

$$P(\alpha) = k_\alpha^\alpha \tag{A.0.1}$$

$$P(\alpha\gamma) = k_\alpha^{\alpha\gamma} \cdot P_\alpha^*(\gamma) \tag{A.0.2}$$

Proof. By Variable Averaging, there exists a $k_\alpha^\gamma > 0$ such that $P(\gamma) = k_\alpha^\gamma \cdot P_\alpha^*(\gamma) + (1 - k_\alpha^\gamma) \cdot P_{\neg\alpha}(\gamma)$. It follows from Lemma A.1 that $P_\alpha^*(\neg\alpha\gamma) = 0$ and hence that

$$P_\alpha^*(\gamma) = P_\alpha^*(\alpha\gamma) + P_\alpha^*(\neg\alpha\gamma) = P_\alpha^*(\alpha\gamma)$$

Hence, by Variable Averaging and Lemma A.1,

$$P(\alpha\gamma) = k_\alpha^{\alpha\gamma} \cdot P_\alpha^*(\alpha\gamma) + (1 - k_\alpha^\gamma) \cdot P_{\neg\alpha}(\alpha\gamma)$$
$$= k_\alpha^{\alpha\gamma} \cdot P_\alpha^*(\gamma).$$

So $P(\alpha\gamma) = k_\alpha^{\alpha\gamma} \cdot P_\alpha^*(\gamma)$. In particular, since, by Lemma A.1, $P_\alpha^*(\alpha) = 1$, $P(\alpha) = k_\alpha^\alpha \cdot P_\alpha^*(\alpha) = k_\alpha^\alpha$. ∎

Theorem A.9. *Let $\Omega = \langle X, \models \rangle$ be an atomless Boolean algebra with P a probability measure defined on it and ⊛ a supposition function that satisfies the axioms of supposition $P^*1 - P^*3$. Let $\mathcal{A} = \{\alpha_i\}$ be a partition of prospects. The following are equivalent:*

1. *Belief Dominance*
2. *Constant Averaging*
3. *For all $\gamma \in X$, $P_\alpha^*(\gamma) = P(\gamma|\alpha)$*

Proof. Constant Averaging clearly implies Belief Dominance. The converse follows from proposition 2 in Mongin (1995). Now assume that, for all $\alpha_i \in \mathcal{A}$, $\gamma \in X$, $P_{\alpha_i}^*(\gamma) = P(\gamma|\alpha_i)$. Then it follows from the Law of Total Probability that $P(\gamma) = \sum_i P(\gamma|\alpha_i) \cdot P(\alpha_i) = P_{\alpha_i}^*(\gamma) \cdot P(\alpha_i)$. Hence Constant Averaging is satisfied with the $k_i = P(\alpha_i)$. Finally, assume Constant Averaging. Then by Lemma A.8, and the fact that the weights are independent of γ, for any $\alpha \in \mathcal{A}$, $P(\alpha\gamma) = k_\alpha \cdot P_\alpha^*(\gamma)$ and, in particular, $P(\alpha) = k_\alpha$. Hence $P(\alpha\gamma) = P(\alpha) \cdot P_\alpha^*(\gamma)$. From which it follows that $P_\alpha^*(\gamma) = P(\gamma|\alpha)$. ∎

Theorem A.10. *Assume Variable Averaging and P^+4 Commutativity. Let ⊛ be a supposition function that satisfies $P^*1 - P^*3$. Then:*

$$(\textit{Constant Averaging}) \; P(\gamma) = \sum_i \alpha_i \cdot P_{\alpha_i}^*(\gamma)$$

Proof. It follows from Variable Averaging that, for any α and γ, there exists a weight $k_\alpha^\gamma > 0$ that depends on P_α^* and γ and is such that $P(\gamma) = k_\alpha^\gamma \cdot P_\alpha^*(\gamma) + (1 - k_\alpha^\gamma) \cdot P_{\neg\alpha}^*(\gamma)$. Note that it follows immediately that $k_{\neg\alpha}^\gamma = 1 - k_\alpha^\gamma$.

Let P, Q and R be probability functions such that $Q = P_{\neg(\alpha\beta)}^*$ and $R = P_\alpha^*$. Now, for any α, β and γ, let k_α^γ be such a weight on the suppositional probability function P_α^* that depends on P, α and γ, let l_α^γ be a weight on the suppositional probability function Q_α^* that depends on Q, α and γ, and let m_β^γ be a weight on the suppositional probability function R_β^* that depends on R, α and γ. Note that it follows from P^+4 that $Q_\alpha^* = (P_{\neg(\alpha\beta)}^*)_\alpha^* = P_{\alpha\neg\beta}^*$, and

$R_\beta^* = (P_\alpha^*)_\beta^* = P_{\alpha\beta}^*$. Now:

$$
\begin{aligned}
P(\alpha\gamma) &= k_{\alpha\beta}^{\alpha\gamma} \cdot P_{\alpha\beta}^*(\alpha\gamma) + (1 - k_{\alpha\beta}^{\alpha\gamma}) \cdot P_{\neg(\alpha\beta)}^*(\alpha\gamma) \\
&= k_{\alpha\beta}^{\alpha\gamma} \cdot P_{\alpha\beta}^*(\gamma) + (1 - k_{\alpha\beta}^{\alpha\gamma}) \cdot [l_\alpha^{\alpha\gamma} Q_\alpha^*(\alpha\gamma) + (1 - l_\alpha^{\alpha\gamma}) Q_{\neg\alpha}^*(\alpha\gamma)] \\
&= k_{\alpha\beta}^{\alpha\gamma} \cdot P_{\alpha\beta}^*(\gamma) + (1 - k_{\alpha\beta}^{\alpha\gamma}) \cdot [l_\alpha^{\alpha\gamma} P_{\alpha\neg\beta}^*(\alpha\gamma) + (1 - l_\alpha^{\alpha\gamma}) P_{\neg\alpha}^*(\alpha\gamma)] \\
&= k_{\alpha\beta}^{\alpha\gamma} \cdot P_{\alpha\beta}^*(\gamma) + (1 - k_{\alpha\beta}^{\alpha\gamma}) \cdot l_\alpha^{\alpha\gamma} P_{\alpha\neg\beta}^*(\gamma)
\end{aligned}
$$

since, by Lemma A.2, $P_{\neg\alpha}^*(\alpha\gamma) = 0$. Similarly, $P^*(\alpha\gamma) = k_\alpha^{\alpha\gamma} \cdot P_\alpha^*(\alpha\gamma) + (1 - k_\alpha^{\alpha\gamma}) \cdot P_{\neg\alpha}^*(\alpha\gamma) = k_\alpha^{\alpha\gamma} \cdot P_\alpha^*(\alpha\gamma)$. So

$$
\begin{aligned}
P^*(\alpha\gamma) &= k_\alpha^{\alpha\gamma} \cdot [m_\beta^{\alpha\gamma} R_\beta^*(\alpha\gamma) + (1 - m_\beta^{\alpha\gamma}) R_{\neg\beta}^*(\alpha\gamma)] \\
&= k_\alpha^{\alpha\gamma} \cdot [m_\beta^{\alpha\gamma} P_{\alpha\beta}^*(\alpha\gamma) + (1 - m_\beta^{\alpha\gamma}) P_{\alpha\neg\beta}^*(\alpha\gamma)] \\
&= k_\alpha^{\alpha\gamma} \cdot m_\beta^{\alpha\gamma} P_{\alpha\beta}^*(\gamma) + k_\alpha^{\alpha\gamma} \cdot (1 - m_\beta^{\alpha\gamma}) P_{\alpha\neg\beta}^*(\gamma)]
\end{aligned}
$$

But this can be so only if, for all $\gamma, P_{\alpha\beta}^*(\gamma) = P_{\alpha\neg\beta}^*(\gamma)$ or $k_{\alpha\beta}^{\alpha\gamma} = k_\alpha^{\alpha\gamma} \cdot m_\beta^{\alpha\gamma}$ and $(1 - k_{\alpha\beta}^{\alpha\gamma}) \cdot l_\alpha^{\alpha\gamma} = k_\alpha^{\alpha\gamma} \cdot (1 - m_\beta^{\alpha\gamma})$. But, by Lemma A.2, $P_{\alpha\beta}^*(\beta) = 1 \neq 0 = P_{\alpha\neg\beta}^*(\beta)$. So

$$
\begin{aligned}
& k_{\alpha\beta}^{\alpha\gamma} = k_\alpha^{\alpha\gamma} \cdot m_\beta^{\alpha\gamma} \\
\Leftrightarrow \quad & \frac{P(\alpha\beta\gamma)}{P_\beta^*(\gamma)} = \frac{P(\alpha\gamma)}{P_\alpha^*(\gamma)} \cdot \frac{P_\alpha^*(\beta\gamma)}{P_{\alpha\beta}^*(\gamma)} \\
\Leftrightarrow \quad & \frac{P_\alpha^*(\beta\gamma)}{P_\alpha^*(\gamma)} = \frac{P(\alpha\beta\gamma)}{P(\alpha\gamma)} = P(\beta|\alpha\gamma) \\
\Leftrightarrow \quad & P_\alpha^*(\beta|\gamma) = P(\beta|\alpha\gamma)
\end{aligned}
$$

In particular, by substitution of α for γ, we obtain $P_\alpha^*(\beta|\alpha) = P(\beta|\alpha)$. Hence, in virtue of Lemma A.2, $P_\alpha^*(\beta) = P(\beta|\alpha)$. ∎

Theorem A.11. *Let* $\Omega = \langle X, \models \rangle$ *be a complete and atomless Boolean algebra and* $\Omega_Y = \langle Y, \models_Y \rangle$ *be a coarsening subalgebra of it. Let* $\mathcal{J} = \langle \Omega, \trianglerighteq, \succsim \rangle$ *be a Joyce structure and* $\langle P, V \rangle$ *a corresponding Jeffrey representation of it. Let* $\trianglerighteq_Y$ *and* $\succsim_Y$ *be the restrictions of* $\trianglerighteq$ *and* $\succsim$ *to* Y. *Let* $\alpha \in X$ *be such that (1)* $P(\alpha) > 0$ *and (2), for all* $\beta_i \in Y, \alpha \wedge \beta_i \neq \bot$. *Suppose that* $\succsim$ *is separable on the subset* $\{\alpha\} \otimes Y$ *of* X. *Then there exists a real number* $a > 0$ *such that* $\forall \beta_i \in Y$:

$$
V(\alpha\beta_i) = a.V(\beta_i) + V(\alpha)
$$

$$
P(\alpha\beta_i) = P(\beta_i) \cdot P(\alpha)
$$

Proof. First we define a probability–desirability pair (V', P'), respectively on Y' and Y, by

$$
V'(\beta_i) := V(\beta_i)
$$

$$
P'(\beta_i) := P(\beta_i)
$$

Note that (V', P') is a Jeffrey representation of $\unrhd_Y$ and $\succsim_Y$ in virtue of the fact that $\langle Y, \models_Y \rangle$ is a coarsening Boolean subalgebra of Ω. Next we define a second desirability–probability pair (V_Y, P_Y), respectively on Y' and Y, by, for all $\beta_i \in Y$,

$$V_Y(\beta_i) := V(\beta_i | \alpha) = V(\alpha \beta_i) - V(\alpha)$$

$$P_Y(\beta_i) := P(\beta_i | \alpha) = \frac{P(\alpha \beta_i)}{P(\alpha)}$$

1. (1) Suppose that α is separable from Y under $\unrhd_Y$. It follows that, for all $\beta_i, \beta_j \in Y$, $\beta_i \unrhd_Y \beta_i$

$$\Leftrightarrow \beta_i \unrhd \beta_j$$

$$\Leftrightarrow \alpha \beta_i \unrhd \alpha \beta_j$$

$$\Leftrightarrow P(\alpha \beta_i) \geq P(\alpha \beta_j)$$

$$\Leftrightarrow P_Y(\beta_i) \geq P_Y(\beta_j)$$

So P_Y is a numerical representation of $\unrhd_Y$. But by Villegas's theorem (Theorem 5.4), $\unrhd_Y$ has a unique numerical representation. So $P_Y(\beta_i) = P'(\beta_i) = P(\beta_i)$. But $P_Y(\beta_i) = P(\beta_i | \alpha)$. So, for all $\beta_i \in Y$,

$$P(\alpha \beta_i) = P(\alpha) \cdot P(\beta_i)$$

2. (2) Now suppose that α is also separable from Y under $\succsim_Y$. It follows that, for all $\beta_i, \beta_j \in Y$, $\beta_i \succsim_Y \beta_i$:

$$\Leftrightarrow \beta_i \succsim \beta_j$$

$$\Leftrightarrow \alpha \beta_i \succsim \alpha \beta_j$$

$$\Leftrightarrow V(\alpha \beta_i) \geq V(\alpha \beta_j)$$

$$\Leftrightarrow V_Y(\beta_i) \geq V_Y(\beta_j)$$

So (V_Y, P_Y) is also a Jeffrey representation of $\unrhd_Y$ and $\succsim_Y$. It then follows, by Joyce's uniqueness theorem, that there exists a real number $a > 0$ such that, for all $\beta_i \in Y$,

$$V_Y(\beta_i) = aV'(\beta_i) = aV(\beta_i)$$

$$P_Y(\beta_i) = P'(\beta_i) = P(\beta_i)$$

But $P_Y(\beta_i) = P(\beta_i | \alpha)$ and $V_Y(\beta_i) = V(\beta_i | \alpha)$. So, for all $\beta_i \in Y$,

$$P(\alpha \beta_i) = P(\alpha) \cdot P(\beta_i)$$

$$V(\alpha \beta_i) = V(\alpha) + aV(\beta_i)$$

∎

Theorem A.12. *Suppose Desirabilism and Probabilism. Then:*

1. *Thesis 7.3 implies Thesis 7.1*
2. *Thesis 7.4 implies Thesis 7.2*
3. *Bradley's Thesis implies Adams' Thesis*

Proof. (1) By Lemma A.1(5a):

$$
\begin{aligned}
P(\alpha \to \beta) &= \frac{-V(\alpha \to \neg\beta)}{V(\alpha \to \beta) - V(\alpha \to \neg\beta)} \\[2mm]
&= \frac{-k_A V^*_\alpha(\neg\beta)}{k_A V^*_\alpha(\beta) - k_A V^*_\alpha(\neg\beta)} \quad \text{[Thesis 7.3]} \\[2mm]
&= \frac{-V^*_\alpha(\neg\beta)}{V^*_\alpha(\beta) - V^*_\alpha(\neg\beta)} \quad \text{[Cancelling } k_A] \\[2mm]
&= P^*_\alpha(\beta) \quad \text{[Lemma A.1(5a)]}
\end{aligned}
$$

(2) and (3) are proved in the same way, substituting $V^+_\alpha(\neg\beta) \cdot P(\alpha)$ (applying Thesis 7.4) and $V(\neg\beta|\alpha) \cdot P(\alpha)$ (applying Bradley's Thesis) for $k_A V^*_\alpha(\neg\beta)$. ∎

Lemma A.13. *Let P and V be a pair of probability and desirability functions defined on a simple conditional algebra of prospects* $\Gamma = \langle Y, \models \rangle$. *Let* $\Gamma_\alpha = \langle Y_\alpha, \models \rangle$ *be the Boolean subalgebra of* Γ, *with* Y_α *the set of simple conditionals with antecedent* α. *Let* α' *be such that* $\alpha(\alpha') = \bot$. *Let the functions P* and V* be defined by*

$$
P^*(\alpha \to \cdot) = \frac{P((\alpha \to \cdot)(\neg\alpha \to \gamma))}{P(\neg\alpha \to \gamma)}
$$

$$
V^*(\alpha \to \cdot) = V((\alpha \to \cdot)(\neg\alpha \to \gamma)) - V(\neg\alpha \to \gamma)
$$

Then P* and V* are, respectively, a probability and desirability on Γ_α.

Proof. Note that by the Boundedness property $(\alpha \to \alpha)(\alpha' \to \gamma) = \alpha' \to \gamma$. Hence $P^*(\alpha \to \alpha) = 1$ and $V^*(\alpha \to \alpha) = 0$. Now, if $\beta\delta = \bot$, then

$$
P^*(\alpha \to (\beta \vee \delta)) = \frac{P((\alpha \to (\beta \vee \delta))(\alpha' \to \gamma))}{P(\alpha' \to \gamma)}
$$

$$
V^*(\alpha \to (\beta \vee \delta)) = V((\alpha \to (\beta \vee \delta))(\alpha' \to \gamma)) - V(\alpha' \to \gamma)
$$

But by Conditional Distributivity and then Distributivity:

$$
\alpha \to ((\beta \vee \delta)(\alpha' \to \gamma)) = (\alpha \to \beta) \vee (\alpha \to \delta)(\alpha' \to \gamma)
$$

$$
= (\alpha \to \beta)(\alpha' \to \gamma) \vee (\alpha \to \delta)(\alpha' \to \gamma)
$$

And, since $\beta\delta = \perp$, by Conditional Distributivity and Boundedness, $(\alpha \to \beta)(\alpha \to \delta) = \alpha \to \beta\delta = \perp$. So, by P2,

$$\frac{P((\alpha \to \beta)(\alpha' \to \gamma) \vee (\alpha \to \delta)(\alpha' \to \gamma))}{P(\alpha' \to \gamma)} = \frac{P((\alpha \to \beta)(\alpha' \to \gamma)) + P((\alpha \to \delta)(\alpha' \to \gamma))}{P(\alpha' \to \gamma)}$$

$$= P^*(\alpha \to \beta) + P^*(\alpha \to \delta)$$

Hence P^* is a probability on Y_α. Similarly, by V1,

$$V((\alpha \to \beta)(\alpha' \to \gamma) \vee (\alpha \to \delta)(\alpha' \to \gamma)) - V(\alpha' \to \gamma)$$

$$= \frac{\begin{array}{c}V((\alpha \to \beta)(\alpha' \to \gamma)) \cdot P((\alpha \to \beta)(\alpha' \to \gamma)) \\ + V((\alpha \to \delta)(\alpha' \to \gamma)) \cdot P^*((\alpha \to \delta)(\alpha' \to \gamma))\end{array}}{P((\alpha \to \beta)(\alpha' \to \gamma) \vee (\alpha \to \delta)(\alpha' \to \gamma))} - V(\alpha' \to \gamma)$$

$$= \frac{\begin{array}{c}[V^*(\alpha \to \beta) + V(\alpha' \to \gamma)] \cdot P^*(\alpha \to \beta) \\ + [V^*(\alpha \to \delta) + V(\alpha' \to \gamma)] \cdot P^*(\alpha \to \delta)\end{array}}{P^*(\alpha \to \beta) + P^*(\alpha \to \delta)} - V(\alpha' \to \gamma)$$

$$= \frac{V^*(\alpha \to \beta) \cdot P^*(\alpha \to \beta) + V^*(\alpha \to \delta) \cdot P^*(\alpha \to \delta)}{P^*(\alpha \to \beta) + P^*(\alpha \to \delta)}$$

So V^* is a desirability function on Y'_α. ∎

Theorem A.14. *(Simple Conditionals) Let $\Gamma = \langle Y, \models \rangle$ be a complete and atomless simple conditional algebra of prospects based on $\Omega = \langle X, \models \rangle$ and $\mathcal{R}^\trianglerighteq$ and $\mathcal{R}^{\succsim}$ be classes of complete and continuous credibility and preference relations, respectively on Ω and Ω'. Let $J = \langle \Gamma, \trianglerighteq, \succsim \rangle$ be a Joyce structure and ⊛ be a supposition function on both $\mathcal{R}^\trianglerighteq$ and $\mathcal{R}^{\succsim}$ that satisfies Coherent Supposition. Then:*

1. *If $\trianglerighteq$ and $\trianglerighteq_\alpha^*$ satisfy the Ramsey property then there exists a unique probability P, on Y, and suppositional probability P_α^*, on X, that numerically represent $\trianglerighteq$ and $\trianglerighteq_\alpha^*$ and which jointly satisfy the Ramsey Test for belief – i.e. are such that, for all $\beta \in X$, $P(\alpha \to \beta) = P_\alpha^*(\beta)$.*
2. *If $\succsim$ and $\succsim_\alpha^*$ also satisfy the Ramsey property then there exists a desirability function V on Y and a suppositional desirability V_α^* on Y', both unique up to linear transformation, that numerically represent $\succsim$ and $\succsim_\alpha^*$ and which jointly satisfy the Ramsey Test for desire – i.e. there exists $k_\alpha > 0$, such that, for all $\beta \in X$, $V(\alpha \to \beta) = k_\alpha \cdot V_\alpha^*(\beta)$.*

Proof. By Theorems 5.6 (Joyce's theorem) and A.5, there exists a unique probability P, on Y, and corresponding suppositional probability P_α^*, on X, that numerically represent $\trianglerighteq$ and $\trianglerighteq_\alpha^*$, and a desirability function V, on Y', and corresponding suppositional desirability V_α^*, on X', both unique up to linear transformation, that numerically represent $\succsim$ and $\succsim_\alpha^*$. Now:

1. Define P_α on X by $P_\alpha(\cdot) = P(\alpha \to \cdot)$. By Lemma A.13, P_α is a probability on X and $P_\alpha(\top) = P(\alpha \to \top) = P(\top)$ in virtue of Conditional Normality. Now:

$$P_\alpha(\beta) \geq P_\alpha(\beta) \Leftrightarrow P(\alpha \to \beta) \geq P(\alpha \to \gamma)$$

$$\Leftrightarrow \alpha \to \beta \trianglerighteq \alpha \to \gamma$$

$$\Leftrightarrow \beta \trianglerighteq^*_\alpha \gamma$$

in virtue of the Ramsey Property. So P_α is a suppositional probability that represents $\trianglerighteq^*_\alpha$. But since P^*_α is unique it follows that $P^*_\alpha = P_\alpha$. Hence $P(\alpha \to \cdot) = P^*_\alpha(\cdot)$ in accordance with the Ramsey Test for belief.

2. Define V_α on X by $V_\alpha(\cdot) = V(\alpha \to \cdot)$. Note that V_α is a desirability on X and that $V_\alpha(\top) = V(\alpha \to \top) = V(\top)$ in virtue of Conditional Normality. Now:

$$V_\alpha(\beta) \geq V_\alpha(\beta) \Leftrightarrow V(\alpha \to \beta) \geq V(\alpha \to \gamma)$$

$$\Leftrightarrow \alpha \to \beta \succsim \alpha \to \gamma$$

$$\Leftrightarrow \beta \succsim^*_\alpha \gamma$$

in virtue of the Ramsey Property. So V_α is a suppositional probability that represents $\succsim^*_\alpha$. But since V^*_α is unique up to linear transformation it follows that, for some $k_\alpha > 0$, $V^*_\alpha = k_\alpha V_\alpha$. Hence $V(\alpha \to \cdot) = k_\alpha V^*_\alpha(\cdot)$ in accordance with the Ramsey Test for desire.

∎

Corollary A.15.   1. *If $\trianglerighteq$ and $\trianglerighteq^*_\alpha$ satisfy Suppositional Rigidity then Adams' Thesis holds.*
2. *If $\succsim$ and $\succsim^*_\alpha$ satisfy Suppositional Rigidity then, for some $k_\alpha > 0$, $V(\alpha \to \beta) = k_\alpha \cdot V(\beta | \alpha)$.*

Proof. (1) By Corollary A.6, if $\trianglerighteq$ and $\trianglerighteq^*_\alpha$ satisfy Suppositional Rigidity, then $P^*_\alpha(\cdot) = P(\cdot | \alpha)$. Hence, by Theorem A.14, $P(\alpha \to \cdot) = P(\cdot | \alpha)$ in accordance with Adams' Thesis. (2) By Corollary A.6, if $\succsim$ and $\succsim^*_\alpha$ satisfy Suppositional Rigidity, then, for some $k_\alpha > 0$, $V^*_\alpha(\cdot) = k_\alpha \cdot V(\cdot | \alpha)$. Hence, by Theorem A.14,

$$V(\alpha \to \beta) = k_\alpha \cdot V(\beta | \alpha)$$

∎

Theorem A.16. *Let $\Gamma = \langle Y, \models \rangle$ be a complete and atomless simple conditional algebra of prospects based on $\Omega = \langle X, \models \rangle$. Let $\trianglerighteq$ and $\succsim$ be continuous credibility and preference orders, respectively on Y and Y', that cohere with one another and let (P, V) be a Jeffrey representation of them unique up to linear transformation of V. Then:*

1. *If* $\unrhd$ *has the Indicative property then* $P(\alpha \to \beta) = P(\beta | \alpha)$
2. *If* $\succsim$ *has the Indicative property then* $V(\alpha \to \beta) = a \cdot V(\beta | \alpha)$ *for some* $a > 0$

Proof. Let $\Gamma_\alpha = \langle Y_\alpha, \models \rangle$ be the subalgebra of Γ based on the set Y_α of simple conditionals with antecedent α. Since this is a coarsening Boolean subalgebra of Γ (see Theorem 5.1), it follows that (P, V) restricted to Γ_α is a Jeffrey representation of $\unrhd$ and $\succsim$ restricted to Γ_α.

1. Define P' on Y_α by $P'(\alpha \to \beta) = P(\beta | \alpha)$. Then:

$$P'(\alpha \;\to\; \beta) \geq P'(\alpha \to \gamma)$$
$$\Longleftrightarrow P(\beta | \alpha) \geq P(\gamma | \alpha)$$
$$\Longleftrightarrow P(\alpha\beta) \geq P(\alpha\gamma)$$
$$\Longleftrightarrow \alpha\beta \unrhd \alpha\gamma$$
$$\Longleftrightarrow \alpha \to \beta \unrhd \alpha \to \gamma$$

in virtue of the Indicative property. So P' is a probability on Γ_α that represents $\unrhd$ restricted to Γ_α. But P uniquely represents $\unrhd$ restricted to Γ_α. So $P(\alpha \to \beta) = P(\beta | \alpha)$, in accordance with Adams' Thesis.

2. Define V' on Y_α by $V'(\alpha \to \beta) = V(\beta | \alpha)$. Then:

$$V'(\alpha \;\to\; \beta) \geq V'(\alpha \to \gamma)$$
$$\Longleftrightarrow V(\beta | \alpha) \geq V(\gamma | \alpha)$$
$$\Longleftrightarrow V(\alpha\beta) - V(\alpha) \geq V(\alpha\gamma) - V(\alpha)$$
$$\Longleftrightarrow \alpha\beta \succsim \alpha\gamma$$
$$\Longleftrightarrow \alpha \to \beta \succsim \alpha \to \gamma$$

in virtue of the Indicative property of $\succsim$. So V' is a desirability on Γ_α that represents $\succsim$ restricted to Γ_α. But V is a unique representation of $\succsim$ restricted to Γ_α up to linear transformation. So there exists $a > 0$ such that $V(\alpha \to \beta) = a \cdot V(\beta | \alpha)$.

■

Corollary A.17. *Suppose that* Γ *is a regular conditional algebra and that* $\succsim$ *satisfies* $\mapsto$-*Betweenness. Then:*

1. *Bradley's Thesis:* $V(\alpha \to \beta) = P(\alpha) \cdot V(\beta | \alpha)$
2. *Restricted Actualism:* $V(\neg\alpha(\alpha \to \beta)) = V(\neg\alpha)$

Proof. By $\mapsto$-Betweenness, $\alpha \to \beta$ lies between $\alpha\beta$ and $\neg\alpha$ in the preference ordering. Hence there exists a real number $k \geq 0$ such that

$$V(\alpha \to \beta) = kV(\alpha\beta) + (1 - k)V(\neg\alpha)$$

So, in virtue of Lemma A.1(5b),

$$a(V(\alpha\beta) - V(\alpha)) = aV(\alpha\beta) + a\frac{P(\neg\alpha)}{P(\alpha)} \cdot V(\neg\alpha)$$

$$= kV(\alpha\beta) + (1-k) \cdot V(\neg\alpha)$$

It follows that

$$1 - a = a\frac{P(\neg\alpha)}{P(\alpha)}$$

$$\Leftrightarrow aP(\neg\alpha) = P(\alpha) - aP(\alpha)$$

$$\Leftrightarrow a = P(\alpha)$$

So $V(\alpha \to \beta) = P(\alpha) \cdot V(\beta|\alpha)$, in accordance with Bradley's Thesis. It then follows, by the definition of conditional desirability and Lemma A.1(1), that $V(\alpha \to \beta) = V(\alpha\beta) \cdot P(\alpha) + V(\neg\alpha) \cdot P(\neg\alpha)$. But, by Desirabilism, $V(\alpha \to \beta)$:

$$= V(\alpha(\alpha \to \beta)) \cdot P(\alpha|\alpha \to \beta) + V(\neg\alpha(\alpha \to \beta)) \cdot P(\neg\alpha|\alpha \to \beta)$$

$$= V(\alpha\beta) \cdot P(\alpha) + V(\neg\alpha(\alpha \to \beta)) \cdot P(\neg\alpha)$$

in virtue of Modus Ponens and Adams' Thesis, a consequence, by Theorem A.12(3), of Bradley's Thesis. It follows that $V(\neg\alpha(\alpha \to \beta)) = V(\neg\alpha)$, in accordance with Restricted Actualism. ∎

Corollary A.18. *Suppose that Γ is a regular conditional algebra and that $\succsim$ satisfies $\mapsto$-Separability. Then:*

1. *For all $\alpha, \beta \in X$, $P(\alpha \to \beta|\alpha' \to \gamma) = P(\alpha \to \beta)$*
2. *If $\succsim$ also satisfies $\mapsto$-Betweenness then, for all $\alpha, \beta \in X$, $V(\alpha \to \beta|\alpha' \to \gamma) = V(\alpha \to \beta)$*

Proof. Since $\succsim$ satisfies $\mapsto$-Separability it follows from Theorem A.11 that, for all $\alpha, \beta, \gamma \in X$,

$$P((\alpha \to \beta)(\alpha' \to \gamma)) = P(\alpha \to \beta) \cdot P(\alpha' \to \gamma)$$

$$V((\alpha \to \beta)(\alpha' \to \gamma)) = V(\alpha \to \beta) + aP(\alpha' \to \gamma)$$

for some $a > 0$. Hence (1) $P(\alpha \to \beta|\alpha' \to \gamma) = P(\alpha \to \beta)$ in accordance with Belief Independence. Now, by Corollary A.17, $V(\alpha \to \beta) = P(\alpha) \cdot V(\beta|\alpha)$. Hence:

$$V((\alpha \to \beta)(\alpha' \to \gamma)) = P(\alpha) \cdot (V(\alpha\beta) - V(\alpha)) + aP(\neg\alpha) \cdot (V(\alpha'\gamma) - V(\neg\alpha))$$

$$= P(\alpha) \cdot V(\alpha\beta) + aP(\neg\alpha) \cdot V(\alpha'\gamma) + (1-a)P(\neg\alpha) \cdot V(\neg\alpha)$$

in virtue of Lemma A.1. But, by $\mapsto$-Betweenness, $(\alpha \to \beta)(\alpha' \to \gamma)$ lies between $\alpha\beta$ and $\neg\alpha\gamma$ in the preference ordering. Hence there exists a real number $k \geq 0$ such that

$$V((\alpha \to \beta)(\alpha' \to \gamma)) = k \cdot V(\alpha\beta) + (1-k) \cdot V(\neg\alpha\gamma) + 0 \cdot V(\neg\alpha)$$

So:

$$aP(\neg\alpha) = 1 - P(\alpha) = P(\neg\alpha)$$

$$(1-a)P(\neg\alpha) = 0$$

Hence $a = 1$ or $P(\alpha) = 1$. If $a = 1$, then, by the definition of conditional desirability,

$$V((\alpha \to \beta)|(\alpha' \to \gamma)) = P(\alpha) \cdot V(\alpha\beta) + P(\neg\alpha) \cdot V(\alpha'\gamma) - P(\neg\alpha) \cdot (V(\alpha'\gamma) - V(\neg\alpha))$$

$$= P(\alpha) \cdot (V(\alpha\beta) - V(\alpha))$$

$$= V(\alpha \to \beta)$$

by Lemma A.1(1) and Corollary A.17. On the other hand, if $P(\alpha) = 1$ then $V((\alpha \to \beta)|(\alpha' \to \gamma)) = P(\alpha) \cdot (V(\alpha\beta) - V(\alpha)) = V(\alpha \to \beta)$. So (2), for all $\alpha, \beta, \gamma \in X$, $V(\alpha \to \beta|\alpha' \to \gamma) = V(\alpha \to \beta)$, in accordance with Value Independence. ∎

Theorem A.19. *Let $\Gamma = \langle Y, \models \rangle$ be an indicative conditional algebra of prospects based on $\Omega = \langle X, \models \rangle$. Let (P, V) be a pair of probability and desirability functions on Γ and $\{\alpha_i\}_{i=1}^n$ an n-fold partition of Y. Then:*

1. *If P satisfies Belief Independence and Adams' Thesis then*

$$P(\bigwedge(\alpha_i \mapsto \beta_i)) = \prod_i P(\beta_i|\alpha_i)$$

2. *If V satisfies Value Independence and Bradley's Thesis then*

$$V(\bigwedge(\alpha_i \mapsto \beta_i)) = \sum_i V(\alpha_i\beta_i) \cdot P(\alpha_i)$$

Proof. Recall that in an indicative conditional algebra $\alpha \mapsto (\beta \mapsto \gamma) = \alpha\beta \mapsto \gamma$. Hence:

1. (1) By theorem 6 of Bradley & Stefánsson (2015), $P(\bigwedge(\alpha_i \mapsto \beta_i)) = \prod_i P(\alpha_i \mapsto \beta_i)$. But, by Adams' Thesis, $P(\alpha_i \mapsto \beta_i) = P(\beta_i|\alpha_i)$. Hence $P(\bigwedge(\alpha_i \mapsto \beta_i)) = \prod_i P(\beta_i|\alpha_i)$.

2. (2) By repeated application of the definition of conditional desirability and Value Independence, $V(\bigwedge(\alpha_i \mapsto \beta_i)) =$

$$V((\alpha_1 \mapsto \beta_1)(\alpha_2 \mapsto \beta_2)...(\alpha_{n-1} \mapsto \beta_{n-1})|\alpha_n \mapsto \beta_n) + V(\alpha_n \mapsto \beta_n)$$

$$= V(\neg\alpha_n \mapsto ((\alpha_1 \mapsto \beta_1)(\alpha_2 \mapsto \beta_2)...(\alpha_{n-1} \mapsto \beta_{n-1}))|\alpha_n \mapsto \beta_n)$$

$$+ V(\alpha_n \mapsto \beta_n)$$

$$= V(\neg\alpha_n \mapsto ((\alpha_1 \mapsto \beta_1)(\alpha_2 \mapsto \beta_2)...(\alpha_{n-1} \mapsto \beta_{n-1}))) + V(\alpha_n \mapsto \beta_n)$$

$$= V((\alpha_1 \mapsto \beta_1)(\alpha_2 \mapsto \beta_2)...(\alpha_{n-1} \mapsto \beta_{n-1})) + V(\alpha_n \mapsto \beta_n)$$

$$= ....$$

$$= V(\alpha_1 \mapsto \beta_1) + V(\alpha_2 \mapsto \beta_2)... + V(\alpha_n \mapsto \beta_n)$$

But, by Bradley's Thesis, $V(\alpha_i \mapsto \beta_i) = V(\beta_i|\alpha_i) \cdot P(\alpha_i)$. Hence $V(\bigwedge(\alpha_i \mapsto \beta_i)) = \sum_i V(\alpha_i\beta_i) \cdot P(\alpha_i)$.

∎

Lemma A.20. *Assume Centring. Let $X \subseteq A$ and $Y, Z \subseteq W$. Then $(X, Y_A, Z_{\bar{A}}) = (X \cap Y, W_A, Z_{\bar{A}})$*

Proof. $(X, Y_A, Z_{\bar{A}}) = \{\langle w_i, w_j, w_k \rangle : w_i \in X, w_j \in Y_A \text{ and } w_k \in Z_{\bar{A}}\}$. Since $X \subseteq A$, it follows from Centring that $\langle w_i, w_j, w_k \rangle \in (X, Y_A, Z_{\bar{A}}) \Leftrightarrow w_i = w_j$. So $(X, Y_A, Z_{\bar{A}}) = \{\langle w_i, w_i, w_k \rangle : w_i \in X \cap Y_A \text{ and } w_k \in Z_{\bar{A}}\}$. Similarly, $(X \cap Y, W_A, Z_{\bar{A}}) = \{\langle w_i, w_i, w_k \rangle : w_i \in X \cap Y \cap W_A \text{ and } w_k \in Z_{\bar{A}}\}$. But $X \cap Y_A = X \cap Y \cap W_A = X \cap Y$. So $(X, Y_A, Z_{\bar{A}}) = (X \cap Y, W_A, Z_{\bar{A}})$. ∎

Theorem A.21. *World Independence is equivalent to Fact–Counterfact Independence*

Proof. World Independence is obtained from Fact–Counterfact Independence by substitution of $\{w_j\}$ for $X \subseteq A$ and $\{w_i\}$ for $Y \subseteq \bar{A}$. Now assume World Independence. Then, for all $X \subseteq W_A$ and $Y_A \subseteq W_A$,

$$P(X, Y_A) = \sum_{w_i \in X} \sum_{w_j \in Y_A} p(w_i, w_j)$$

$$= \sum_{w_i \in X} \sum_{w_j \in Y_A} p(w_i) \cdot p_A^*(w_j)$$

$$= \sum_{w_i \in X} p(w_i) \sum_{w_j \in Y_A} p_A^*(w_j)$$

$$= \sum_{w_i \in X} p(w_i) \cdot P_A^*(Y_A)$$

$$= P(X) \cdot P_A^*(Y_A)$$

in accordance with Fact–Counterfact Independence. ∎

Theorem A.22. *World Independence and World Actualism are jointly equivalent to Fact–Counterfact Independence and Prospect Actualism*

Proof. World Actualism is obtained from Prospect Actualism by substitution of $\{w_j\}$ for $X \subseteq \bar{A}$. Then, in view of Theorem A.21, we need only establish that World Independence and World Actualism jointly imply Prospect Actualism.

Assume World Independence and World Actualism. Then by V2, the axiom of averaging, for all $X \subseteq \bar{A}$,

$$V(X, Y_A) = \frac{\sum_{w_i \in X} \sum_{w_j \in Y_A} u(w_i, w_j) \cdot p(w_i, w_j)}{P(X, Y_A)}$$

$$= \frac{\sum_{w_i \in X} \sum_{w_j \in Y_A} u(w_i) \cdot p(w_i) \cdot p_A(w_j)}{P(X, Y_A)}$$

$$= \frac{P_A^*(Y_A)}{P(X) \cdot P_A^*(Y_A)} \sum_{w_i \in X} u(w_i) \cdot p(w_i)$$

in virtue of the fact that $\sum_{w_j \in Y_A} p_A(w_j) = P_A^*(Y_A)$ and that, by Theorem A.3, World Independence implies that $P(X, Y_A) = P(X) \cdot P_A^*(Y_A)$. But then:

$$V(X, Y_A) = \frac{\sum_{w_i \in X} u(w_i) \cdot p(w_i)}{P(X)} = V(X)$$

in accordance with Prospect Actualism. ∎

Theorem A.23. *Given Centring, Counterfact Independence is equivalent to the condition that, for all $w_A \in A, w_{\bar{A}} \in \bar{A}$, $p(\langle w_A, w_{\bar{A}}\rangle) = p_A(w_A) \times p_{\bar{A}}(w_{\bar{A}})$ and $u(\langle w_A, w_{\bar{A}}\rangle) = u_A(w_A) + u_{\bar{A}}(w_{\bar{A}})$.*

Proof. Clearly, Counterfact Independence implies that, for all $w_A \in A, w_{\bar{A}} \in \bar{A}$, $u(\langle w_A, w_{\bar{A}}\rangle) = u_A(w_A) + u_{\bar{A}}(w_{\bar{A}})$. By Theorem A.4, Counterfact Independence implies that, for all $X \subseteq W_A$ and $Y_A \subseteq W_A$, $P(X_A, Y_{\bar{A}}) = P_A^*(X_A) \cdot P_{\bar{A}}^*(Y_{\bar{A}})$. So it follows that, for all $w_A \in A, w_{\bar{A}} \in \bar{A}$, $p(\langle w_A, w_{\bar{A}}\rangle) = p_A(w_A) \times p_{\bar{A}}(w_{\bar{A}})$. Now assume that, for all $w_A \in A, w_{\bar{A}} \in \bar{A}$, $p(\langle w_A, w_{\bar{A}}\rangle) = p_A(w_A) \times p_{\bar{A}}(w_{\bar{A}})$ and $u(\langle w_A, w_{\bar{A}}\rangle) = u_A(w_A) + u_{\bar{A}}(w_{\bar{A}})$. Assume Centring. Then by V2, the axiom of averaging, for all $X \subseteq W_A$ and $Y_A \subseteq W_A$, $V(X_A, Y_{\bar{A}})$

$$= \frac{\sum_{w_i \in X_A} \sum_{w_j \in Y_{\bar{A}}} u(w_i, w_j) \cdot p(w_i, w_j)}{P(X_A, Y_{\bar{A}})}$$

$$= \frac{\sum_{w_i \in X_A} \sum_{w_j \in Y_{\bar{A}}} [u_A(w_i) + u_{\bar{A}}(w_j)][p_A(w_i) \cdot p_{\bar{A}}(w_j)]}{P(X_A, Y_{\bar{A}})}$$

$$= \frac{\sum_{w_i \in X_A} \sum_{w_j \in Y_{\bar{A}}} [p_A(w_i) \cdot u_A(w_i)] \cdot p_{\bar{A}}(w_j) + [u_{\bar{A}}(w_j) \cdot p_{\bar{A}}(w_j)] \cdot p_A(w_i)}{P_A^*(X_A) \cdot P_{\bar{A}}^*(Y_{\bar{A}})}$$

$$= \frac{P_{\bar{A}}^*(Y_{\bar{A}})}{P_A^*(X_A) \cdot P_{\bar{A}}^*(Y_{\bar{A}})} \sum_{w_i \in X_A} p_A(w_i) \cdot u_A(w_i) + \frac{P_A^*(X_A)}{P_A^*(X_A) \cdot P_{\bar{A}}^*(Y_{\bar{A}})} \sum_{w_j \in Y_{\bar{A}}} u_{\bar{A}}(w_j) \cdot p_{\bar{A}}(w_j)$$

$$= V_A^*(X_A) + V_{\bar{A}}^*(Y_{\bar{A}})$$

in virtue of the fact that $\sum_{w_j \in Y_{\bar{A}}} p_{\bar{A}}(w_j) = P_{\bar{A}}^*(Y_{\bar{A}})$ and $\sum_{w_i \in X_A} p_A(w_i) = P_A^*(X_A)$, and that, by Theorem A.4, Counterfact Independence implies that $P(X, Y_A) = P(X) \cdot P_A^*(Y_A)$. But then:

$$\begin{aligned} V(X_A, Y_{\bar{A}}) &= \sum_{w_i \in X_A} \frac{p_A(w_i) \cdot u_A(w_i)}{P_A^*(X_A)} + \frac{u_{\bar{A}}(w_j) \cdot p_{\bar{A}}(w_j)}{P_{\bar{A}}^*(Y_{\bar{A}})} \\ &= V_A^*(X_A) + V_{\bar{A}}^*(Y_{\bar{A}}) \\ &= V(X_A) + V(Y_{\bar{A}}) \end{aligned}$$

in accordance with Counterfact Independence. ∎

Theorem A.24. *(Imprecise Probabilism) Let $\trianglerighteq$ be a relation on a complete atomless Boolean algebra of prospects $\Omega = \langle X, \models \rangle$ that has a minimal coherent extension to a continuous credibility relation on X. Then there exists a maximal set of probability functions $S = \{P_1, ..., P_n\}$ that explains or rationalises $\trianglerighteq$ in the sense that, for all $\alpha, \beta \in \Omega$,*

$$\alpha \trianglerighteq \beta \Leftrightarrow \forall P_i \in S, \ P_i(\alpha) \ge P_i(\beta)$$

Proof. Let $R = \{\trianglerighteq_i\}$ be the set of all continuous credibility relations on Ω that are coherent minimal extensions of $\trianglerighteq$. By assumption R is non-empty. Since each $\trianglerighteq_i$ is a continuous and complete credibility relation on Ω, it follows by Theorem 5.4 (Villegas's theorem) that each $\trianglerighteq_i$ is uniquely represented by a probability function P_i such that, for all $\alpha, \beta \in \Omega$, $\alpha \trianglerighteq_i \beta \Leftrightarrow P_i(\alpha) \ge P_i(\beta)$. Suppose that $\alpha \trianglerighteq \beta$. Then, for all $\trianglerighteq_i$, $\alpha \trianglerighteq_i \beta$ and hence $\forall P_i \in S, \ P_i(\alpha) \ge P_i(\beta)$. Suppose S is not maximal. Then there exists a probability $P^* \notin S$ and associated continuous and complete credibility relation $\trianglerighteq^*$ such that $\alpha \trianglerighteq \beta \Rightarrow \alpha \trianglerighteq^* \beta$. But then $\trianglerighteq^*$ is a coherent minimal extension of $\trianglerighteq$, hence $\trianglerighteq^* \in R$. So $P^* \in S$, contrary to what we supposed. It follows that S is maximal. ∎

Theorem A.25. *(Imprecise Bayesianism) Let $\Omega = \langle X, \models \rangle$ be a complete, atomless Boolean algebra of prospects. Let $\succsim$ be a weak relation on Ω that has a minimal coherent extension to a continuous and impartial preference order. Then there exists a maximal set of pairs of probability and desirability functions $S = \{\langle P_i, V_i \rangle\}$ on Ω that explains or rationalises the preference relation $\succsim$ in the sense, that for all $\alpha, \beta \in \Omega$,*

$$\alpha \succsim \beta \Leftrightarrow \forall V_i \in S, \ V_i(\alpha) \ge V_i(\beta)$$

Proof. Let $R = \{\succsim_i\}$ be the set of all continuous and impartial preference orders that are minimal coherent extensions of $\succsim$. By Bolker's representation theorem (Theorem 5.5) each $\succsim_i$ is represented by a pair of probability and desirability functions (P_i, V_i) such that, for all $\alpha, \beta \in \Omega$, $\alpha \succsim_i \beta \Leftrightarrow V_i(\alpha) \ge V_i(\beta)$. Let S_i be the set of all such pairs and $S = \bigcup(S_i)$. Suppose that $\alpha \succsim \beta$. Then, for all

$\succsim_i$, $\alpha \succsim_i \beta$ and hence $\forall V_i \in S$, $V_i(\alpha) \geq V_i(\beta)$. Suppose S is not maximal. Then there exists a pair of probability and desirability functions (P_i^*, V_i^*) and associated complete preference relation $\succsim^*$ such that $\alpha \succsim \beta \Rightarrow \alpha \succsim^* \beta$. But then $\succsim^*$ is a coherent extension of $\succsim$. Hence $\succsim^* \in R$ and $(P_i^*, V_i^*) \in S_i$, contrary to what we supposed. It follows that S is maximal. ∎

Bibliography

Adams, Ernest W. 1975. *The Logic of Conditionals: An Application of Probability to Deductive Logic.* Dordrecht: D. Reidel Publishing Company.

Ahmed, Arif. 2014. *Evidence, Decision and Causality.* Cambridge: Cambridge University Press.

Al-Najjar, Nabil I., & Weinstein, Jonathan. 2009. The ambiguity aversion literature: A critical assessment. *Economics and Philosophy*, 25(3), 249–284.

Allais, Maurice. 1953. Le comportement de l'homme rationnel devant le risque: Critique des postulats et axiomes de l'ecole américaine. *Econometrica: Journal of the Econometric Society*, 21(4), 503–546.

Allais, Maurice. 1979. The foundations of a positive theory of choice involving risk and a criticism of the postulates and axioms of the american school. In Allais, Maurice, & Hagen, Ole (eds.), *Expected Utility Theory and the Allais Paradox: Contemporary Discussions of Decisions under Uncertainty with Allais' Rejoinder*, 47–73. Berlin: Springer. Reidel.

Anscombe, Francis J., & Aumann, Robert J. 1963. A definition of subjective probability. *Annals of Mathematical Statistics*, 34(1), 199–205.

Arrow, Kenneth J. 1959. Rational choice functions and orderings. *Economica*, 26, 121–127.

Arrow, Kenneth J., & Fisher, Anthony C. 1974. Environmental preservation, uncertainty, and irreversibility. *Quarterly Journal of Economics*, 88(2), 312–319.

Aumann, Robert J. 1962. Utility theory without the completeness axiom. *Econometrica: Journal of the Econometric Society*, 30(3), 445–462.

Ben-Haim, Yakov. 2006. *Info-Gap Decision Theory: Decisions under Severe Uncertainty*, 2nd edn. Oxford: Academic Press.

Bewley, Truman F. 2002. Knightian decision theory: Part I. *Decisions in Economics and Finance*, 25(2), 79–110.

Binmore, Ken. 2008. *Rational Decisions.* Princeton, NJ: Princeton University Press.

Bolker, Ethan D. 1966. Functions resembling quotients of measures. *Transactions of the American Mathematical Society*, 124(2), 292–312.

Bossert, Walter, & Suzumura, Kōtarō. 2010. *Consistency, Choice, and Rationality.* Cambridge, MA: Harvard University Press.

Bradley, Richard. 1998. A representation theorem for a decision theory with conditionals. *Synthese*, 116(2), 187–229.

Bradley, Richard. 1999. Conditional desirability. *Theory and Decision*, 47(1), 23–55.

Bradley, Richard. 2000. A preservation condition for conditionals. *Analysis*, 60(3), 219–222.

Bradley, Richard. 2002. Indicative conditionals. *Erkenntnis*, 56(3), 345–378.

Bradley, Richard. 2004. Ramsey's representation theorem. *Dialectica*, 58(4), 483–497.

Bradley, Richard. 2005a. Bayesian utilitarianism and probability homogeneity. *Social Choice and Welfare*, 24(2), 221–251.

Bradley, Richard. 2005b. Radical probabilism and Bayesian conditioning. *Philosophy of Science*, 72(2), 342–364.

Bradley, Richard. 2007a. The kinematics of belief and desire. *Synthese*, 56(3), 513–535.

Bradley, Richard. 2007b. Reaching a consensus. *Social Choice and Welfare*, 29(4), 609–632.

Bradley, Richard. 2007c. A unified Bayesian decision theory. *Theory and Decision*, 63(3), 233–263.

Bradley, Richard. 2008. Comparing evaluations. *Proceedings of the Aristotelian Society*, 108(1), 85–100.

Bradley, Richard. 2009. Becker's thesis and three models of preference change. *Politics, Philosophy and Economics*, 8(2), 223–242.

Bradley, Richard. 2011. Conditionals and supposition-based reasoning. *Topoi*, 30(1), 39–45.

Bradley, Richard. 2012. Multidimensional possible-world semantics for conditionals. *Philosophical Review*, 121(4), 539–571.

Bradley, Richard. 2016. Ellsberg's paradox and the value of chances. *Economics and Philosophy*, 32(2), 231–248.

Bradley, Richard, & Drechsler, Mareile. 2014. Types of uncertainty. *Erkenntnis*, 79(6), 1225–1248.

Bradley, Richard, & Stefánsson, H. Orri. 2016. Desire, expectation, and invariance. *Mind*, 125(499), 691–725.

Bradley, Richard, & Stefánsson, H. Orri. 2017. Counterfactual desirability. *British Journal for the Philosophy of Science*, 68(2), 485–533.

Broome, John. 1990. Bolker–Jeffrey expected utility theory and axiomatic utilitarianism. *Review of Economic Studies*, 57(3), 477–502.

Broome, John. 1991. *Weighing Goods: Equality, Uncertainty and Time*. Oxford: Blackwell.

Broome, John. 1999. Can a Humean be moderate? In *Ethics out of Economics*, 68–87. Cambridge: Cambridge University Press.

Broome, John. 2012. *Climate Matters: Ethics in a Warming World*. New York: W. W. Norton.

Buchak, Lara. 2013. *Risk and Rationality*. Oxford: Oxford University Press.

Chateauneuf, Alain, & Faro, José Heleno. 2009. Ambiguity through confidence functions. *Journal of Mathematical Economics*, 45(9), 535–558.

Davey, Brian A., & Priestley, Hilary A. 2002. *Introduction to Lattices and Order*, 2nd edn. Cambridge: Cambridge University Press.

De Finetti, Bruno. 1937. La prévision: Ses lois logiques, ses sources subjectives. *Annales de l'institut Henri Poincaré*, 7(1), 1–68.

Dekel, Eddie, Lipman, Barton L., & Rustichini, Aldo. 1998. Standard state-space models preclude unawareness. *Econometrica: Journal of the Econometric Society*, 66(1), 159–173.

Dennett, Daniel C. 1971. Intentional systems. *Journal of Philosophy*, 68(4), 87–106.

Diaconis, Persi, & Zabell, Sandy L. 1982. Updating subjective probability. *Journal of the American Statistical Association*, 77(380), 822–830.

Dietrich, Franz, & List, Christian. 2013a. A reason-based theory of rational choice. *Nous*, 47(1), 104–134.

Dietrich, Franz, & List, Christian. 2013b. Where do preferences come from? *International Journal of Game Theory*, 42(3), 613–637.

Dietrich, Franz, List, Christian, & Bradley, Richard. 2016. Belief revision generalized: A joint characterization of Bayes' and Jeffrey's rules. *Journal of Economic Theory*, 162, 352–371.

Döring, Frank. 1999. Why Bayesian psychology is incomplete. *Philosophy of Science*, 66(3), 379–389.

Douven, Igor. 2007. On Bradley's preservation condition for conditionals. *Erkenntnis*, 67(1), 111–118.

Douven, Igor, & Romeijn, Jan-Willem. 2011. A new resolution of the Judy Benjamin problem. *Mind*, 120(479), 637–670.

Drèze, Jacques H. 1990. *Essays on Economic Decisions under Uncertainty*. Cambridge: Cambridge University Press.

Earman, John. 1992. *Bayes or Bust? A Critical Examination of Bayesian Confirmation Theory*. Cambridge: MA: MIT Press.

Edgington, Dorothy. 1995. On conditionals. *Mind*, 104(414), 113–128.

Eells, Ellery. 1982. *Rational Decision and Causality*. Cambridge: Cambridge University Press.

Elga, Adam. 2010. Subjective probabilities should be sharp. *Philosophers' Imprint*, 10(5), 1–11.

Ellsberg, Daniel. 1961. Risk, ambiguity, and the Savage axioms. *Quarterly Journal of Economics*, 75(4), 643–669.

Elster, Jon. 1985. *Sour Grapes: Studies in the Subversion of Rationality*. Cambridge: Cambridge University Press.

Evren, Özgür, & Ok, Efe A. 2011. On the multi-utility representation of preference relations. *Journal of Mathematical Economics*, 47(4), 554–563.

Fagin, Ronald, & Halpern, Joseph Y. 1987. Belief, awareness, and limited reasoning. *Artificial Intelligence*, 34(1), 39–76.

Fishburn, Peter C. 1973. A mixture-set axiomatization of conditional subjective expected utility. *Econometrica: Journal of the Econometric Society*, 41(1), 1–25.

Gärdenfors, Peter. 1988. *Knowledge in Flux: Modeling the Dynamics of Epistemic States*. Cambridge, MA: MIT Press.

Gärdenfors, Peter, & Sahlin, Nils-Eric. 1982. Unreliable probabilities, risk taking, and decision making. *Synthese*, 53(3), 361–386.

Genest, Christian, & Zidek, James V. 1986. Combining probability distributions: A critique and an annotated bibliography. *Statistical Science*, 4(1), 114–135.

Ghirardato, Paolo, Maccheroni, Fabio, & Marinacci, Massimo. 2004. Differentiating ambiguity and ambiguity attitude. *Journal of Economic Theory*, 118(2), 133–173.

Gibbard, Allan, & Harper, William L. 1981. Counterfactuals and two kinds of expected utility. In Harper, William L., Stalnaker, Robert, & Pearce, Glenn (eds.), *Ifs: Conditionals, Belief, Decision, Chance, and Time*, 153–190. Dordrecht: Springer.

Gilboa, Itzhak. 2009. *Theory of Decision under Uncertainty*. Cambridge: Cambridge University Press.

Gilboa, Itzhak, & Schmeidler, David. 1989. Maxmin expected utility with non-unique prior. *Journal of Mathematical Economics*, 18(2), 141–153.

Glymour, Clark N. 1980. *Theory and Evidence*. Princeton, NJ: Princeton University Press.

Gold, Natalie, & List, Christian. 2004. Framing as path dependence. *Economics and Philosophy*, 20(2), 253–277.

Good, Irving John. 1983. *Good Thinking: The Foundations of Probability and Its Applications*. Minneapolis: University of Minnesota Press.

Grove, Adam J., & Halpern, Joseph Y. 1998. Updating sets of probabilities. In Cooper, Gregory F., & Moral, Serafín (eds.), *Proceedings of the Fourteenth Conference on Uncertainty in Artificial Intelligence*, 173–182. San Francisco: Morgan Kaufmann.

Hacking, Ian. 2006. *The Emergence of Probability: A Philosophical Study of Early Ideas about Probability, Induction and Statistical Inference*. Cambridge: Cambridge University Press.

Hall, Ned. 2004. Two mistakes about credence and chance. *Australasian Journal of Philosophy*, 82(1), 93–111.

Halpern, Joseph Y. 2001. Alternative semantics for unawareness. *Games and Economic Behavior*, 37(2), 321–339.

Halpern, Joseph Y. 2003. *Reasoning about Uncertainty*. Vol. 21. Cambridge, MA: MIT Press.

Hansen, Lars Peter, & Sargent, Thomas J. 2001. Robust control and model uncertainty. *American Economic Review*, 91(2), 60–66.

Hansson, Sven Ove. 1995. Changes in preference. *Theory and Decision*, 38(1), 1–28.

Hansson, Sven Ove. 1996a. Decision making under great uncertainty. *Philosophy of the Social Sciences*, 26(3), 369–386.

Hansson, Sven Ove. 1996b. What is philosophy of risk? *Theoria*, 62(1–2), 169–186.

Hansson, Sven Ove. 2001. *The Structure of Values and Norms*. Cambridge: Cambridge University Press.

Hansson, Sven Ove. 2004. Great uncertainty about small things. *Techné: Research in Philosophy and Technology*, 8(2), 26–35.

Hansson, Sven Ove. 2009. Replacement: A Sheffer stroke for belief change. *Journal of Philosophical Logic*, 38(2), 127–149.

Hansson, Sven Ove. 2016. Evaluating the uncertainties. In Hansson, Sven Ove, & Hirsch Hadorn, Gertrude (eds.), *The Argumentative Turn in Policy Analysis: Reasoning about Uncertainty*, 79–104. Cham, Switzerland: Springer.

Hausman, Daniel M. 2011a. Mistakes about preferences in the social sciences. *Philosophy of the Social Sciences*, 41(1), 3–25.

Hausman, Daniel M. 2011b. *Preference, Value, Choice, and Welfare*. Cambridge: Cambridge University Press.

Hill, Brian. 2013. Confidence and decision. *Games and Economic Behavior*, 82, 675–692.

Howson, Colin, & Urbach, Peter. 2006. *Scientific Reasoning: The Bayesian Approach.* Chicago: Open Court Publishing.

Jackson, Frank. 1979. On assertion and indicative conditionals. *Philosophical Review,* 88(4), 565–589.

James, William. 1897. *The Will to Believe, and Other Essays in Popular Psychology.* New York: Longmans, Green.

Jaynes, Edwin T. 1968. Prior probabilities. *IEEE Transactions on Systems Science and Cybernetics,* 4(3), 227–241.

Jaynes, Edwin T. 2003. *Probability Theory: The Logic of Science.* Cambridge: Cambridge University Press.

Jeffrey, Richard. 1990/1965. *The Logic of Decision,* 2nd edn. Chicago: University of Chicago Press.

Jeffrey, Richard. 1992. *Probability and the Art of Judgment.* Cambridge: Cambridge University Press.

Jeffrey, Richard. 2004. *Subjective Probability: The Real Thing.* Cambridge: Cambridge University Press.

Joyce, James M. 1998. A nonpragmatic vindication of probabilism. *Philosophy of Science,* 65(4), 575–603.

Joyce, James M. 1999. *The Foundations of Causal Decision Theory.* Cambridge: Cambridge University Press.

Joyce, James M. 2005. How probabilities reflect evidence. *Philosophical Perspectives,* 19(1), 153–178.

Joyce, James M. 2010. A defense of imprecise credences in inference and decision making. *Philosophical Perspectives,* 24(1), 281–323.

Joyce, James M. 2011. The development of subjective Bayesianism. In Gabbay, Gov M., Hartmann, Stephen, & Woods, John (eds.), *Handbook of the History of Logic,* vol. X, *Inductive Logic,* 415–475. Oxford: North Holland.

Kaplan, Mark. 1998. *Decision Theory as Philosophy.* Cambridge: Cambridge University Press.

Karni, Edi. 1985. *Decision Making under Uncertainty: The Case of State-Dependent Preferences.* Cambridge, MA: Harvard University Press.

Karni, Edi, & Vierø, Marie-Louise. 2013. Reverse Bayesianism: A choice-based theory of growing awareness. *American Economic Review,* 103(7), 2790–2810.

Keynes, John Maynard. 1937. The general theory of employment. *Quarterly Journal of Economics,* 51(2), 209–223.

Keynes, John Maynard. 1973/1921. *A Treatise on Probability, vol. VIII, The Collected Writings of John Maynard Keynes.* London: Macmillan.

Klibanoff, Peter, Marinacci, Massimo, & Mukerji, Sujoy. 2005. A smooth model of decision making under ambiguity. *Econometrica: Journal of the Econometric Society,* 73(6), 1849–1892.

Krantz, David, Luce, R. Duncan, Suppes, Patrick, & Tversky, Amos. 1971. *Foundations of Measurement, vol. I, Additive and Polynomial Representations.* Cambridge, MA: Academic Press.

Kreps, David M. 1988. *Notes on the Theory of Choice.* Boulder, CO: Westview Press.

Kreps, David M., & Porteus, Evan L. 1978. Temporal resolution of uncertainty and dynamic choice theory. *Econometrica: Journal of the Econometric Society,* 46(1), 185–200.

Lance, Mark. 1991. Probabilistic dependence among conditionals. *Philosophical Review*, 100(2), 269–276.

Landes, Jürgen, & Williamson, Jon. 2013. Objective Bayesianism and the maximum entropy principle. *Entropy*, 15(9), 3528–3591.

Lange, Marc. 2004. Is Jeffrey conditionalization defective by virtue of being non-commutative? Remarks on the sameness of sensory experiences. *Synthese*, 123(3), 393–403.

Lehrer, Keith. 1976. When rational disagreement is impossible. *Noûs*, 10, 327–332.

Lehrer, Keith. 1983. Rationality as weighted averaging. *Synthese*, 57(3), 283–295.

Lehrer, Keith, & Wagner, Carl. 1981. *Rational Consensus in Science and Society: A Philosophical and Mathematical Study*. Berlin: Springer.

Leitgeb, Hannes, & Pettigrew, Richard. 2010a. An objective justification of Bayesianism I: Measuring inaccuracy. *Philosophy of Science*, 77(2), 201–235.

Leitgeb, Hannes, & Pettigrew, Richard. 2010b. An objective justification of Bayesianism II: The consequences of minimizing inaccuracy. *Philosophy of Science*, 77(2), 236–272.

Levi, Isaac. 1978. On indeterminate probabilities. In Hooker, Cliff A., Leach, Jim J., & McClennen, Edward F. (eds.), *Foundations and Applications of Decision Theory*. 233–261. Berlin: Springer.

Levi, Isaac. 1990. *Hard Choices: Decision Making under Unresolved Conflict*. Cambridge: Cambridge University Press.

Levi, Isaac. 1997. *The Covenant of Reason: Rationality and the Commitments of Thought*. Cambridge: Cambridge University Press.

Lewis, David. 1973. *Counterfactuals*. Oxford: Blackwell.

Lewis, David. 1976. Probabilities of conditionals and conditional probabilities. *Philosophical Review*, 85(3), 297–315.

Lewis, David. 1981. Causal decision theory. *Australasian Journal of Philosophy*, 59(1), 5–30.

Lewis, David. 1988. Desire as belief. *Mind*, 97(387), 323–332.

Lewis, David. 1996. Desire as belief II. *Mind*, 105(418), 303–313.

List, Christian, & Puppe, Clemens. 2009. Judgment aggregation. In Anand, Paul, Pattanaik, Prasanta K., & Puppe, Clemens (eds.), *Oxford Handbook of Rational and Social Choice*, 457–482. Oxford: Oxford University Pres.

Loomes, Graham, & Sugden, Robert. 1982. Regret theory: An alternative theory of rational choice under risk. *Economic Journal*, 92(368), 805–824.

Luce, R. Duncan, & Krantz, David H. 1971. Conditional expected utility. *Econometrica: Journal of the Econometric Society*, 39(2), 253–271.

Machina, Mark J. 1982. 'Expected utility' analysis without the independence axiom. *Econometrica: Journal of the Econometric Society*, 50(2), 277–323.

McDermott, Michael. 1996. On the truth conditions of certain 'if'-sentences. *Philosophical Review*, 105(1), 1–37.

McGee, Vann. 1989. Conditional probabilities and compounds of conditionals. *Philosophical Review*, 98(4), 485–541.

Meacham, Christopher J. G., & Weisberg, Jonathan. 2011. Representation theorems and the foundations of decision theory. *Australasian Journal of Philosophy*, 89(4), 641–663.

Milne, Peter. 1997. Bruno de Finetti and the logic of conditional events. *British Journal for the Philosophy of Science*, **48**(2), 195–232.

Modica, Salvatore, & Rustichini, Aldo. 1999. Unawareness and partitional information structures. *Games and Economic Behavior*, **27**(2), 265–298.

Mongin, Philippe. 1995. Consistent Bayesian aggregation. *Journal of Economic Theory*, **66**(2), 313–351.

Mongin, Philippe. 2016. Spurious unanimity and the Pareto principle. *Economics and Philosophy*, **32**(3), 511–532.

Nagel, Thomas. 1979. The fragmentation of value. In *Mortal Questions* 128–141. Cambridge: Cambridge University Press.

Nehring, Klaus. 2009a. Coping rationally with ambiguity: Robustness versus ambiguity-aversion. *Economics and Philosophy*, **25**(3), 303–334.

Nehring, Klaus. 2009b. Imprecise probabilistic beliefs as a context for decision-making under ambiguity. *Journal of Economic Theory*, **144**(3), 1054–1091.

Norton, John D. 2008. Ignorance and indifference. *Philosophy of Science*, **75**(1), 45–68.

Over, David E., & Evans, Jonathan St B. T. 2003. The probability of conditionals: The psychological evidence. *Mind and Language*, **18**(4), 340–358.

Parfit, Derek. 2013. *On What Matters*, vol. I. Oxford: Oxford University Press.

Paris, Jeff B. 2006. *The uncertain reasoner's companion: a mathematical perspective.* Cambridge: Cambridge University Press.

Pearl, Judea. 2009. *Causality: Models, Reasoning, and Inference*, 2nd edn. Cambridge: Cambridge University Press.

Pedersen, Arthur Paul, & Wheeler, Gregory. 2014. Demystifying dilation. *Erkenntnis*, **79**(6), 1305–1342.

Popper, Karl R. 1959/1934. *The Logic of Scientific Discovery*. London: Hutchinson.

Quiggin, John. 2012. *Generalized Expected Utility Theory: The Rank-Dependent Model*. Berlin: Springer.

Ramsey, Frank P. 1990/1926. Truth and probability. In Mellor, David H. (ed.), *F. P. Ramsey: Philosophical Papers*, 52–94. Cambridge: Cambridge University Press.

Ramsey, Frank P. 1990/1929. General propositions and causality. In Mellor, David H. (ed.), *F. P. Ramsey: Philosophical Papers*, 145–163. Cambridge: Cambridge University Press.

Rényi, Alfréd. 1955. On a new axiomatic theory of probability. *Acta Mathematica Hungarica*, **6**(3), 285–335.

Samson, A. (ed.) 2014. *The Behavioral Economics Guide 2014.* Retrieved from www.behavioraleconomics.com.

Sartre, Jean-Paul, & Elkaïm-Sartre, Arlette. 2007. *Existentialism Is a Humanism*. New Haven, CT: Yale University Press.

Savage, Leonard J. 1974/1954. *The Foundations of Statistics*, 2nd edn rev. New York: Dover.

Schmeidler, David. 1989. Subjective probability and expected utility without additivity. *Econometrica: Journal of the Econometric Society*, **57**(3), 571–587.

Seidenfeld, Teddy, Schervish, Mark J., and Kadane, Joseph B. 1995. A representation of partially ordered preferences. *Annals of Statistics*, **23**(6), 2168–2217.

Seidenfeld, Teddy, & Wasserman, Larry. 1993. Dilation for sets of probabilities. *Annals of Statistics*, **21**(3), 1139–1154.

Sen, Amartya. 1970. The impossibility of a Paretian liberal. *Journal of Political Economy*, 78(1), 152–157.

Sen, Amartya K. 1971. Choice functions and revealed preference. *Review of Economic Studies*, 38(3), 307–317.

Sen, Amartya K. 1977. Rational fools: A critique of the behavioral foundations of economic theory. *Philosophy and Public Affairs*, 6(4), 317–344.

Simon, Herbert A. 1957. *Models of Man, Social and Rational: Mathematical Essays on Rational Human Behavior in a Social Setting*. New York: Wiley.

Simon, Herbert A. 1986. Rationality in psychology and economics. *Journal of Business*, 59(4), S209–S224.

Simon, Herbert A. 1990. Bounded rationality. In John Eatwell, Murray Milgate & Peter Newman (eds.), *Utility and Probability*, 15–18. New York: Norton.

Skyrms, Brian. 1977. Resiliency, propensities, and causal necessity. *Journal of Philosophy*, 74(11), 704–713.

Skyrms, Brian. 1981. The prior propensity account of subjunctive conditionals. In Harper, William L., Stalnaker, Robert, & Pearce, Glenn (eds.), *Ifs: Conditionals, Belief, Decision, Chance, and Time*, 259–265. Dordrecht: Springer.

Skyrms, Brian. 1987. Dynamic coherence and probability kinematics. *Philosophy of Science*, 54(1), 1–20.

Spirtes, Peter, Glymour, Clark, & Scheines, Richard. 2000. *Causation, Prediction, and Search*, 2nd edn Cambridge, MA: MIT Press.

Stalnaker, Robert. 1981a. Letter to David Lewis. In Harper, William L., Stalnaker, Robert, & Pearce, Glenn (eds.), *Ifs: Conditionals, Belief, Decision, Chance, and Time*, 151–152. New York: Springer.

Stalnaker, Robert. 1984. *Inquiry*. Cambridge, MA: MIT Press.

Stalnaker, Robert C. 1981b. Probability and conditionals. In Harper, William L., Stalnaker, Robert, & Pearce, Glenn (eds.), *Ifs: Conditionals, Belief, Decision, Chance, and Time*, 107–128. Dordrecht: Springer.

Stalnaker, Robert, & Jeffrey, Richard. 1994. Conditionals as random variables. In Eells, Ellery, & Skyrms, Brian (eds.), *Probability and Conditionals: Belief Revision and Rational Decision*, 31–46. Cambridge: Cambridge University Press.

Starmer, Chris. 2000. Developments in non-expected utility theory: The hunt for a descriptive theory of choice under risk. *Journal of Economic Literature*, 38(2), 332–382.

Stefánsson, H. Orri, & Bradley, Richard. 2015. How valuable are chances? *Philosophy of Science*, 82(4), 602–625.

Suppes, Patrick. 2002. *Representation and Invariance of Scientific Structures*. Stanford, CA: CSLI Publications.

Suppes, Patrick, & Zanotti, Mario. 1976. Necessary and sufficient conditions for existence of a unique measure strictly agreeing with a qualitative probability ordering. *Journal of Philosophical Logic*, 5(3), 431–438.

Teller, Paul. 1973. Conditionalization and observation. *Synthese*, 26(2), 218–258.

Titelbaum, Michael G. 2012. *Quitting Certainties: A Bayesian Framework Modeling Degrees of Belief*. Oxford: Oxford University Press.

Trautmann, Stefan T., & van de Kuilen, Gijs. 2016. Ambiguity attitudes. In Keren, Gideon, & Wu, George (eds.), *The Wiley Blackwell Handbook of Judgment and Decision Making*, vol. I, 89–116. Chichester, UK: John Wiley.

Van Benthem, Johan, & Liu, Fenrong. 2007. Dynamic logic of preference upgrade. *Journal of Applied Non-Classical Logics*, 17(2), 157–182.

Van Fraassen, Bas C. 1981. A problem for relative information minimizers in probability kinematics. *British Journal for the Philosophy of Science*, 32(4), 375–379.

Van Fraassen, Bas C. 1989. *Laws and Symmetries*. Oxford: Oxford University Press.

Villegas, Carlos. 1964. On qualitative probability/sigma-algebras. *Annals of Mathematical Statistics*, 35(4), 1787–1796.

Von Neumann, John, & Morgenstern, Oskar. 2007/1944. *Theory of Games and Economic Behavior*. Princeton, NJ: Princeton University Press.

Wagner, Carl G. 2002. Probability kinematics and commutativity. *Philosophy of Science*, 69(2), 266–278.

Wakker, Peter. 2010. *Prospect Theory: For Risk and Uncertainty*. Cambridge: Cambridge University Press.

Walker, Oliver, & Dietz, Simon 2011. A representation result for choice under conscious unawareness, Working Paper no. 68. Leeds: Centre for Climate Change Economics and Policy.

Walley, Peter. 1991. *Statistical Reasoning with Imprecise Probabilities*. London: Chapman and Hall.

Weirich, Paul. 2004. *Realistic Decision Theory: Rules for Nonideal Agents in Nonideal Circumstances*. Oxford: Oxford University Press.

Weisberg, Jonathan. 2009. Commutativity or holism? A dilemma for conditionalizers. *British Journal for the Philosophy of Science*, 60(4), 793–812.

Weiss, Paul, & Hartshorne, Charles (eds.). 1932. *Collected Papers of Charles Sanders Peirce*, vol. II, *Elements of Logic*. Cambridge, MA: Harvard University Press.

Wenmackers, Sylvia, & Romeijn, Jan-Willem. 2016. New theory about old evidence. *Synthese*, 193(4), 1225–1250.

White, Roger. 2009. Evidential symmetry and mushy credence. In Gendler, T. Szabo, & Hawthorne, J. (eds.), *Oxford Studies in Epistemology*, vol. III, 161–186. New York: Oxford University Press.

Williamson, Jon. 2007a. Inductive influence. *British Journal for the Philosophy of Science*, 58(4), 689–708.

Williamson, Jon. 2007b. Motivating objective Bayesianism: From empirical constraints to objective probabilities. In Harper, William, & Wheeler, Gregory (eds.), *Probability and Inference: Essays in Honour of Henry E. Kyburg Jr*, 151–179. London: College Publications.

Williamson, Jon. 2011. Objective Bayesianism, Bayesian conditionalisation and voluntarism. *Synthese*, 178(1), 67–85.

Zynda, Lyle. 2000. Representation theorems and realism about degrees of belief. *Philosophy of Science*, 67(1), 45–69.

Index

Printed in the United States
By Bookmasters